**The Power of Artificial Intelligence
for the Next-Generation Oil
and Gas Industry**

The Power of Artificial Intelligence for the Next-Generation Oil and Gas Industry

Envisaging AI-Inspired Intelligent
Energy Systems and Environments

Pethuru Raj Chelliah
Edge AI Division
Reliance Jio Platforms Ltd
Bangalore, India

Venkatraman Jayasankar
Leading Oil and Gas Company
Bangalore, India

Mats Agerstam
Principal Engineer, Intel's Network and Edge Group
Portland, OR, USA

B. Sundaravadivazhagan
University of Technology and Applied Sciences - Al Mussanah
Oman

Robin Cyriac
Federation University
Brisbane-Campus
Australia

IEEE Press Series on Power and Energy Systems
Ganesh Kumar Venayagamoorthy, Series Editor

IEEE PRESS
WILEY

Published by John Wiley & Sons, Inc., Hoboken, New Jersey.
Published simultaneously in Canada.

For general information on our other products and services or for technical support, please contact our Customer Care Department within the United States at (800) 762-2974, outside the United States at (317) 572-3993 or fax (317) 572-4002.

Wiley also publishes its books in a variety of electronic formats. Some content that appears in print may not be available in electronic formats. For more information about Wiley products, visit our web site at www.wiley.com.

Library of Congress Cataloging-in-Publication Data Applied for:
Hardback: 9781119985587

Cover Design: Wiley
Cover Image: © Zen Rial/Getty Images

Set in 9.5/12.5pt STIXTwoText by Straive, Pondicherry, India

Contents

About the Authors

Pethuru Raj Chelliah – Working at Reliance Jio Platforms Ltd. (JPL) Bangalore. Previously He has worked at IBM Global Cloud Center of Excellence (CoE), Wipro Consulting Services (WCS), and Robert Bosch Corporate Research (CR). In total, he has gained more than 22 years of IT industry experience and 8 years of research experience. He finished the CSIR-sponsored PhD degree at Anna University, Chennai, and continued with the UGC-sponsored postdoctoral research in the Department of Computer Science and Automation, Indian Institute of Science (IISc), Bangalore. Thereafter, he was granted a couple of international research fellowships (Japan Society for the Promotion of Science (JSPS) and Japan Science and Technology Agency (JST)) to work as a research scientist for 3.5 years in two leading Japanese universities. He has been an ACM and IEEE professional member. He focus is on some of the digital transformation technologies such as the Internet of Things (IoT), artificial intelligence (AI), streaming data analytics, blockchain, digital twins, cloud-native computing, edge and serverless computing, reliability engineering, microservices architecture (MSA), event-driven architecture (EDA), and 5G/6G.

Venkatraman Jayasankar (Venkat) is an experienced IT professional with over 25 years of experience in different business domains like financial sector, manufacturing, healthcare, and energy. He has worked in multiple IT domains, industries, and technologies. He is a TOGAF (The Open Group Architecture Framework) certified architect and has worked extensively in architecture, designing solutions to complex business problems including IoT, edge computing, and asset management and digitalization. He has worked in different consulting organizations for clients in financial services, manufacturing, retail, healthcare, and energy. He is working for a leading oil and gas major as a Senior Principal Data Engineer focusing on all things related to data. His primary job is to make sure information and data are organized and governed for seamless analytics. He worked on architecture of digital twin, IoT, etc.

Mats Agerstam is an experienced system and software architect and researcher with over 20 years of professional experience in the information technology and software industry. Mats has a long background in security and wireless communication and protocols, particularly in technologies in the unlicensed spectrum such as WiFi, Bluetooth, and IEEE 802.15.4. He has been a key contributor in standardization bodies such as the Open Connectivity Foundation (OCF) and its open source reference implementation. He brings an extensive experience and expertise in the IoT space from multiple vertical segments such as industrial, retail, and enterprise. Mats holds a Master of Science degree in Information Technology Engineering from Uppsala University in Sweden and holds over 35 patents.

Dr. B. Sundaravadivazhagan is an experienced researcher and educator in the field of Information and Communication Engineering. He has more than 22 years of experience in teaching and research, and he earned his PhD in Information and Communication Engineering from Anna University in Chennai in 2016. He is a member of various professional bodies such as IEEE, ISACA, ISTE, and ACM, and has published over 40 research articles in SCI and Scopus journals. He has also served as a resource person, keynote speaker, and advisory committee member in more than 20 international and national conferences. He has received two research grants from the Ministry of Higher Education, Research and Innovation, The Research Council (TRC), Oman. His research interests include IoT, AI and machine learning, deep learning, cloud computing, networks and security, wireless networks, and MANET. He is a reputable journal editor and reviewer and serves on the Doctoral Committee in International Committee Member–Amrita University, Bangalore, as well as an Adjunct Faculty in Saveetha School of Engineering, Chennai.

Dr. Robin Cyriac is a distinguished computer science lecturer whose career has left an indelible mark on both academia and industry. With a PhD in Computer Science from a renowned institution, he embarked on a journey to share his profound knowledge and passion for the subject with eager minds. Dr. Robin's engaging teaching style, coupled with his ability to simplify complex concepts, has earned him a reputation as a beloved educator among his students. Beyond his academic pursuits, Robin is known for his dedication to promoting diversity and inclusion in technology, striving to create an inclusive learning environment that empowers students from all backgrounds. He continues to inspire the next generation of computer scientists through his innovative curriculum and commitment to fostering critical thinking and problem-solving skills in his students. His commitment to fostering innovation and critical thinking is evident through his engaging lectures and mentorship. Beyond the lecture halls, Robin has contributed significantly to research in IoT, security, and cloud technologies.

Foreword

As vice president of Architecture and Data for a major oil and gas company, I have had the privilege of witnessing firsthand the transformative power of artificial intelligence (AI) in the energy sector. The rapid evolution of digital technologies, combined with the explosion of data generated by the oil and gas industry, has presented a unique opportunity to leverage AI and advanced analytics to drive innovation and operational excellence.

Prior to joining Shell, I studied computer science and philosophy at university, where I became fascinated by the intersection of these two fields in the realm of AI. Despite the initial promise of AI, progress was hindered by the "frame problem," which refers to the challenge of determining which (of the infinite number of) aspects of a changing environment should be considered when planning actions, a task that the human brain can handle well, but computers cannot. While there have been advancements in processing power, data handling, and algorithms, game-changing AI technology remained elusive.

Today, the oil and gas industry is entering a new phase in which energy transition and digitalization are converging, creating an urgent need to embrace the advantages offered by AI. In their book, Pethuru Raj, Venkat, Mats, Sundar, and Robin provide an intriguing overview of the breadth and depth of AI and digital technology applications in the oil and gas industry. Their case studies and examples span the entire oil and gas value chain, from exploration and production to downstream asset management, showcasing how the applications of IoT, cloud, robotics, and digital twin technologies have a significant impact on safety and efficiency.

The book also provides insights into cybersecurity, which has become a growing concern as the integration of information technology (IT) and operational technology (OT) increases, and AI is used more widely across both domains. Anyone interested in these trends and the need for the oil and gas industry to transform and decarbonize while ensuring future energy security should read this book.

This book on AI in the oil and gas industry is a timely contribution to the growing body of knowledge in this area. The editors and contributors have brought together a diverse range of perspectives and expertise, reflecting the multidisciplinary nature of AI in the energy sector. From upstream exploration and production to downstream refining and managing assets, this book provides insights into how AI is transforming every aspect of the oil and gas value chain.

The case studies and examples presented in this book illustrate how AI is already delivering significant benefits to the industry, including improved efficiency, cost reduction, and enhanced safety. However, the potential of AI goes beyond operational optimization. As the energy transition accelerates, AI can help companies to decarbonize their operations, reduce their environmental footprint, and develop new business models that align with the changing energy landscape.

I commend the editors and contributors for their efforts in compiling this valuable resource for the oil and gas industry. I am confident that this book will serve as a catalyst for further innovation and collaboration in the application of AI to energy operations.

Nils Kappeyne
The Hague, The Netherlands
24 April 2023

Preface

Business behemoths and start-ups across the globe are keenly embracing the disruptive and transformative power of digital technologies and tools to produce and deliver strategically sound and sophisticated applications to their customers and consumers. Besides embarking on technology-driven process optimization, enterprises are meticulously working in assimilating trend-setting architectural patterns such as microservices and event-driven architectures (MSA and EDA) to be agile and adaptive in their operations, offerings, and outputs. Further on, the infrastructure optimization primarily induced through the paradigm of cloud computing is another grandiose initiative of worldwide organizations to be right and relevant to their business partners, employees, and end-users. Breakthrough digital technologies such as artificial intelligence (AI) are being leveraged by enterprising businesses to visualise and realize state-of-the-art and sophisticated systems to enhance operational efficiency and the much-desired customer delight.

With the widespread adoption of digitization and edge technologies, every kind of physical, mechanical, and electrical systems get digitized. With high-bandwidth and highly reliable communication technologies, digitized entities connect with one another in the vicinity and also with cloud-based software applications, enablement platforms, middleware solutions, and databases. Electronic devices are intrinsically being instrumented to be connected and cognitive in their activities and assignment.

Leading market analysts and researchers have predicted that there will be billions of connected devices and trillions of digitized elements on the planet earth in the years to unfurl. Resultantly, there will be a massive amount of multi-structured digital data. It is therefore imperative to make sense out of digital data mountains. There are integrated data analytics platforms to collect, cleanse, and crunch digital data to emit actionable insights. For any enterprise to march ahead in the right direction with all the confidence and clarity, all kinds of internal and external data have to be meticulously gleaned and subjected to a variety of deeper investigations to discover hidden knowledge. The discovered knowledge then gets disseminated to the concerned applications and devices to exhibit intelligent behavior.

Artificial intelligence (AI) is the latest popular paradigm to transition data into information and into knowledge. AI is being methodically supported in its vision of making intelligent systems by a host of machine and deep learning algorithms and models. Precisely speaking, AI is the flexible and futuristic paradigm for making sense out of data heaps. In the digital era, it is paramount and pertinent for product, solution, and service providers to extract actionable insights out of data to envisage next-generation offerings to retain their customers and to attract fresh consumers. AI has laid down a stimulating and sparkling foundation for worldwide enterprises to explore fresh avenues to enhance their revenues.

This book is dedicated to articulate and accentuate how the growing power of AI helps immensely in transforming the oil and gas industry to meet up the fast-emerging and evolving business, technical, user, and sustainability requirements. The book chapters are prepared and presented to understand the challenges and concerns of the oil and gas industry and how they are being addressed by leveraging the distinct capabilities of AI algorithms, models, frameworks, and toolsets.

The AI technology is being positioned and primed as the path-breaking paradigm for the entire human society as it has the innate wherewithal to visualize and realize plenty of intelligent systems and environments across industry verticals. AI is becoming famous and feverish as it has the required competency to replicate human brain capabilities into our everyday devices, business workloads, and IT services. AI-attached software products, solutions, and services can inherently exhibit data-driven insights and insights-driven decisions. There is less intervention, interpretation, and involvement of human beings in operating and managing AI-inspired systems situated in our living, relaxing, and working environments. The AI power grows and glows as there are many technological innovations and disruptions in the IT space. Essentially, we are tending toward the big data era with the exponential growth of different and distributed data sources.

Besides articulating and accentuating the AI power in strategically and significantly empowering the oil and gas industry, we have prepared and presented several chapters leveraging other related technologies such as the Industrial Internet of Things (IIoT), cyber-physical systems (CPS), digital twins, blockchain, edge computing, 5G communication, etc. Further on, we have dealt with new fuels and how they are going to sustain the human society. How intelligent drones and robots automate, accelerate, and augment various operations such as exploration, extraction, transportation, refinement, and retailing are vividly illustrated in this book. Finally, we have incorporated a chapter on explainable AI (XA) in order to insist the importance of trust and transparency in AI decisions, recommendations, predictions, and outcomes. One chapter exclusively talks about the blockchain technology, which is being portrayed as the way forward to ensure the tightest security for IoT edge devices and data.

In this book, we would like to focus on the oil and gas industry and how to make it smarter in its operations, outputs, and offerings through the leverage of the distinct advancements happening in the AI space. There are several problems and needs for leveraging the AI power to drastically and deftly empower oil and gas systems. The goal is to arrive at the digitally transformed oil and gas industry. This book will start with the challenges and concerns of the oil and gas industry. Then we will focus on the unique capabilities of the AI paradigm. Finally, we dig deeper and deal with how AI comes handy in empowering the oil and gas industry significantly. This book is comprehensive yet compact in illustrating how digitization and digitalization technologies blend together to bring forth a bevy of innovations, disruptions, and transformations for the oil and gas industry.

Pethuru Raj Chelliah
Edge AI, Reliance Jio Platforms Ltd.
Bangalore
24 April 2023

1

A Perspective of the Oil and Gas Industry

Oil was always used for commercial purposes like lighting and heating for a long time. In 1859, the first commercial oil well was drilled for the purpose of finding oil and using it for industrial purposes

What does crude oil contain?

- Crude oil is a naturally occurring fossil fuel – meaning it comes from the remains of dead organisms like algae
- Crude oil is made up of a mixture of hydrocarbons – hydrogen and carbon atoms, methane, ethane, propane, and butane exist as gases, while pentane exists as liquids
- An oil well predominantly produces crude oil, which is mixed a bit with natural gas, mainly methane
- A gas well produces natural gas predominantly with very little crude
- Crude oil is often referred to as petroleum. This is because petroleum includes both the unrefined crude oil as well as refined petroleum products.

The oil and gas industry has the following stages of oil extraction:

Upstream – Exploration and Production
Midstream – Transportation
Downstream – Refining and Marketing

The Power of Artificial Intelligence for the Next-Generation Oil and Gas Industry: Envisaging AI-Inspired Intelligent Energy Systems and Environments, First Edition. Pethuru Raj Chelliah, Venkatraman Jayasankar, Mats Agerstam, B. Sundaravadivazhagan, and Robin Cyriac.
© 2024 The Institute of Electrical and Electronics Engineers, Inc.
Published 2024 by John Wiley & Sons, Inc.

1.1 Exploration and Production

Oil and gas exploration encompasses the processes and methods involved in locating potential sites for oil and gas drilling and extraction. Early oil and gas explorers relied upon surface signs like natural oil seeps, but developments in science and technology have made oil and gas exploration more efficient.

In the past, surface features such as tar seeps or marks provided initial clues to the location of shallow hydrocarbon deposits. Today, accurate geological surveys are conducted using various means for exploration. Rock or sand surveys are done to see if there can be any possible deposits under the surface. Seismic surveys are also conducted by geologists to find oil deposits under the surface. If a site seems to have oil, an exploratory well is drilled and if there are deposits worth of value, then full-fledged development wells are drilled to extract the oil

Once the prospective reserve is found, companies will start drilling using mobile offshore drilling units (MODU). Once the drilling units find oil, the company will replace it with a more permanent oil production rig to capture oil.

Exploration is of high risk and expensive. The cost of a basic exploration, such as one that involves deep seismic studies, can cost $5–$20 million per exploration site, and in some cases, much more. However, when an exploration site is successful and oil and gas extraction is productive, exploration costs are recovered and are significantly less in comparison to other production costs.

Proven reserves measure the extent to which a company thinks it can produce economically recoverable oil and gas in place, as of a certain point in time, using existing technology.

Once a company identifies where oil or gas is located, plans begin for drilling, and the drilling methods vary depending on the type of oil or gas and the geology of the location.

To drill a well, it is necessary to simultaneously carry out the following drilling process.

- Crush the rocks under the earth to small pieces so that liquid can flow and the drill can travel down to get to the oil or hydrocarbons
- Remove the rock debris and continue drilling
- To make sure the holes do not cave in, preventing the drilling
- Prevent the fluids contained in the drilled formations from entering the well.

This can be achieved by using rotary drilling rigs, which are the ones operating today in the field of hydrocarbon exploration and production. The drills have a conveyor belt which continuously removes the debris that the drill digs out. These modern drills operate with great efficiency, but they can also cause damages to the areas around the drill site if they are not operated according to the conditions. The specialized equipment and complex technology make drilling of oil wells extremely expensive and of high risk.

1.1.1 Onshore

In onshore drilling facilities, the wells are grouped together in a field, ranging from half an acre per well for heavy crude oil to 80 acres per well for natural gas. The group of wells are connected by steel tubes, which send the oil and gas to a production and processing facility where the oil and gas are treated through a chemical and heating process. Onshore production companies can turn on and off rigs more easily than offshore rigs to respond to market conditions due to the need that offshore rigs need to be visited by engineers to inspect and close.

1.1.2 Offshore

Offshore drilling uses a single platform that is either fixed (bottom supported) or mobile (floating secured with anchors). Offshore drilling is more expensive than onshore drilling, and fixed rigs are more expensive than mobile rigs. Most production facilities are located on coastal shores near offshore rigs to reduce expenses and safety issues.

1.1.3 Hydraulic Fracturing

Fracking, or hydraulic fracturing, is a technique where a high-pressure liquid is injected to form cracks or fractures on the rocks to extract oil or gas. The use of fracking has led to recovering gas, and the oil extracted by this method is called shale oil.

Once a prospective reserve is found, companies will drill highly regulated exploration wells with Mobile Offshore Drilling Units (MODUs).

The MODU's job is to drill down into the ocean floor and find oil and natural gas reserves.

Once a well is found, the company switches to a more viable model of production. All the legal entities are created, and in many cases the drill that was used for initial explorations will be used for the production as well. The part of the well that allows the drill bit to drill without interruptions, like a pipe or casing, is called riser. A drilling riser may terminate at the sea floor or may extend slightly into the earth to prevent water infiltration.

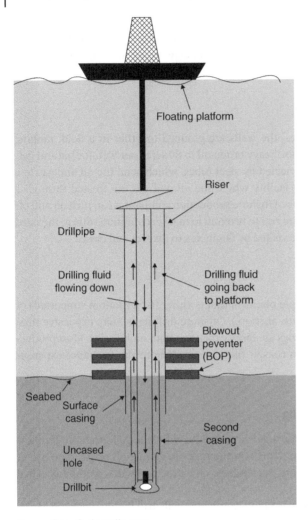

Drillpipe

Floating platform

Riser

Drilling fluid
flowing down

Drilling fluid
going back
to platform

Blowout
peventer
(BOP)

Seabed
Surface
casing

Second
casing

Uncased
hole

Drillbit

Source: EnggCyclopedia.com.

1.2 Midstream Transportation

The midstream sector involves the transportation (by pipeline, rail, barge, oil tanker, or truck), storage, and wholesale marketing of crude or refined petroleum products. Pipelines and other transport systems can be used to transport crude oil from production sites to refineries and deliver the various refined products to downstream distributors. Natural gas pipeline networks aggregate gas from natural gas purification plants and deliver it to downstream customers, such as local utilities.

Midstream activity starts after oil and gas is extracted. Once the gas is extracted from the wells, it must be refined. Midstream activity consists of the different process steps to transport and refine the oil.

Modes of transportation include the following:

Oil tankers: A tank vessel as one that is constructed or adapted to carry oil or hazardous material in bulk as cargo or cargo residue – as defined by the US Coast Guard. There are various types of tankers like oil, parcel tanker, combination, and barges. Many oil companies work on upstream, downstream, midstream and hence considered integrated. In many countries, midstream business does not exist as a separate business and is combined with upstream business.

LNG tankers: High-pressure possibility of explosions make it difficult to transport natural gas. For this reason, natural gas is liquefied at extremely low temperatures and transported as LNG via liquefied natural gas (LNG) tankers. LNG tankers are specially designed with double hulls to allow extra ballast water because LNG is lighter than gasoline.

Pipelines: Pipelines can do the work of transporting oil and gas to gathering systems (wellhead to processing facilities), transmission lines (supply areas to markets), or distribution pipelines (most commonly to transport natural gas to medium or small consumer units). Pipelines play a very critical role in the transportation process because most of the oil moves through pipelines for at least part of the route. After the crude oil is separated from natural gas, pipelines transport the oil to another carrier or directly to a refinery. The only challenge is laying of pipelines to get the oil transported.

Strategic planning involves determining the shortest and most economical routes where pipelines are built, the number of pumping stations and natural gas compression stations along the line, and terminal storage facilities so that oil from almost any field can be shipped to any refinery on demand.

Railroad/tank trucks: Historically, before the introduction of pipelines, railroads were used to transport petroleum. Today, railroads compete with pipelines. The existing railroad infrastructure creates a more flexible, alternative route when pipelines are at capacity. For instance, many a times, a railroad exists; however, laying a pipeline becomes a huge challenge. The railroad would have been laid before the actual development or roadblocks came up due to growth in the area, and it would now become a challenge to create new pipelines. Also, there are many chances of pipelines breaking or getting damaged. Reaching a damaged pipeline maybe a difficult task for engineers, and it is time-consuming. However, transporting oil by railroads has its own set of challenges.

Storage of oil and natural gas helps smooth out supply and demand discrepancies. Companies store more when the prices are lower than they would like and withdraw when prices are high. The oil storage is also done by different countries and is known as strategic oil reserves. Strategic oil reserves usually help a country when there are issues in oil supply due to natural calamities, man-made issues, etc. The cheapest storage method is underground tanks, such as depleted reservoirs. This method is primarily used for natural gas. Aboveground tanks are used for crude and refined oil, finished oil products, and natural gas. At retail locations, like gas stations, tanks are stored underground for safety reasons.

1.3 Downstream – Refining and Marketing

Downstream covers refining and marketing, or in some organizations it is trading and supply. The goal of refining is straightforward, to take crude oil, which is virtually unusable in its natural state, and transform it into petroleum products used for a variety of purposes and then sell it. Downstream depends on upstream for a steady supply of crude oil to help refine the oil and supply downstream.

One of the key processes involved in refining depending on the product needed is hydrotreating.

1.3.1 Hydrotreating

Hydrotreating or hydrodesulfurization refers to a set of operations that remove sulfur and other impurities. During hydrotreating, crude oil is made to react with hydrogen in the presence of a catalyst at relatively high temperatures and moderate pressure.

Hydrotreating helps petroleum refineries transform crude oil into useful fuels and products while satisfying government safety requirements. One of the most common issues is nickel catalyst poisoning by sulfur, which is present in the crude form extracted from catalytic beds.

Marketing is the wholesale and retail distribution of refined petroleum products to business, industry, government, and public consumers. Marketing is also known as trading and supply in some organizations.

Gasoline service stations handle the bulk of public consumer sales, and oil companies sell their petroleum products directly to factories, power plants, and transportation-related industries. Natural gas sales are almost evenly divided between industrial consumers, electrical providers, and residential and commercial heating.

In simpler terms, upstream operations include oil and gas production, midstream includes storage and transportation, and downstream includes distribution and retail outlets.

1.4 Meaning of Different Terms of Products Produced by the Oil and Gas Industry

1.4.1 Natural Gas

Natural gas is a fossil fuel. Natural gas contains many different compounds. The components of natural gas include methane, natural gas liquids (NGLs, which are also hydrocarbon gas liquids), and nonhydrocarbon gases, such as carbon dioxide and water vapor. We use natural gas as a fuel and to produce materials and chemicals. Natural gas is the cleanest burning hydrocarbon. A natural gas-powered station takes lesser time to start and stop compared to a coal-powered station. The main natural gases are methane, ethane, butane, and propane.

It is used for many things, especially in the home. Some common examples are as follows:

- Home heating through furnaces
- Warming water in water heaters
- Cooking food on barbecues and gas-burning stoves
- Operating gas fired fireplaces

1.4.2 Extraction

Natural gas, being an unconventional gas, must be extracted from deeper areas below the surface of the Earth. It consists of fracking or fracturing the rocks, passing water in high pressure to simulate the gas to flow via the pipes to the surface.

1.4.3 Advantages and Disadvantages

Natural gas has a high energy density and can be used flexibly for multiple applications, which make it a popular fuel. People advocating using natural gas often point to it as the cleanest burning fossil fuel. Even as a cleanest burning fossil fuel, natural gas is still composed of hydrocarbons, and burning it releases CO_2 and other pollutants (NOx being a problem specifically). Natural gas use is often an improvement over that of coal; however, its combustion still contributes to air pollution and climate change.

With advances in fracking techniques, natural gas reserves are expected to last a long time.

Natural gas power plants generate electricity by burning natural gas as their fuel. There are many types of natural gas power plants which all generate electricity but serve different purposes. All natural gas plants use a gas turbine; natural gas is added, along with a stream of air, which combusts and expands through this turbine, causing a generator to spin a magnet, producing electricity.

Natural gas power plants are cheap and quick to build. They also have very high thermodynamic efficiencies compared to other power plants. Burning of natural gas produces fewer pollutants like NO_x, SO_x and particulate matter than coal and oil.

Despite the improved air quality, natural gas plants significantly contribute to climate change, and that contribution is increasing. Natural gas power plants produce considerable carbon dioxide, although less than coal plants do. On the other hand, the process of getting natural gas from where it is mined to the power plants leads to considerable release of methane (natural gas leaks into the atmosphere). If natural gas plants are used to produce electricity, their emissions will continue to warm the planet in dangerous ways.

1.4.4 Types

The use of natural gas accounts for around 23% of the world's electricity generation.

This is second only to coal, and the fraction that is natural gas is expected to grow in coming years. This means that the contribution of natural gas to climate change will continue to increase.

1.4.5 Types of Natural Gas Deposits

Natural gas can be contained in a variety of different types of deposits that must be accessed if the natural gas is to be used. According to the Canadian Association of Petroleum Producers (CAPP), Canada has a natural gas reserve of between 700 and 1300 trillion cubic feet. While a little over 15% of that natural gas has been recovered, the rest is contained in four types of deposits: conventional and unconventional deposits: shale gas deposit, tight gas deposit, and coal bed methane.

Natural gas has been extracted from conventional natural gas deposits for a long time; the unconventional resources are resources that are being extracted using substantially new techniques.

1.4.6 Conventional Natural Gas Deposits

Conventional resources are gas contained within relatively porous rock, and they are the most easily mined. While hydraulic fracturing has allowed for more expansive access to these deposits, they can be mined without its use.

1.4.7 Coal Bed Methane

Coal bed methane is natural gas consisting mostly of methane, which is trapped inside coal deposits under the surface. This is extracted while the coal is being mined, as the diminishing pressure in the coal seam allows the gas to flow out of the seam and into a wellbore, where it is extracted.

1.4.8 Shale Gas

Shale gas is natural gas found inside a fine-grained sedimentary rock called shale. Shale is porous (there are lots of tiny spaces inside it), but it is non-permeable, which means the gas cannot flow through it. Shale gas requires the use of hydraulic fracturing for extraction.

1.4.9 Tight Gas

Tight gas is like shale gas, in that it is trapped inside a porous, non-permeable reservoir rock. The only differentiation between the two is that the term tight gas includes natural gas trapped inside reservoir rocks that are not shale.

1.4.10 Environmental Impacts of Natural Gas

1.4.10.1 Global Warming Emissions

Natural gas is a fossil fuel at the end of the day, which means there will be environmental impacts starting from drilling for gas, emissions during combustion, etc., though the global warming emissions from its combustion are much lower than those from coal or oil.

Natural gas emits 50–60% less greenhouse gases when combusted in a new, natural gas power plant compared with emissions from a typical new coal plant. Considering only tailpipe emissions, natural gas also emits 15–20% less heat-trapping gases than gasoline when burned in today's typical vehicles.

Drilling and extraction of natural gas from wells and its transportation in pipelines results in the leakage of methane, the primary component of natural gas, which is many times more stronger than CO_2 at trapping heat, which results in environmental hazards.

The emissions created by natural gas usually depend on the assumed leakage of methane, which is eventually a major reason for trapping the heat. One study found that methane losses must be kept below 3.2% for natural gas power plants to have lower life cycle emissions than new coal plants over short time frames of 20 years or lesser. This means implementation of more tighter controls and better systems to make sure the losses are monitored and reduced as per need. Similarly, vehicles burning natural gases must also keep methane losses much below 1% and 1.6% compared with those burning diesel fuel and gasoline, respectively. Technologies are available to reduce much of the leaking methane, but deploying such technology would require new policies and investments.

1.4.10.2 Air Pollution

Combustion of natural gas produces negligible amounts of sulfur, mercury, and other particulate matter. Burning natural gas does produce nitrogen oxides (NOx), which are precursors to smog, but at lower levels than gasoline and diesel used for motor vehicles. DOE analyses indicate that every 10,000 U.S. homes powered with natural gas instead of coal avoids annual emissions of 1900 tons of NO_x, 3900 tons of SO_2, and 5,200 tons of particulate matter. Reductions in these emissions translate into public health benefits.

However, despite these benefits, unconventional gas development can affect local and regional air quality. Some areas where drilling occurs have experienced increases in concentrations of hazardous air pollutants. Exposure to elevated levels of these air pollutants can lead to adverse health outcomes, including respiratory problems, cardiovascular diseases, and cancer. Another study found that residents living less than half a mile from gas well sites were at greater risk of health effects from air pollution from natural gas development than those living farther from the well sites.

1.4.10.3 Land Use and Wildlife

The construction and land required for oil and gas drilling can alter land use and harm local ecosystems by causing erosion and fragmenting wildlife habitats and migration patterns. When oil and gas operators clear a site to build a well pad, pipelines, and access roads, the construction process can cause erosion of dirt, minerals, and other harmful pollutants into nearby streams.

1.4.10.4 Water Use and Pollution

Unconventional oil and gas development can produce health risks by contaminating water and making people around fall sick. The hazardous chemicals which were once deep under the surface are drilled, and some of them get mixed with the water, causing serious contamination of water resources. There are many instances where clean water gets completely contaminated by chemicals as a result of deep drilling for oil and gas. Radioactive chemicals and methane can pose major health risks when leaked into the water supplies by carelessness or leaks or both.

The large volumes of water used in unconventional oil and gas development also raise water availability concerns in some communities.

1.4.10.5 Groundwater

There have been multiple documented instances of ground water getting contaminated with fracking or other fluids that are used to explore shale oil. One of the key challenges when ground water gets contaminated is that there are little options to clean up the mess that is created, and there must be a massive people/livestock relocation due to safety concerns. One major cause of gas contamination is improperly constructed or failing wells that allow gas to leak from the well into groundwater.

Another potential avenue for groundwater contamination is natural or man-made fractures in the subsurface, which could allow stray gas to move directly between an oil and gas formation and groundwater supplies.

In addition to gases, groundwater can become contaminated with hydraulic fracturing fluid. In several cases, groundwater was contaminated from surface leaks and spills of fracturing fluid.

1.4.10.6 Surface Water

Unconventional oil and gas development also poses contamination risks to surface waters through spills and leaks of chemical additives, spills and leaks of diesel or other fluids from equipment on-site, and leaks of wastewater from facilities for storage, treatment, and disposal. Unlike groundwater contamination risks, surface water contamination risks are mostly related to land management and to on- and off-site chemical and wastewater management.

There has been more than 1000 chemical additives identified that are used for hydraulic fracturing, including acids (notably hydrochloric acid), bactericides, scale removers, and friction-reducing agents. Large quantities – tens of thousands of gallons for each well – of the chemical additives are trucked to and stored on a well pad. If not managed properly, the chemicals could leak or spill out of faulty storage containers or during transport.

Drilling debris, diesel, and other material/fluids that are used when drilling can also spill at the surface, creating temporary or permanent damage. Improper management of flowback or produced wastewater can cause leaks and spills. There is also risk to surface water contamination from improper disposal. This could lead to areas around the drill site getting impacted: earthquakes around the drill site, water issues, and agriculture getting impacted and many more.

1.4.10.7 Water Use

The growth of hydraulic fracturing and its use of huge volumes of water per well may strain local ground and surface water supplies, particularly in water-scarce areas. The amount of water used for hydraulically fracturing a well can vary because of differences in formation geology, well construction, and the type of hydraulic fracturing process used The EPA estimates that 70–140 billion gallons of water were used in 2011 for fracturing an estimated 35,000 wells. This water is not sea water or ground water, which can get years to be replenished. Unlike other energy-related water withdrawals, which are commonly returned to rivers and lakes, most of the water used for unconventional oil and gas development is not recoverable. Depending on the type of well along with its depth and location, a single well with horizontal drilling can require 3–12 million gallons of water when it is first fractured – dozens of times more than what is used in conventional vertical wells. Similar vast volumes of water are needed each time a well undergoes a "work over," or additional fracturing later in its life to maintain well pressure and gas production.

1.4.11 The Future of Natural Gas

Replacing coal with natural gas in the electricity sector is not an effective long-term climate strategy.

Low natural gas prices and recent increases in the cost of generating electricity from coal have resulted in a significant shift from coal to natural gas over the past few years. With sufficient regulatory oversight, burning natural gas instead of coal could help reduce air pollution, providing immediate public health and environmental benefits. And because natural gas generators can be ramped up and down quickly, they could support the integration of wind and solar energy, provide increased flexibility to the electricity system, and continue to be used to meet peak demand.

Natural gas can also play an important role in meeting peak electricity demand and fueling cogeneration plants that generate both heat and power – which are up to twice as efficient as plants that only generate electricity using highly efficient technologies that provide both heat and power in the commercial and industrial sectors. The greenhouse emissions produced by natural gas are half the amount of emissions produced by coal when it is burnt.

However, natural gas is by no means a silver bullet solution for the environmental problems caused by our energy use. There is broad agreement among climate scientists that carbon reductions of at least 80% by 2050 will be needed to avert the worst effects of climate change. Simply switching to natural gas from coal and oil will not ultimately bring about the necessary reductions as natural gas pollutes the atmosphere and emits greenhouse gases when burnt, and in addition, the development of our newly discovered shale gas resource will disturb areas previously untouched by oil and gas exploration and raise serious water management, contamination, and other disturbances for the habitat.

Also, using natural gas as the primary source of electricity will motivate companies to find more sources, which means more contamination and polluting of the surroundings. The better option would be going in for a good mix of generating electricity by getting energy from renewable resources along with other sources. This mix coupled with renewables can provide an important hedge against future natural gas price increase.

Thus, while natural gas has a role to play in our future electric mix, a natural gas-centered energy pathway would also carry significant economic, environmental, and public health risks. Instead, a diversified electricity system – with amplified roles for renewable energy and energy efficiency and a modest role for natural gas – would both limit the threat of climate change and mitigate the risks of an overdependence on natural gas.

1.4.11.1 Liquefied Natural Gas (LNG)

Decades ago, the fossil fuel industry figured out how to transport gas by ships, in its quest to open markets beyond the domestic pipeline network. Its trick – liquefying natural gas – was a boon for energy companies. However, there were issues in that too. Liquefying huge volumes of natural gas to transport them to other places (that cannot be connected by pipelines) is no less carbon-intensive. The energy/electricity needed to chill, ship, and regasify the natural gas makes it far more carbon-intensive and increases the potential for leakage of dangerous methane. In addition, once there is a way to store huge volumes of gas through massive long-term infrastructure projects, it would mean more hazards to the environment, and this could make it impossible to limit global warming to 1.5°C.

What Was the Issue and Why Was Liquefaction Invented? The fossil fuel known as "natural gas" or "fracked gas" – the stuff frackers are after – exists in a gaseous state at room temperature. Natural gas is part of unconventional gas since it mostly needs to be drilled from deep under the surface from shale rock. The conventional oil is drilled from far lesser deeper rock like limestone or sandstone.

This presents a problem for transport. Since gases are made up of substantial empty space with just a few molecules bouncing around, it would take a massive container to ship a useful amount of gas by road or rail. That leaves pipelines as the primary means to transport gas at room temperature. This huge amount of new gas needs to be transported, and that needs to be done quickly without the need for laying of pipelines.

Enter liquefied natural gas (LNG). By cooling natural gas to −259°F – a very energy-intensive undertaking—processors can transform it into a denser liquid. (For context, the coldest temperature ever recorded outdoors on Earth is −135.8°F.) This liquid is then loaded into specialized oceangoing containers, which look like gigantic domes rising from the ship's deck.

Shippers have a trick to keep the LNG cold and in a liquid state during its long maritime journey. The product is loaded onto the ship at its condensation point of −259°F. As the storage container absorbs heat from the outside air, that energy goes toward converting small amounts of the LNG back into its gaseous form. The regasified molecules are directed out of the storage container and into the ship's engines, where they power the ship. Most of the LNG, however, remains in its liquid state until it reaches its destination, where it is regasified and transported through existing pipelines to consumers.

What Is Liquefied Natural Gas Used For? Once LNG is regasified at room temperature, it is used the same way as any natural gas – it is burned to generate electricity, heat, and cooking fuel in homes and businesses, as well as processed for plastics or other petrochemical products. LNG can also be used as a vehicle fuel without being returned to its gaseous form. But, since holding the liquid at −259°F is difficult, LNG's transport use is limited

Power plants also use LNG as a backup fuel. By cooling gas, utilities are able to store LNG on-site in cryogenic tanks. When demand peaks, or supply drops due to limited pipeline availability, the utility returns the LNG to its gaseous state and burns it to generate electricity.

What Are the Differences Between Raw, Compressed, and Liquefied Natural Gas? Raw natural gas is the name for unprocessed natural gas that has just been pumped from the ground. Because raw natural gas is unprocessed, it contains several components (like water, nitrogen, sand particles, rocks, and carbon dioxide) that must be removed.

Once the natural gas has been separated into its useful parts, there are two ways to make it dense enough to transport: compressing it or liquefying it.

Compressed natural gas (CNG) is gas that has been physically stuffed into a chamber, which is then gradually shrunk. CNG has 100 times as much energy as the same volume of uncompressed gas and can be stored at room temperature.

Liquefied natural gas, as mentioned, is chilled and liquefied gas held at very cold temperatures. Although LNG comes with storage and transport challenges, it is much more energy-dense than CNG – and about 600 times more energy-dense than ordinary gas.

LNG is an especially problematic form of natural gas for the climate. Chilling gas to incredibly cold temperatures uses much energy. Holding it at that temperature also uses energy. Transporting it by ship, rail, and truck uses energy. So, liquefying natural gas causes additional emissions and energy use to even get the natural gas to its usable state. Warming it back up also uses a lot of energy. When you add all of that up, LNG is responsible for almost twice as much greenhouse gas as ordinary natural gas, including from gas leaks, flaring, or intentional venting (for example, when operators release gas into the atmosphere to allow for maintenance on a pipe) during production and transport.

1.4.11.2 Compressed Natural Gas (CNG)

CNG comprises

- Mostly methane gas which, like gasoline, produces engine power when mixed with air and fed into your engine's combustion chamber.
- When CNG reaches the combustion chamber, it mixes with air, is ignited by a spark, and the energy from the explosion moves the vehicle.

- CNG is compressed so that enough fuel can be stored in your vehicle to extend the driving range, much like the gasoline tank in vehicle.
- Although vehicles can use natural gas as either a liquid or a gas, most vehicles use the gaseous form compressed to 3000 psi. CNG is produced by compressing natural gas to less than 1% of its volume at standard atmospheric pressure.

Due to the high production cost of CNG, it must be stored in cryogenic tanks, and hence CNG use in commercial applications has been limited.

Hydrocarbon resources are resources that contain both hydrogen and carbon. Hydrocarbon resources are often known as fossil fuels (natural gas, oil, and coal) since hydrocarbons are the primary constituents in these.

1.4.12 CNG vs. Liquid Fuels

- CNG is one of the most viable alternatives to traditional liquid fuels for vehicles.
- CNG is called as clean fuel because it is free from lead and sulfur and reduces harmful emissions. It is also lighter than air; hence, if there are leaks, it rises up, disperses into the atmosphere, and mixes easily and evenly
- CNG is approximately one-fifth the price of gasoline, resulting in substantial savings in fuel costs.
- CNG reduces maintenance costs since it contains no additives and burns cleanly, leaving no by-products of combustion to contaminate your spark plugs and engine oil.
- The combustion chamber parts function at peak output for longer periods before requiring service. The engine oil also remains clean, which minimizes engine wear and requires less frequent changes.
- CNG is more environment-friendly, and CNG engines are much quieter due to the higher octane rating of CNG over gasoline.
- CNG produces less exhaust emissions, and as a result, harmful emissions such as carbon monoxide (CO), carbon dioxide (CO_2), and nitrous oxide (N_2O) are generally reduced by as much as 95% when compared to gasoline-powered vehicles.
- CNG is the safe bet as its components are designed and made to international standards and are monitored to ensure safe operation.
- CNG fuel systems are also sealed, which prevents any spill or evaporation losses.

The extraction and processing cost of CNG makes it a very expensive proposition to use as fuel.

1.4.13 Unconventional Oil and Gas

Unconventional resources are generally oil or natural gas resources that do not appear in traditional formations and must use specialized extraction or production techniques to obtain. Fracking or hydraulic fracturing is increasing the viability of the extraction of these resources.

Examples of unconventional deposits are shale oil, oil sands (refer to a mixture of sand, water, clay, and bitumen), shale gas, tight gas (type of natural gas), and coal bed methane (natural gas stored in coal beds).

The reservoirs described earlier are called conventional sources of oil and gas. As demand increases, prices increase, and new conventional resources become economically viable. At the same time, production of oil and gas from unconventional sources becomes more attractive. These unconventional sources include very heavy crudes, oil sands, oil shale, gas, and synthetic crude from coal, coal bed methane, methane hydrates, and biofuels.

Products produced by the oil and gas industry are as follows:

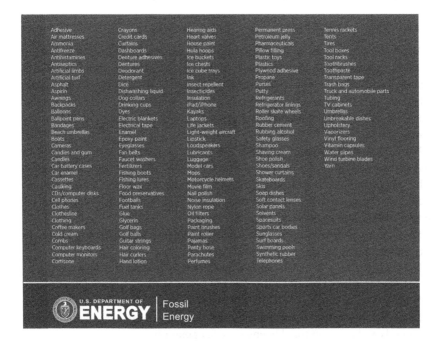

1.5 Oil and Gas Pricing

Oil and gas pricing is determined very carefully as the oil price has political ramifications. Oil prices are based on barrels of oil as the unit. A barrel of oil is equivalent to 42 US gallons or 159l.

Price benchmarks are used in the oil and gas industry to give buyers a way to value the commodity based on quality and locations. The main benchmarks used in this industry are as follows:

- **Brent blend** is the most common, internationally used oil benchmark. It is based in London, traded on the InterContinental Exchange (ICE), and consists of light, sweet crude oil from offshore drilling in the North Sea.

- **West Texas intermediate (WTI)** is used for light and sweet oil in the United States, specifically crude oil that comes from land-locked wells in Oklahoma.
- **Dubai/Oman** is used for heavier oil, with a higher sulfur content from the Persian Gulf to the Asian market.
- **Henry Hub** is the benchmark for North American natural gas and global lique-fied natural gas (LNG), based off of the Henry Hub natural gas pipeline in Louisiana

1.6 A Note on Renewable Energy Sources

What is renewable energy?
Renewable energy, as the term indicates, is the energy that can be renewed naturally. It is energy from sources that can be naturally replenished

The major types of renewable energy sources are as follows:

- Biomass
- Hydropower
- Geothermal
- Wind
- Solar

1.6.1 Biomass

Biomass is obtained from plants and animals. Biomass contains stored chemical energy from the Sun. Biomass can be burned directly for heat or converted to renewable liquid and gaseous fuels through various processes.

Biomass sources for energy include the following:

- **Wood and wood processing wastes**: firewood, wood pellets, and wood chips, lumber and furniture mill sawdust and waste, and black liquor from pulp and paper mills
- **Agricultural crops and waste materials**: corn, soybeans, sugar cane, switch-grass, woody plants and algae, and crop and food processing residues, mostly to produce biofuels
- **Biogenic materials in municipal solid waste**: paper, cotton, and wool products and food, yard, and wood wastes
- Animal manure and human sewage for producing biogas/renewable natural gas.

1.6.2 Hydropower

Hydropower is generated by fast-moving water in a large river or rapidly descending water from a high point and converts the force of that water into electricity by spinning of a generator's turbine blades. Hydropower is used in almost all countries where there is forceful waterflow.

1.6.3 Geothermal

The Earth's core is about as hot as the Sun's surface, due to the slow decay of radioactive particles in rocks at the center of the planet. The water that has seeped into the Earth's crust heats up and comes out as steam. This steam is captured and used for powering a turbine, which in turn will run a generator.

1.6.4 Wind

Many parts of the world have strong wind speeds, but the best locations for generating wind power are sometimes remote ones. Offshore wind power offers tremendous potential.

Wind energy is obtained by harnessing the kinetic energy of moving air by using large wind turbines located on land (onshore) or in sea or freshwater (offshore). Large windmills are installed in the areas where huge winds are expected, and the blades when turned run a turbine, which in turn produces power.

Though average wind speeds vary considerably by location, the world's technical potential for wind energy exceeds global electricity production, and ample potential exists in most regions of the world to enable significant wind energy deployment.

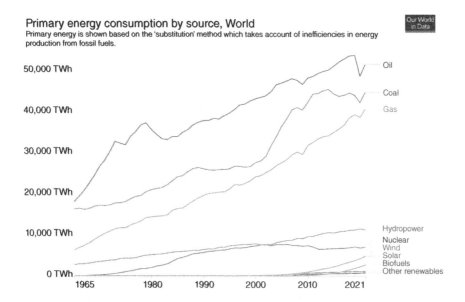

Primary energy consumption by source, World
Primary energy is shown based on the 'substitution' method which takes account of inefficiencies in energy production from fossil fuels.

1.6.5   A Note About Hydrogen

Hydrogen, looked up as an energy hydrogen, is a clean fuel that, when consumed in a fuel cell, produces only water. Hydrogen can be produced from a variety of domestic resources, such as natural gas, nuclear power, biomass, and renewable power like solar and wind. These qualities make it an attractive fuel option for transportation and electricity generation applications. It can be used in cars, in houses, for portable power, and in many more applications.

1.6.6   How is Hydrogen Produced

There are four main sources that can help in extracting or producing hydrogen: natural gas, oil, coal, and electrolysis. The predominant composition of oil is hydrocarbon, which is a mix of hydrogen and carbon, and extracting hydrogen is one possible way of producing hydrogen.

1.6.7   Production of Hydrogen

Today, hydrogen fuel can be produced through several methods. The most common methods today are natural gas reforming (a thermal process) and electrolysis. Other methods include solar-driven and biological processes.

Synthesis Gas: Natural gas is reacted with high-temperature steam, which creates a mixture of carbon monoxide, hydrogen, and carbon dioxide. The carbon monoxide is reacted with water to produce more hydrogen.

1.6.8   Electrolysis

An electric current splits water into hydrogen and oxygen. If the electricity required to split the water comes from renewable resources, then the hydrogen is considered a fully renewable resource.

1.6.9   Biological Processes

Biological processes use microbes such as bacteria and microalgae and can produce hydrogen through biological reactions. In microbial biomass conversion, the microbes break down organic matter like biomass or wastewater to produce hydrogen.

1.6.10   Converting Hydrogen to Hydrogen-Based Fuels

Hydrogen has low energy density, which makes it more challenging to store and transport than fossil fuels. However, it can be converted into hydrogen-based fuels such as synthetic methane, synthetic liquid fuels, and ammonia, which can make use of the existing infrastructure for their transport, storage, and distribution. This can reduce the costs of reaching end users. Some of the synthetic hydrocarbons produced from hydrogen can be direct substitutes for their fossil equivalents.

1.7 Environmental Impact

As of 2020, most hydrogen is produced from fossil fuels, resulting in carbon dioxide emissions. This is often referred to as **gray hydrogen** when emissions are released to the atmosphere, and **blue hydrogen** when emissions are captured through carbon capture and storage (CCS). Blue hydrogen has been estimated to have a greenhouse gas footprint 20% greater than that of burning gas or coal for heat and 60% greater when compared to burning diesel for heat.

Hydrogen produced using the newer, non-polluting technology like methane pyrolysis is often referred to as **turquoise hydrogen**. High-quality hydrogen is produced directly from natural gas by splitting methane (from in natural gas) into hydrogen and carbon.

Hydrogen produced from renewable energy sources is often referred to as **green hydrogen**. There are two ways of producing hydrogen from renewable energy sources. One is to use power to gas, in which electric power is used to produce hydrogen from electrolysis of water, and the other is to use landfill gas to produce hydrogen in a steam reformer. Hydrogen fuel, when produced by renewable sources of energy like wind or solar power, is a renewable fuel.

Turquoise hydrogen has been touted as a game changer.

One major factor here is that the methane pyrolysis process does not result in any CO_2 emissions, so there is no need to invest in carbon capture facilities and no need to worry about where to store it.

Especially while electrolyzer capacity worldwide is building up, methane pyrolysis could provide another effective route toward generating carbon-free hydrogen, helping build the hydrogen value chain.

1.8 Uses of Hydrogen

So why is there so much buzz about the benefits of hydrogen as an energy transition fuel?

One of its key advantages is that it is the perfect complement to renewables. Wind, solar, and other renewables remain vital for the global energy transition, and they can be complemented by on-demand power generation when the output from renewable sources cannot meet all the electricity demands. Part of the reason is hydrogen's potential to help decarbonize sectors like hard-to-abate industries, mobility, and power generation.

1.8.1 Challenges in Using Hydrogen as a Fuel

1.8.1.1 Old Product, Old Problems

This is not the first time hydrogen has been hailed as the fuel of the future. It powered the first internal combustion engine in 1806. In 1970, the term hydrogen economy was coined after a report by a US academic Lawrence Jones started hydrogen hype in the country, which eventually died down in the 1980s.

Today, hydrogen produced from fossil fuels, without capturing the carbon, is used mainly in the chemicals and refining industries. It is responsible to 830 million tons of CO_2 emissions per year, equivalent to the annual emissions of the UK and Indonesia combined.

While green hydrogen will avoid this pollution, many old problems remain, most notably how to store it safely. While less toxic and more readily dispersed than natural gas, hydrogen has a wide range of flammable concentrations in air and lower ignition energy than petrol or natural gas, which means it can ignite more easily.

Currently, hydrogen storage requires extremely high pressure and is therefore too expensive and inefficient for widespread use in vehicles.

With technologies advancing and other forms of fuel becoming easier to produce, maintain, and renew, it is only a matter of time before the world finds a replacement for fossil fuels as a renewable way to produce energy

Most of the air pollution occurs due to incomplete burning of fossil fuels. Coal and oil are burnt for generating electricity, fuel for transportation, etc. Inhaling air induced with pollutants due to the burning of natural gas and fossil fuel reduces the heart's ability to pump enough oxygen, hence causing one to suffer from various respiratory and heart illnesses. One of the solutions that is being explored is to use nuclear-fired power plants instead of coal-fired power plants.

Nuclear power is the use of nuclear reactions to produce electricity. Nuclear power can be obtained from nuclear fission, nuclear decay, and nuclear fusion reactions.

However, disposal of nuclear waste, health hazard in case of accident within the power plants, and many more challenges are faced by commissioning a nuclear power plant. While a nuclear power plant may not be ideal, it is one of the least polluting in the current scenario.

Bibliography

1 Oil and gas industry: a research guide. https://guides.loc.gov/oil-and-gas-industry/upstream#:~:text=Oil%20and%20gas%20exploration%20encompasses,and%20gas%20exploration%20more%20efficient (accessed 14 August 2022).

2 The basics of oil and gas. https://www.noia.org/basics-offshore-oil-gas/ (accessed 14 August 2022).

3 Nisbet, E.G., Fisher, R.E., Lowry, D., et al. (2020). Methane mitigation: methods to reduce emissions, on the path to the Paris agreement. https://doi.org/10.1029/2019RG000675.

4 Rajput, S. and Thakur, N.K. The road ahead and other thoughts. Geological Controls for Gas Hydrate Formations and Unconventionals. https://www.sciencedirect.com/topics/earth-and-planetary-sciences/tight-gas#:~:text=Tight%20gas%20refers%20to%20natural,underground%20formation%20extremely%20%E2%80%9Ctight%E2%80%9D (accessed 13 September 2022).

5 Environmental impacts of natural gas (2014). Updated May 9, 2023. https://www.ucsusa.org/resources/environmental-impacts-natural-gas(accessed 14 December 2022).

6 Gürsan, C. and de Gooyert, V. The systemic impact of a transition fuel: does natural gas help or hinder the energy transition? https://www.sciencedirect.com/science/article/pii/S1364032120308364 (accessed 13 September 2022).

7 McDonald, M. (2015). What is holding the green revolution back?, CDT. https://oilprice.com/Alternative-Energy/Renewable-Energy/What-Is-Holding-The-Green-Revolution-Back.html (accessed 15 December 2022).

8 Renewable energy – powering a safer future. https://www.un.org/en/climatechange/raising-ambition/renewable-energy#:~:text=The%20science%20is%20clear%3A%20to,affordable%2C%20sustainable%2C%20and%20reliable (accessed 14 August 2022).

9 Palmer, B. (2022). Liquefied Natural Gas 101. What is it? Why is it? And what does it mean for the climate? https://www.nrdc.org/stories/liquefied-natural-gas-101 (accessed 15 December 2022).

10 Oil and gas industry: a research guide – oil and gas pricing. https://guides.loc.gov/oil-and-gas-industry/pricing#note5 (accessed 14 August 2022).

11 Ritchie, H., Roser, M., and Rosado, P. (2022). Energy. https://ourworldindata.org/energy (accessed 15 December 2022).

12 Naujokaitytė, G. (2021). Clean hydrogen: smoke screen or the future of energy? https://sciencebusiness.net/climate-news/news/clean-hydrogen-smoke-screen-or-future-energy (accessed 15 December 2022).

13 https://guides.loc.gov/oil-and-gas-industry/upstream#:~:text=Oil%20and%20gas%20exploration%20encompasses,and%20gas%20exploration%20more%20efficient.

14 https://www.noia.org/basics-offshore-oil-gas/

2

Artificial Intelligence (AI) for the Future of the Oil and Gas (O&G) Industry

2.1   Introduction

The oil and gas sector is being continuously enabled through a bevy of innovations and transformations with the faster evolution and stability of digital technologies and tools. The power of digitization technologies (the Internet of Things (IoT) and edge technologies, 5G, etc.) and digitalization technologies (cloud-native, server-less and edge computing paradigms, data analytics, blockchain, cybersecurity, microservice, and event-driven architectures, AI, digital twins, etc.) have brought in a massive disruption for the oil and gas industry. In short, the much-expected digital transformation is being precipitated through the intrinsic power of digital technologies and tools across industry verticals. The oil and gas industry also cannot escape to the transitions happening around it. As the entire world continues to depend on the organic sources of energy, there is an insistence for pioneering technologies to tackle micro and macro challenges that are affecting the oil and gas industry.

The oil and gas industry faces many critical challenges in its operational processes and practices. The operating environments are situated in risky, rough, and tough locations. The oil and gas industry assets are not fully digitized and hence unconnected. Therefore, remote monitoring, measurement, and management and maintenance of oil rigs and other machineries are difficult. Also, market prices are hugely volatile, with fluctuating demands. This chapter is primarily dedicated to articulating the everyday challenges and concerns and how they can be surmounted through artificial intelligence (AI) technologies and tools.

The Power of Artificial Intelligence for the Next-Generation Oil and Gas Industry: Envisaging AI-Inspired Intelligent Energy Systems and Environments, First Edition. Pethuru Raj Chelliah, Venkatraman Jayasankar, Mats Agerstam, B. Sundaravadivazhagan, and Robin Cyriac.
© 2024 The Institute of Electrical and Electronics Engineers, Inc.
Published 2024 by John Wiley & Sons, Inc.

2.2 The Emergence of Digitization Technologies and Tools

Without an iota of doubt, the prime focus is to sharply improve various upstream, midstream, and downstream processes and practices using groundbreaking digital technologies, which get typically subdivided into two major categories: digitization and digitalization. Digitization technologies such as the Internet of Things (IoT), scores of edge technologies, and connectivity methods such as 5G and Wi-Fi 6 and 7 empower ordinary assets to become digitized artifacts. Digitized elements can gain the power to find and interact with one another purposefully in the vicinity and get integrated with cloud-based software applications and data stores through the internet communication infrastructure. The typical process is to attach edge technologies such as minuscule sensors, actuators, microcontrollers, RFID tags, barcodes, stickers, beacons, LED lights, and specks on ordinary items. Such an attachment can make any tangible product to perform and contribute as digitized entities. Digitized elements can be further empowered to be computational, communicative, sensitive, perceptive, vision-enabled, decision-making, responsive, and active artifacts through additional modules.

By getting linked up with cloud-based resources and recipes, digitized elements can be enabled to do improved things. When digitized assets start interacting with one another, there is a huge possibility for generating a large amount of multistructured digital data. In short, the concept of digitization is rapidly spreading into every domain of substance. Every object becomes a digitized entity. Market watchers, researchers, and analysts have forecasted that there will be trillions of digitized entities in the years to come. All kinds of physical, mechanical, and electrical systems are being prepared and presented as digitized elements through the transformative power of the IoT technologies and tools. There is a related concept of cyber-physical systems (CPSs). This is all about empowering physical assets at the ground level to get integrated with cyber-resources such as cloud-based software packages. Such a linkage instantaneously empowers physical devices to be blessed with additional capabilities. This transition helps physical systems to join in the mainstream computing arena to contribute for producing context-aware software solutions and services.

The idea of digital twins is blossoming these days. As mentioned above, in the digital era, all kinds of physical systems are methodically digitized to remarkably contribute the long-standing goal of computing everywhere all the time. Furthermore, digitized artifacts can interact with cloud-based business and IT services. Digital twins are the virtual/logical/cyber/software version of physical assets, which are at the ground. Digital twins hugely simplify engineering complicated physical systems. That is, for a physical product and process to be produced and deployed with all the confidence and clarity, establishing and leveraging a corresponding digital twin for the product are termed as mandatory activities.

With such an arrangement gaining momentum, there are many business, technical, and user benefits. Through digital twins, it is quite easy to understand the use cases, distinct capabilities, hidden risks, improvements to be made, operational difficulties, etc. of a complex physical system. Thus, as products and processes become complex and intertwined, having appropriate digital twins comes handy in moderating the rising complexity.

In short, every tangible item becomes digitized methodically to be constructive, contributing immensely for the ensuing digital era.

2.3 Demystifying Digitalization Technologies and Tools

In the abovementioned section, we discussed about digitization technologies. Herein, let us concentrate on digitalization technologies. As described above, digitization results in an enormous number of multifaceted digitized systems and sensors, which, when interacting for fulfilling business processes, generate a lot of digital data. Just generating data is not sufficient anymore. Every bit of data getting generated must be collected, cleansed, and crunched to emit out actionable insights hidden inside data collection, thus deriving useful information out of data through a bevy of digitalization technologies such as integrated data analytics methods and platforms, AI algorithms and models, software-defined cloud infrastructures, visualization dashboards, databases, data warehouses and lakes, message brokers, hubs, and queues. Thus, transitioning data to information and to knowledge is the key goal of digitalization technologies and tools. In short, digital data becomes digital intelligence. Unraveling insights out of data volumes is being simplified through the correct application of digitalization techniques.

AI is being recognized as the prime digitalization technology for extracting predictive and prescriptive insights out of data mountains. AI is primarily used in applications such as quality control, asset inspection, and predictive maintenance in the upstream, midstream, and downstream domains. There are two AI implementation techniques: machine and deep learning (ML/DL) algorithms. These algorithms help create AI models, which can learn from data in an automated manner and emit out actionable insights in time. Deep learning algorithms enable the creation of powerful deep neural networks (DNNs) to learn from data.

2.4 Briefing the Potentials of Artificial Intelligence (AI)

Without an iota of doubt, oil and gas exploration and production are an expensive process. There are complicated and costly equipment, which are being operated by professionals and trained technicians. Exploration involves seismic surveys, which produce substantial data along with the related environment data. The traditional analytics methods and platforms fail to get deep and decisive insights. Thus, researchers and data scientists must embark on leveraging AI algorithms to make sense out of the Big Data. AI can emit out highly right and relevant patterns and associations instantaneously and could train on the data produced by active wells and their exploration details to bring forth viable models, which are then utilized to predict about the details about new oil reserves. AI can also prescribe the optimal ways and means to access known resources and project the total lifetime yield from them.

Artificial intelligence (AI) is proving to be a cost-saving, thought-provoking, and game-changing phenomenon for the entire society. AI is famous and fabulous in solving some of the important personal, professional, and social problems such as prediction through classification, regression, and clustering algorithms and recognition, detection, translation, etc. through deep neural networks (DNNs). In short, AI facilitates the transition from raw data to information and to knowledge. That is, knowledge discovery and dissemination are the primary targets for leveraging AI technologies and tools. Knowledge discovered from data gets disseminated to appropriate actuation systems, which, in turn, exhibit adaptivity and adroitness in their operations, outputs, and offerings. Precisely speaking, replicating human brain capabilities into networked embedded systems and software applications is the foremost goal for AI methods. Transitioning all sorts of devices in our everyday environments (homes, hotels, hospitals, etc.) into AI-enabled devices to be intelligent in their decisions, deals, and deeds is the primary goal of AI. Not only that, all kinds of software applications (business workloads, IT services, technology, and device-centric applications, etc.) will be AI-inspired to be cognitive in their service deliveries. Thus, for establishing intelligent environments, establishments, and enterprises, AI is the way forward without an iota of doubt.

AI is being tried across multiple industries. There are several use cases with positive results. AI is being projected as the next-generation technological paradigm to bring in genuine business transformation. AI is being ably supported by machine and deep learning (ML/DL) algorithms. Finding useful patterns in datasets and extracting actionable insights out of data are being facilitated through ML and DL algorithms. In other words, these algorithms come handy in empowering computing systems to learn something new from data and express/explain/expose it in an automated manner. Thus, turning data into intelligence is the primary duty of ML and DL algorithms. There are myriads of enabling frameworks, toolkits, and libraries and integrated platforms for accelerating and augmenting the complicated process of transforming data into highly beneficial insights. Additionally, there are AI-specific processors and accelerators. In short, AI is the state-of-the-art technology for implementing digitally transformed homes, cities, hotels, retail stores, supply chain, industrial and manufacturing environments, etc. Due to the extreme speed of computers, the much-expected targets of setting up and sustaining self-learning, configuring, healing, scaling, defending, evolving, and managing systems are being speeded up. Thus, it is not an exaggeration to state that AI is the giant killer and game-changer.

Data are the greatest asset for any start-ups and enterprises to proceed with all the confidence, clarity, and alacrity. It is a universally accepted fact that with more data, the prediction accuracy of AI algorithms increases sharply. The oil and gas industry generates substantial data, and hence AI turns out to be the most suitable technology partner for the oil and gas industry to visualize and realize competent and cognitive solutions.

AI comes handy in several aspects. AI helps avoid accidents and production disruptions through the predictive and prescriptive capabilities of AI models. AI anticipates equipment failures and dangerous operating conditions, and then it articulates them in advance to the concerned. All the noteworthy advancements in the AI space are being immaculately used for fulfilling the mantra of more with less. The much-needed choice, convenience, comfort, and care are being elegantly accomplished through AI.

Fuels are notified as the essential items for human beings to lead a cool and civilized life in this planet Earth. There are sources spread across the world for procuring and processing oil and gas in enormous quantities. There are huge investments being made by national governments and private organizations to produce and reserve a huge stock of oil and gas to fulfil the growing energy requirements. However, the oil and gas industry is facing a number of challenges internally as well as externally. To overcome them, like other industries, the oil and gas (O&G) sector also is destined to embark on digital transformation by smartly leveraging promising and potential digital technologies. In the subsequent sections, we will focus on digital transformation methods.

2.5 AI for the Oil and Gas (O&G) Industry

Considering the growing operational and environmental challenges in the oil and gas industry, pundits and proponents have strongly recommended the smarter leverage of the predictive and prescriptive powers of AI algorithms, models, frameworks, libraries, toolkits, and accelerators to surmount the identified and hidden challenges. Additionally, several newer possibilities and opportunities are being conceived and concretized through the glowing AI paradigm. This section is fully dedicated to articulate and accentuate a variety of futuristic and fabulous use cases.

Oil companies increasingly leverage the AI capabilities to drill and mine raw hydrocarbons and other products to produce fuel. There are suitable AI algorithms and models providing accurate and actionable insights to expertly guide drills on water and land. Precision drilling helps in remarkably reducing the risks of fire accident, oil spillage, etc. AI helps the oil companies in optimizing the oil production by eliminating the areas of inefficiency.

2.5.1 Automate and Optimize Inspection

Automation comes handy in saving of time, treasure, and talent. The oil and gas (O&G) industry is highly labor-intensive, and the employees usually work in rough and tough locations. AI helps in automating all the repetitive tasks while eliminating the redundancy. Therefore, the people and property safety and security are fully ensured, while the operational efficiency is bound to increase remarkably.

For an example, during inspection, AI helps in detecting any perceptible anomalies that may ensure the risk-free, reliable, and continuous operations of oil and gas (O&G) assets. Also, O&G companies are trying to bring in robust optimization in inspection and maintenance procedures and processes through AI. Without appropriate optimization, operational, management, and maintenance costs are bound to increase significantly. There are many proven ML and DL algorithms for implementing the use case of anomaly detection. Besides prediction, prescriptive insights are also borne out through AI-enabled analytics. Precisely speaking, AI-inspired automation, orchestration, and optimization competencies do several good things for the struggling O&G industry. Process excellence is being achieved through AI, and hence the adaptivity, adroitness, and affordability of oil and gas systems and services are eternally guaranteed.

2.5.2 Defect Detection and Enhance Quality Assurance

The oil and gas (O&G) industry is heavily using big machineries and long pipelines in its day-to-day operations. Therefore, identifying improper threading in pipelines is one serious requirement for the correct functioning of the system. Further on, defects on products and processes of pipelines must be proactively and pre-emptively identified to avoid costly slowdowns and breakdowns. AI has shown its incredibility in precisely pointing out defects and verifying the production quality with precision and perfection. In short, AI-powered defect detection in time is being seen as a breakthrough solution for the O&G industry.

Defect detection is a crucial factor for the oil and gas industry to survive and thrive. Detecting incorrect threading in pipelines or any fault in processors in time saves a lot. AI comes handy in validating production quality and provides deep insights into defects. By applying deep learning (DL) algorithms on video streams captured and transmitted on-site cameras, it is possible to detect a variety of things. Deep learning models are efficient and elegant in recognizing and articulating distinct patterns.

Timely recognition of any inappropriate threading in pipelines and faults in mechanisms that may go wrong paves the way to minimize production losses. If we go for a computer-vision-based system, it is a futuristic move because a CV system simplifies pinpointing defects and deficiencies in time. Thus, AI-enabled quality check empowers the industry remarkably. The quality processes also can be optimized and adopted to do better and bigger things.

2.5.3 Monitoring

As accentuated above, the O&G industry is stuffed with a variety of assets and state-of-the-art machineries. There are oil fields with rigs, gas stations, trucks, reservoirs, and refineries with a variety of equipment and appliances. There is a need for employee, enterprise, and equipment safety. Any incident and accident can be minutely monitored through AI-powered cameras, robots, drones, etc. to reduce the extent of potential and physical damage. Frequent inspection of equipment and risk assessment will help companies in taking predictive measures to avoid unforeseen circumstances. Briefly, the aspects of monitoring and observability, which is getting greater attention for ensuring the continuous availability and high reliability of cloud applications, infrastructure services, and data stores, are also becoming a valuable tool for the intended success of the oil and gas industry.

Object detection can be used for security and alerting on rigs. The monitoring and control team is alerted whenever some unwarranted equipment movements or if there is a trespass into the site. There are pre-trained, curated, and refined self-learning models for object detection and tracking, activity recognition, and image captioning. These work together to bring timely and correct alerts.

2.5.4 Reduce Production and Maintenance Costs

The environmental and seasonal changes could result in degradation and corrosion in oil and gas equipment and instruments. Such an unavoidable thing can break the system if not found and corrected in time. Again, AI technologies and tools are found to be hugely effective in pinpointing the signs and symptoms of corrosion. The prediction capability being exhibited by AI systems and the domain-specific knowledge graph combine well to detect the commencement of corrosion in particular spots. Such a critical discovery comes handy in sharply reducing operational costs.

The production optimization is another beautiful use case amid frequent changes in oil prices. Further on, the life of oil wells must be extended as much as possible. There are several independent variables such as flow rate and pressure, etc. As this industry is primely data-intensive, the application of AI technologies and tools is seen as a huge boost in solving a variety of everyday problems.

2.5.5 Accurate Decision-Making

The oil and gas industry generates a massive amount of multistructured data due to the participation and contribution of multiple components. However, the conventional business analytics methods are not competent enough to supply accurate decisions. However, AI has proved decisive in decision-making with Big Data. AI also contributes well with streaming data emanating from different and distributed sensors in oil fields to emit out real-time insights. Unplanned downtimes cost a lot to offshore oil and gas platforms in the event of catastrophic asset failures. Thus, it is not an exaggeration to state that for achieving the expected success, AI is the way forward.

2.5.6 Improve Supply Chain and Logistics Efficiency

Typically, the supply chain for the oil and gas industry is complicated with several intertwined activities and decisions such as the crude oil purchase, the price for the purchase, the crude oil transportation to the refinery, activating the refining and gantry operations, and then initiating the retail sale of fuels. In the upstream business, AI helps in synchronizing the operations team with the warehouse team to ensure the timely availability of crucial parts.

Demand forecasting, price prediction for crude oil and finished products, planning, scheduling, and route optimization, etc. are activated through the distinct power of AI. There are other contributions such as setting up and sustaining a smart warehouse, inventory management, shipping operations management, risk hedging, timely delivery, and cost optimization through the growing power of AI algorithms and models in conjunction with historical and real-time datasets.

With the entire world trying to fulfil the sustainability goal, oil and gas companies have a big task at hand. They must strategize and plan decarbonization efforts with clarity and confidence. They must meticulously optimize their end-to-end operations to achieve production cut by gaining the full visibility into their direct and indirect emissions. AI-enabled data analytics is pertinent and paramount for identifying the sources and drivers of emissions, for reducing energy consumption and heat dissipation, etc. AI-based processing of all sorts of operational data gives important cues and clues to victoriously embark on their long and arduous journey of decarbonization goals.

2.5.7 Geoscience Data Analytics

Without an iota of doubt, oil and gas companies are associated with a large amount of seismic data sets. Seismic data are structured data (time-series) to simplify forecasting. The decision trees (DT) algorithm, one of the popular machine learning (ML) algorithms, can tackle this kind of data comfortably and to make sense out of data quickly and reliably. DTs could classify subsurface rock facies by analyzing oil well log data.

Big oil and gas companies put up and operate their plants in multiple locations across the globe. For insight-driven management, AI plays a vital role here. Data access, assessment, and leverage are important. There is a need for data aggregation to give a consolidated view to data scientists to proceed with their analytics activity. There are AI-centric automated tools extending a helping hand to oil and gas companies to digitize their records and automate data analysis.

The continued advancements in computer vision, a prime application area of deep learning, can squeeze out unparalleled insights out of unstructured data such as rock images. As experts point out, deep learning excels at unraveling useful patterns in unstructured datasets. Deep learning systems can finish tedious and time-consuming works quickly in reduced cost with high accuracy. Object detection using deep neural networks (DNNs) has proved useful in many ways. A DNN examines electron microscope images of rock slices to detect and measure these fractures.

Further on, a physics model for rock porosity got improved using deep learning. The model's accuracy was improved by an autoencoder, which is a special type of DNN to automatically select the decision-enabling features in images. The identified features are then fed into a deep learning model to get better accuracy.

AI is proving to be a gold mine for O&G exploration leaders, quite literally. For example, ExxonMobil aims to use the deep-sea AI robot to enhance its natural seep detection capabilities. ExxonMobil's AI-powered robots are capable enough to detect these oil seeps, which will eventually reduce exploration risk and subsequent lesser harm to marine life.

2.5.8 Predictive Models for Oil Field Development

Oil field development is a complex affair. Multiple decisions must be correctly taken for developing and sustaining oil fields and to arrive at an optimized workflow to cut costs. The other associated aspects include learning the optimum number of wells, best locations to drill, and efficient drilling sequences. Predictive models using machine and deep learning algorithms come handy in assisting in taking right decisions and subsequent actions. Reinforcement learning (RL) is an ML algorithm for learning how to meet up a goal while reducing cost. Deep RL, which uses DNNs, is used to create optimized field plans based on reservoir parameters, rock, and fluid properties.

Engineers run many reservoir simulations with structural and fluid properties that look like the real field. At each step, the RL agent randomly decides whether to drill or not. If it decides to, it chooses a location. Once a virtual well gets ready. the system then simulates a two-phase oil and water flow there. The agent then calculates the target value for the entire field. This process is repeated till the maximum value is obtained. More details can be found on this page (https://www. width.ai/post/machine-learning-in-oil-and-gas).

2.5.9 Predictive Analytics for Reservoir Engineering

Without an iota of doubt, the reservoir is the core part of oil and gas production. Due to its criticality and centrality, it must be continuously maintained and consistently optimized. AI is vital here in making sense out of all sorts of data getting emitted by reservoirs and their equipment.

Oil reservoir management is turning out to be a complicated thing because of the involvement of multiple factors, including seismic interpretations, geology, and oil production. AI-enabled prediction and optimization do a lot for the reservoir management aspects. Therefore, AI systems are being trained on the reservoir data to help in the field surveillance, reservoir engineering, and maintenance cost reduction. AI can be programmed for achieving a variety of desired outcomes (reducing greenhouse gas emissions sharply, modifying maintenance schedules on need basis, maximizing the life of an oil reserve, etc.)

Thus, AI-emitted insights are vital for the future of reservoirs. Recovery factor is a key metric for a reservoir. It is the total oil that can be recovered from a reservoir. Based on multiple independent variables, predictive analytics estimates the recovery value. This helps not to drill in an uneconomical location. For a data-led refinery, AI models can independently propose ways and means to make the most out of a refinery. Event AI can be specifically programmed to discover reserves.

Thus, machine learning can also bring in necessary optimizations to get enhanced oil recovery. Deep reinforcement learning (Deep RL) is being used to select optimal values for important parameters such as water injection rates to remarkably reduce costs and to insightfully increase profits. It involves a 2D reservoir simulator that simulates the two-phase flow of oil and water. At each step, the deep RL agent chooses a different water injection rate to infer the simulator's behavior. Then, it calculates the net present value and changes the next step's injection rate with the intention of obtaining a better net present value.

AI can identify reserves and provide a good estimate about the ease with which the extraction process can be done. AI software can make recommendations on the modifications to be made in operating conditions to succulently increase safety for assets. Also, AI comes handy in prolonging the lifetime of assets, thereby the total cost of ownership decreases significantly. The asset utilization efficiency is also being enhanced using industrial AI. Thus, AI is primed for any data-centric domain. Several advantages are being derived out of AI competencies. In reducing capital, management, operational, and maintenance costs, AI plays a vital role. As the oil and gas industry is capital-intensive, the contributions of AI are being well-applauded.

In summary, the paradigm of AI is proclaimed as the game-changer for technologically strengthening and solidifying the oil and gas industry.

2.5.10 AI for Oil and Gas Production

Machine and deep learning methods contribute better for oil production. There are multiple challenges and concerns being expressed and elegantly addressed through the state-of-the-art AI methods. As the oil and gas industry is lucrative, advanced and accomplished technologies are being lustrously used to bring in required acceleration and automation. In this section, we are to deal with various manual or semi-automated activities and how they can be automated and orchestrated to completely automate complex business processes.

One key problem associated with production is flow metering. But hardware flow meters are expensive, and hence the companies have turned toward leveraging machine learning-enabled virtual flow meters. This can estimate flow rates just from pressure, temperature, and choke data.

There is a research paper. (https://www.sciencedirect.com/science/article/pii/S2405896318307067) proposing an ML approach for virtual flow metering and forecasting. The authors have demonstrated that deep neural networks (DNNs) are good at learning how things can change with time. Also, it is proven that DNNs are good at forecasting flow rates with high accuracy. Generative adversarial network (GAN) is another AI method capable of predicting the production rate of an EOR operation. The GAN involves two ML systems. One tries to generate better data, while the other one tries to get better at predicting the generated data.

We have read and heard a lot about the use case of predictive maintenance. With the flourishing of AI techniques, preventive maintenance has advanced to predictive maintenance. This tells which asset needs some repair and rest. Thus, the significance of AI is growing rapidly. There are proven and potential AI algorithms for anomaly detection, failure prediction, root cause analysis (RCA), correlation analytics, etc. Recommender systems (https://www.width.ai/post/recommender-systems-recommendation-systems) can recommend which equipment and spars to buy to prevent downtime.

2.5.11 AI in Midstream

AI has become the dominant technology for automating several manual activities across the streams (down, middle, and up). Pipelines' predictive maintenance and transport optimization are being seen as viable and verified AI-driven use cases in the midstream operations. For example, machine learning can predict ships' propulsion power and estimate their efficiency. A ship's power is an indicator of its efficiency and emissions. There are multiple factors/features such as ship speeds and wind speed to infer propulsion power.

2.5.12 AI in Downstream

Oil prices are oscillating due to assorted reasons. But, data scientists can help in forecasting oil prices. Usually, regression algorithms are good at forecasting something useful. For the oil price forecasting by evaluating the oil market, nonlinear regression models are preferred. The forecasts are being made based on historical and current data. The key independent variables for accurate forecasting include stock indices, currency exchange rates, past fuel prices, and future contracts. From this data collection, a nonlinear model using the proven support vector machine (SVM) algorithm is generated to predict oil prices.

Another downstream problem is scheduling a refinery's output. Crude oil quality normally varies daily. Therefore, you must plan the refinery's output accordingly to remain profitable. Here too, machine learning can help for improving the scheduling. Deep belief networks are graph neural networks (GNNs) that can model task scheduling accurately by automatically learning parameters from historical data. There is a research article titled as "Refinery scheduling with varying crude: a deep belief network classification and multi-model approach" talking about how a deep belief network accurately classifies the incoming crude oil's composition in real-time and suggests a suitable schedule for the day.

The industry, with a deeper understanding of the path-breaking abilities of AI tools, has already commenced its AI-first formula to be right and relevant to their constituents, customers, and consumers. As described above, computer vision (CV) and natural language processing (NLP) are the top two applications of AI. In the sections below, we will focus on how CV- and NLP-enabled systems contribute for the prominent tasks such as optimization, prediction, automation, and monitoring.

2.6 Computer Vision (CV)-Enabled Use Cases

Oil and gas companies must collect and unleash studies on the production of a variety of data before, during, and after drilling into the Earth. There are numerous analytical methods and platforms for facilitating anomaly detection and correlation insights. The advancements in data analytics through AI algorithms truly help in the target of data-driven insights and insight-driven decisions. The day-to-day operations are efficiently performed to enhance productivity. The insights extracted come handy for executives and experts to decide the best course of actions. The exploration and production problems are being insightfully addressed. Root and secondary causes can be proactively identified and leveraged to surmount any operational and management issues. Thus, AI is being perceived and presented as the cutting-edge technology to enhance the oil and gas industry altogether. Especially, computer vision (CV) is being looked up as one of the prime contributors for the AI success.

Computer vision is turning out to be an important application area of AI through a host of deep learning (DL) algorithms such as the convolutional neural network (CNN), which is a deep neural network (DNN). Networked embedded systems are being fit with cameras to capture images and videos. Through CNN architectures, it is possible for embedded devices to readily and rewardingly analyze captured image and video data to extract actionable insights in time. Thus, increasingly vision-enabled machines and devices are being across environments to perform some exquisite and extraordinary jobs.

A vision-based system can verify the product quality and provide useful insights of defects through visual data analytics. Thus, AI-powered defect detection solutions are flourishing everywhere. Computer vision (CV) contributes to production optimization to analyze seismic and subsurface data quickly. CV enables predictive maintenance of equipment, reservoir understanding, and modeling.

Further on, employees working in oil plants work under extreme temperatures and immense pressure. They are also sometimes exposed to toxic fumes. The manual processes of monitoring employees through camera feeds are found to be ineffective. The current solutions ensure that every employee is wearing personal protection equipment (PPE) at the point of entry. However, an AI-powered vision solution can immaculately monitor the entire work environment to ensure workers to comply with all the mandated safety measures. Any deviation or deficiency can be proactively pinpointed and avoided through the unique power of AI models.

In summary, CV is another application area of AI. There are many machine and deep learning algorithms and pre-trained models to facilitate the complicated concept of CV. Further on, there are AI model optimization and compression techniques such as pruning, quantization, knowledge transfer, and federated learning. Unsupervised machine learning algorithms are being used on aerial images to identify and map the extent of an oil spill. Unsupervised learning techniques like clustering are used to separate pixels that contain clean water from pixels that contain oil pollutants.

2.7   Natural Language Processing (NLP) Use Cases

NLP is a prominent application area of artificial intelligence. A growing number of machine and deep learning methods have emerged and evolving fast to simplify and streamline NLP use cases. There are a large of number of research contributions for the betterment of NLP. NLP systems can learn and understand textual content as we do. NLP methods are powerful enough to classify incident reports and extract useful data from reports. There are document classification and information extraction models, which can infer characteristics like risk type, incident type, and consequence type from text content. Speech recognition, text translation, summarization, etc. are the key contributions of NLP techniques. The research advancements in the NLP space are being explored and experimented in the O&G industry to bring in augmentation, acceleration, and automation capabilities.

The NLP-assisted virtual assistant can interact with customers and advise and sell a range of products, including lubricants, waxes, greases, and even petrol. In the recent past, there have been intelligent chatbots and conversational interfaces. Such an AI-backed facility ensures the highest customer satisfaction, reduces event/incident reduction, proactively understands and offers premium services, and detects frauds and fake products pre-emptively. With 5G communication, edge computing, and AI flourishing together, newer metaverse applications are being rolled out for empowering employees and enterprises. There are other noteworthy accomplishments such as human–machine interactions (HMIs) and machine-to-machine (M2M) communication.

2.8   Robots in the Oil and Gas Industry

People working on the oilfields are liable for several risks as they must handle non-covered rotary equipment. Their work environment involves high pressure and high-temperature operations. Robots (https://www.getac.com/intl/industries/oil-gas/how-ai-is-changing-the-use-of-robotics-in-the-energy-sector/) are increasingly empowered through AI models to be hugely beneficial across industry verticals. With AI in place, robots are becoming autonomous machines. Such AI-powered robots can perform context-aware physical activities and are readily replacing humans in risky and rough places. They do drills and contribute to disposing nuclear power wastes. The AI-enhanced robots are penetrating other mission-critical sectors. Additional features, functionalities, and facilities are being accomplished through the leverage of AI-enabled robots.

AI-enabled mobile robots are extremely beneficial in offshore oil and gas in deep waters. They also help in finding hydrocarbons efficiently. Similarly, fixed-location robots are casually found in manufacturing floors. Especially, car and vehicle manufacturing spots are filled up with such robots. The AI-enabled Iron Roughneck (https://www.nov.com/products/iron-roughnecks) is designed for every onshore and offshore drilling application. This includes other capabilities and options. This robot minimizes rig floor hazards and streamlines the make-up and break-out process. Lumada video insights transform video into insights, combined with the Internet of things (IoT) and business data to gain comprehensive intelligence for exploration operations (https://www.hitachivantara.com/en-us/products/industrial-data-iot-management/video-intelligence/lumada-video-insights.html). This helps in optimizing the performance of its rigs.

Such AI-enabled robot systems crunch the data of past drilling operations by using machine and deep learning algorithms and models to create a digital twin for the drilling site. The drilling machines are attached with a variety of sensors and actuators. All the data are collected and transmitted to the digital twin to understand what to do next. The benefits of automated drilling by robots are many, including faster and precise drilling operations, heightened safety due to unmanned operations, and minimal costs for installation and management.

ExxonMobil's AI-powered robots can detect oil seeps to reduce exploration risks. Several established companies, having understood the unique functionalities, are keen on using multifaceted and feature-rich robots to reduce manual activities and to assist humans with all the perfection and precision. The offshore oil and gas production in future is to be activated and accomplished through unmanned platforms, which are to be operated, inspected, and maintained collectively by multiple AI-enhanced robots. AI-enhanced robots improve oil exploration, production efficiency, etc. And, as I mentioned in the beginning, with a deeper and decisive understanding, companies across the world are realizing and reaping the originally expressed benefits of AI, which is being consciously proclaimed as the savior for many of the industry's pressing and perpetual problems. AI can shepherd decision-makers in safeguarding oil plants. AI can exquisitely unlock efficiencies by articulating and accentuating scores of innovative ways toward simplified, secure, and streamlined exploration, development, production, marketing, etc. AI has the intrinsic potential to accelerate, automate, and augment processes associated with the oil and gas industry.

2.9 Drones in the Oil and Gas Industry

Drones or unmanned flying machines have entered the oil and gas industry. These are primarily used for asset inspection, as a flexible and cost-effective way to conduct inspections. AI-enabled drones speed up inspections and investigations to throw actionable insights. Drones ensure the safety of field workers, check whether there is any damage in pipelines, monitor mission-critical assets continuously, and reduce the operational costs drastically. Thus, drones automate several processes associated with the oil field operations.

Drones are also utilized for indoor and confined space inspections. They guarantee increased safety for human resources and industry assets, come out with predictive insights to reduce equipment downtime, save time and costs, enhance productivity, etc. Drones are fabulous and famous for asset inspection, remote monitoring, detecting leaks, emergency response, handling and moving materials, providing 360° view of objects under surveillance, etc. Further on, drone-captured real-time pictures and videos can be subjected through a host of video analytics methods to detect and track the development of oil spills and fires. Real-time image, audio, and video analytics can also assist in mapping oil spills or fire incidents. Drones aid in predictive maintenance of important equipment used in oil fields and plants.

As the nations across the globe are strategizing and executing a variety of actions to drastically minimize the risks associated with climate change, greenhouse emission, and global warming, oil and gas companies are exploring the possibility of using drones to reduce methane emission from operations. Drones come handy in minimizing any damage to people and properties in the event of an industrial accident or a natural disaster.

2.9.1 Drones in Upstream Activities

Drones contribute to the identification of potential geographical sites, drilling and exploration, and production of crude oil and gas. Drones automate such time-consuming, error-prone, and manual processes.

2.9.2 Drones in Midstream Activities

Here, crude oil and gas procured through the upstream activities is transferred through pipelines, barges, or oil tankers for the downstream process of refining and storage. Drones monitor and ensure there is no seepage and pilferage.

2.9.3 Drones in Downstream Activities

Herein, procured crude oil and raw natural gas are refined to create required end products like petrol, jet fuel, heating oil, natural gas, LPG, and other petrochemicals. Drones monitor refineries and various industrial assets to ensure their continuous and accurate functioning, thereby any kind of untoward incident can be pre-emptively nullified.

We have incorporated a separate chapter on leveraging AI-enabled robots and drones to accelerate and automate different tasks connected with oil exploration, production, transmission, transformation, and selling.

2.10   AI Applications for the Oil and Gas (O&G) Industry

With the widespread acceptance and adoption of the Internet of Things (IoT) technologies and tools, ordinary items become extraordinary and dumb objects become dynamic. Further on, there are so many path-breaking edge technologies to transform any concrete asset into a digitized artifact. Thus, machineries and tools at oil plants are also getting empowered to be digitized and distinct in their power and contributions. Now, such enabled objects generate a lot of digital data. With the growing power of AI algorithms, it is possible to extract hidden patterns and insights out of digital data. That is, the aspect of digital intelligence is gaining momentum. AI directly assists in knowledge discovery out of data mountains. Such knowledge, if applied properly, can result in a variety of optimizations and savings. There are several case studies and blogs, illustrating how companies earn and yearn a lot by applying AI models on IoT data.

2.10.1   Optimizing Production and Scheduling

The cost and schedule overruns are perennial problems for offshore oil projects. This is due to weather variations, resource shortage, and incorrect schedules. The production complexity increases with many intertwined activities during oil field development. Thus, intelligent planning and scheduling need to be derived and put into action. Both cloud and edge AI platforms can analyze incoming data comprehensively and quickly to emit out actionable insights. Logs, metrics, and traces are being meticulously analyzed to identify any deficiency and deviation. Root causes are being found and articulated. Correlation analytics through machine learning algorithms throws a lot of useful information for executives and engineers to do course correction if needed. The life of expensive machineries can be increased through right and relevant AI-inspired analytics.

2.10.2   Asset Tracking and Maintenance Through AI-enabled Digital Twins (DT)

Digital twins turn out to be an important ingredient for adroitly producing and managing oil field machineries and processes. Assets deployed in oil plants are being minutely monitored, measured, managed, and maintained through the power of AI-enabled digital twins. DT predicts whether its physical twin needs rest and repair. Thus, the lifetime of mission-critical assets gets solidly increased.

After extraction, oil and gas are stored in a central repository. From there, it gets dispersed through pipelines, which must be guarded against any kind of corrosion, damage, and debilitation. By integrating AI solutions, the industry can detect early signs of any such damage by analyzing the data recorded for various parameters. In association with knowledge graphs and predictive analytics, any degrading thing can be proactively found and addressed in time. Therefore, there is no slowdown and breakdown. The brand value can be preserved.

AI is emerging as a real enabler for the oil and gas sector, offering a bevy of advancements. By employing computer vision to decisively analyze seismic and subsurface data, it is possible to achieve the much-needed production optimization. Through predictive insights, plant and equipment downtime can be avoided. AI helps have a better understanding about reservoir and its resources. Through appropriate modeling, it is possible to predict oil corrosion risks, and this capability, in turn, aids in reducing maintenance costs.

2.10.3 AI-led Cybersecurity

There would be hundreds, even thousands, of IoT devices, equipment, appliances, wares, and instruments in an oil plant. Due to their connectedness, edge devices are bound to be attacked through external resources. With our everyday environments being stuffed with networked-embedded systems, there is a huge rise in cyberattacks. AI-backed cybersecurity solutions are highly demanded to ensure the tightest security for all the mission-critical assets. Security-related data get captured and crunched using appropriate AI models to pre-emptively tackle cyberterrorism nuisances.

Leveraging an AI-based solution can alleviate some of the challenges in O&G procurement by helping firms in understanding major procurement spend categories; automate purchase-to-pay; identify critical and noncritical supply chain bottlenecks; and gain visibility into planned and actual figures by the supplier, material, geography, and other company-specific dimensions, to name a few.

Undoubtedly, AI is a promising technology slated to play a key role in the future of the oil and gas sector. Sensor-rich assets generate a lot. By deftly applying the sophisticated capabilities of AI models on the captured asset data, it is possible to transition the raw asset data into actionable insights. Oil companies can bring in numerous sophistications in their everyday decisions, deals, and deeds. Oil companies can bring in premium and ground-breaking services through AI-enabled data processing. Business, technical, and user advantages can be obtained through the immaculate use of digital technologies.

The oil and gas environment is highly dangerous for field workers, who are constantly exposed to extreme temperatures and fumes emanating from it. By incorporating and leveraging distinct AI powers, site and spot monitoring is being automated to ensure the highest safety for people. AI-embedded CCTV cameras can autonomously do the job with all the perfection and precisions.

2.11 Better Decision-Making Using AI

As described above, the oil and gas (O&G) industry is one of the critical industries for the world. It supplies fuel for electricity production and transportation. However, there are several challenges getting related to the oil and gas exploration and production. Therefore, pioneering digital technologies are being increasingly used to overcome the identified as well as unidentified challenges. One noteworthy point here is that this industry generates a massive amount of multistructured data. Thus, the aspect of data-driven insights and insight-driven activation and accomplishment is gaining momentum. That is, to automate, accelerate, and augment downstream, midstream, and upstream processes, innovative technologies in conjunction with state-of-the-art infrastructures are being cognitively used. This section elaborates on the leverage of AI algorithms.

We all would have noticed that frequent changes of crude oil prices have resulted in fuel price variations. Not only fuel prices but also food, electricity, and other consumer goods prices change indiscriminately. For meeting the continuously growing demand for fuel supplies, geology specialists and experts meticulously collect and process satellite images to identify rock type and any extra formation and structure that could potentially give oil or gas. With this macrolevel detection of features of interest, there is a deeper and detailed mapping using energy or seismic waves. These waves are reflected off at interfaces where rock type properties change.

Similarly, if there are nearby wells and their log data are available, geologists, then, can utilize the data to build geological models. Geophysical sensors, which were lowered down in the drill hole, generate a lot of useful data to strengthen the geological models. All these come handy for geologists to conjecture the possible location of oil and gas reservoirs. Thus, oil and gas exploration and production are beset with numerous challenges. There are well-intended strategies and solutions articulated by researchers and experts for confidently approaching the persisting problems. Now with the flourishing of AI algorithms (machine and deep learning), the oil and gas industry is poised to see new twists and transformations.

There are open-source and commercial-grade platforms (AutoML) for creating machine learning (ML) models in an automated manner. H_2O Driverless AI (https://h2o.ai/platform/ai-cloud/make/h2o-driverless-ai/) is one such automated platform empowering data scientists to accomplish key machine learning tasks in just minutes or hours. As per the platform page, this AutoML platform delivers automatic feature engineering, model validation, model tuning, and model selection and deployment. In addition, the distinct capability of machine learning interpretability is being provided by this platform. There are other relevant functionalities for simplifying and streamlining the workloads of data scientists. This Driverless AI (DAI) platform is being leveraged for multiple use cases across industry verticals. There are blogs and tutorials explaining the steps to be done for generating competent ML models from input data. With the surge in AI model optimization and compression techniques and frameworks, models are subsequently made lightweight without sacrificing the much-insisted prediction accuracy.

In short, AI is turning out to be a silver bullet for various industries. The oil and gas industry is also immensely benefiting out of all the noteworthy advancements in the AI space. AI algorithms help learn from data to arrive at a highly accurate prediction. That is, the transition of raw data into usable knowledge is the essence of the AI paradigm. Machine and deep learning algorithms fulfil the vision and mission of the AI technology. Machine learning empowers machines to find patterns in data. Identified patterns can make viable and venerable predictions. With more data and training, the decisions and conclusions being supplied by AI models can become more accurate.

Machine learning is being increasingly and insightfully used in the oil and gas industry. AI experts have articulated that a multitude of processes are getting accelerated and automated through the power of AI algorithms. In this section, we will discuss some of the advancements being brought in AI.

2.11.1 Predictive Maintenance

This is a common AI-inspired use case. Industrial assets ought to be closely monitored and managed to keep up their availability and reliability. Predictive maintenance is a vital requirement for that. AI models are being built and used to predict when an equipment or instrument needs some rest or repair. Currently, we are following the reactive approach. But with AI, the much-needed proactive and preemptive approach will become the new normal. By modeling sensor data, it is possible to pinpoint problematic equipment.

Machine learning can help companies make the transition to predictive maintenance by modeling sensor data to find problematic equipment. If any instrument is working outside its parameters, then the dataset will through an anomaly. This helps operators plunge into activation mode before the instrument is damaged.

2.11.2 Identifying Optimal Operating Condition

AI comes handy here. For example, consider steam injection. Steam is being injected into an oil reservoir to heat it. This makes the oil less viscous. Engineers must determine the flow rate and temperature of that steam. They must know which combination of steam flow rate and temperature bring the most oil to the surface. The optimal combination may change daily. This is a time-consuming and repetitive process. Through AI methods, it is possible to model the reservoir to calculate the optimal combination.

2.11.3 Well Logging

Oil wells produce a lot of log data, which are being captured by lowering sensors (measurement probes) to the wellbores of oil wells. These logs convey more about the surrounding environment such as the depth of the well and the amount of oil in the reservoir.

Consider a location, wherein no wells had been drilled before. Then, an exploratory well is used to see if there is oil or gas in the area. If the exploratory well is found to have oil or gas, then more wells are to be drilled. If an oil well is not producing any profit, then the operator may abandon the well. AI has the innate strength to predict the production potential of a well. Then, the loss for operators can be minimized. Also, the number of abandoned wells decreases significantly, thereby the amount of emission is low. Thus, AI is a savior for the fragile environment.

2.11.4 Detecting Contaminant Concentrations

Contaminants in oil wells pose a huge problem as they can plug the wells. This may lead to the slowdown in oil production. AI can help in identifying the root cause of this problem.

2.11.5 Ensuring the People and Property Safety

As widely known, oil wells and refineries are rough, risky, and hazardous. Thus, it is indispensable and uncompromisable in timely identification of any impending risks for workers and to mitigate it pre-emptively. AI algorithms and models minutely monitor video footages, see, and understand what is happening and then articulate what countermeasures ought to be taken to safeguard assets and people in time.

2.11.6 Energy Efficiency

There is a wastage of energy during oil and gas exploration and production activities. Several things associated with the production need expensive energy. With more energy consumption, the heat dissipation is also bound to increase, bringing in irreparable damage for our environment. Thus, energy efficiency is being insisted everywhere.

AI can help reduce energy wastage by pinpointing deficiencies. For example, the sources of energy waste, such as heat from an oil reservoir not being used in the oil treating process, can be identified through AI models. Such identification helps correct and control undue energy consumption.

2.11.7 Equipment Inspection

There are multiple equipment and electronics. Every asset must be inspected to ensure its perfect condition and functioning. However, it is neither straightforward nor simple. For example, the inspection of pipelines is time-consuming. An AI-inspired drone could inspect the pipeline in less time.

Going forward, oil or gas getting extracted through oil rigs is stored in a central repository and then gets distributed across pipelines. Due to varying temperature levels and environmental conditions, oil and gas may get degraded or corroded. AI solutions can prevent these by detecting signs/symptoms of corrosion by analyzing various parameters using knowledge graphs. AI assists oil companies in discovering new oil and gas reserves and enhances production from existing facilities.

2.12 Cloud AI vs. Edge AI for the Oil and Gas Industry

The oil and gas industry generates a lot of polystructured data. There are integrated AI platforms, frameworks, accelerators, libraries, toolkits, etc. for accurately and timely analytics of operational, log, performance, health condition, and security data for pinpointing the root cause for any problem, to detect anomalies/outliers, to facilitate failure detection with the intention of stopping its spread forthwith across system components, to avoid service degradation, to ensure business continuity through auto-remediation, etc. As widely known, cloud environments give an illusion of infinite capacities and capabilities. Due to the heaviness of AI models, cloud infrastructures and platforms are presented to be the appropriate one for training, refining, optimizing, deploying, observing, and maintaining AI models for a variety of business problems.

Fortunately, we have digital twins for all kinds of complex physical, mechanical, electrical, and electronics systems at the ground. Further on, the cloud environment boasts of large-scale data lakes for keeping and querying a massive amount of multistructured data. AI toolkits, accelerators, frameworks, packages, libraries, platforms, and other enablers are readily made available in cloud environments (private and public). Thus, AI model engineering, storage, and management are fairly accomplished through the unique power of cloud environments. Not only creating AI models but also consistently updating them to be on par with the changing business sentiments are being elegantly performed through cloud-based software solutions. As AI model generation is data- and process-intensive, software-defined cloud is the way forward. With the flourishing of cloud centers across the world, the centralized AI model creation and refinement are being recommended.

For AI model engineering, there are integrated platforms for automating and accelerating a variety of related tasks. For example, there are databases, data warehouses, and data lakes being made to run in cloud servers. In the recent past, we had heard about data lakehouse. Thus, data management and storage management concepts are maturing and stabilizing fast with the faster adoption and adaptation of the cloud idea. There are abundant data integration, visualization, and virtualization tools. Leading market watchers and analysts have estimated that more than 75 percent of enterprise applications will reside in cloud environments by the year 2025. AI-enablement libraries are already deployed in cloud servers to speed up the process of AI model building. Thus, data plus the dazzling array of AI algorithms are being seen as game-changing phenomena for the ensuing era of knowledge. Thus, the aspect of cloudification is a real game-changer for the AI world. Today, several business problems are being tackled through the innate strength of AI. Clouds are the most compatible and competent IT environments for creating and running AI models.

Let us move to edge AI now. We have seen the centralization, consolidation, compartmentalization (virtualization and containerization), sharing, and automation aspects of the flourishing cloud paradigm come handy in producing and executing AI models without any hitch and hurdle. The surge in the interest in embracing cloud-native computing and serverless computing models also contributes for the mesmerizing success of the emerging domain of cloud AI. Serverless computing brings in additional and deeper automation for some of the difficult activities of cloud operations. Thus, the much-celebrated cloud paradigm is in the fast tracking process. As most of the software and hardware systems are all set to be cloud-hosted and AI-driven, the concept of cloud AI is bound to flourish in the days to unfurl.

However, there are certain limitations with cloud AI. As AI is data-intensive, collecting and transmitting all kinds of edge data to online, off-premises, and on-demand cloud environments is to waste a lot of network bandwidth resources. Further on, there will be bigger network latency. Thereby, real-time data analytics, decision-making, and action become a tough assignment. Today, the standard procedure is to train and create AI models in clouds and then get deployed in connected edge devices.

IoT edge devices (alternatively termed as networked embedded systems) are plentifully installed in our personal, professional, and social environments. For example, homes in advanced countries are being stuffed with several consumer electronics, appliances, kitchen utensils, wares, Wi-Fi gateway, computing, communication, sensing and perception devices, digital assistants, etc. Similarly, there are hotels, hospitals, retail stores, manufacturing floors, warehouse, airports, railway stations, auditoriums, etc. stuffed with a growing array of multifaceted and eye-catching gadgets and gizmos. Robots and drones will become the new normal in mission-critical environments.

The IoT edge devices are gradually becoming resource-intensive. So, the era of edge computing is dawning upon us. That is, the computing moves to where the data reside. Edge devices and sensors collect a lot of useful data. Edge data explosion is exponentially growing. Every bit of data is paramount, pertinent, and productive. For any industry to march ahead in its long and arduous journey, every data, whether internally or externally sourced, must be meticulously gleaned and used for producing viable insights in time. Thus, edge environment data must be gathered with all the care to be subjected to a variety of deeper investigations to untangle hidden patterns.

Now with the concept of containerization, there is a new twist in clubbing together different and distributed edge devices to form *ad hoc*, dynamic, open, purpose-specific, and transient edge clouds. That is, the device heterogeneity is being tackled through the layer of containerization. Containers fulfil the long-term aim of device portability. Edge-native applications are being visualized and realized. There are several containerization-enablement and container orchestration platforms emerging and evolving fast to catch up with the fast-changing business and personal sentiments. With the ready availability of lightweight versions of the Kubernetes platform, the formation of edge device clouds is getting accelerated. With devices getting integrated with one another, the feasibility of performing complex operations through devices is becoming higher. That is, device clusters and clouds will become the casual and common thing across environments to do real-time data capture, cleansing, and crunching.

In short, the long-standing goal of edge data analytics is understanding something clearly at last. Such an analytical phenomenon is to fulfil real-time analytics, which throws real-time insights, which is the fundamental thing for envisaging real-time services and applications. Thus, edge computing and analytics have set the ball rolling for producing and sustaining real-time intelligent enterprises.

Finally, with the mesmerizing speed and sagacity with which the AI paradigm is metamorphizing, the days of edge AI is beckoning upon us. That is, AI models are being increasingly deployed in edge devices. Such a transformation has laid down a stimulating environment for visualizing a bevy of next-generation business-aware, process-centric, event-driven, service-oriented, cloud-hosted, insight-filled, mission-critical, and edge-native services and applications. As indicated above, the current practice is to derive AI models through public or private clouds and run them in edge devices and clusters. The problem with this approach is the lack of real-time self-learning. There is a communication cost involved when edge data travel to remote cloud platforms and databases to be analyzed purposefully. The network latency is seen as a barrier. The idea is to train and create AI models directly in edge devices by directly leveraging edge data. Such a thing is the future for the knowledge-driven world. Whenever there is a new edge data, edge devices swing into action to receive the newly captured and gathered data to optimize AI model inferences and decisions.

For the oil and gas industry, as noted above, there are several complicated equipment and gadgets. There are other important assets contributing for the success of the oil and industry sector. With the faster maturity of digitization and edge technologies, all kinds of assets are being methodically transformed into digital entities. Digital elements join in the mainstream computing, thereby physical and context-aware services are being designed and delivered. Oil companies, intermediaries, and consumers get immense benefits with the deployment of edge AI devices supporting on-device data processing capabilities. For real dynamism and proximate processing to extract actionable insights in time, edge AI is the way forward. There are AI model optimization techniques such as pruning, quantization, knowledge distillation, and federated learning that are gaining traction so that their usage becomes risk-free and rewarding.

2.13 AI Model Optimization Techniques

AI model engineering platforms are acquiring special significance across industry verticals. Due to the leverage of automated AI model building platforms, the number of features being used for model generation is numerous, and hence the resulting AI models are bulky. Therefore, AI researchers have come out with several powerful optimization techniques and tools to create lightweight AI models. This section explores a suite of pioneering methods for achieving AI model compression. Pruning and quantization techniques are being used in combination to reduce the size of complex AI model architectures and make them optimized and performant so that they can be easily deployed in the IoT edge devices.

In the fields of computer vision (CV) and natural language processing (NLP), deep neural networks (DNNs) are being used in innovative and impactful ways. However, the computational resources required for implementing DNNs (realized through machine and deep learning (ML/DL) algorithms) are on the higher side. Further on, the energy consumption of these AI models is also high, while the heat dissipated by them into our fragile environment is hugely damaging. Additionally, deploying such feature-rich DNNs on the Internet of Things (IoT) edge devices is beset with a number of technical challenges and concerns. Thus, AI models ought to be optimized to reduce the usage of huge computational resources. The optimization typically involves the identification and elimination of redundant and irrelevant features. Especially, there is a surge in edge devices, robots, drones, instruments, equipment, appliances, wares, machineries, and gadgets. There is a strong need for producing and deploying edge-native intelligent applications. Highly optimized AI models are therefore insisted for intrinsically empowering edge devices with all the data-driven insights.

The choice of AI algorithms used in edge devices is crucial as these devices lack large power sources. Selecting the right ML algorithm will help edge devices process copious amounts of data and significantly decrease the complexity of the network. These choices will cut down on parameters such as inference latency, power consumption, and memory use. This will also improve performance parameters of the chosen model. By doing so, we will reduce the number of redundant calculations in neural networks during the training phase. Quantization and pruning are the two common techniques using which ML algorithms can be made to work well in edge devices. Quantization involves converting data from floating point 32 bits to a less precise format, such as integer 8 bits, and then using integer 8 bits to do all the convolution operations. In the last stage, the output with lower precision is changed back to the output with higher precision in floating point 32. This process allows models to be represented in a compact manner and allows vectorized operations to run quickly on a wide range of hardware platforms.

Pruning is a technique used to remove unwanted and unnecessary computation from a network. It works in two ways: structured pruning and unstructured pruning (also known as weight pruning and unit pruning (of neurons)). The weight matrix is set to 0 in weight pruning or unstructured pruning. This allows the removal of connections. To obtain k percent sparsity, arrange the weights in order of their size and set them all to 0 to create a weight matrix. In unit pruning, the focused area will be made up of mapped neurons that are not needed or are not being used enough.

2.14 Conclusion

As the oil and gas industry becomes critical for the entire world, oil and gas companies are actively seeking innovative and transformative solutions to be efficient in their operations through simplification and rationalization initiatives and implementations. Optimization and automation across all the layers and levels have become the hallmarks for the corporates to be on the right side all the time. Technologically assisted and abled exploration for fresh avenues and revenues is picking up fast. Digitization and digitalization aspects have become the new normal for these organizations to proceed in fully fulfilling peoples' aspirations.

AI is being touted as the best-in-class technology to envisage a suite of sophistications. Thus, AI-enabled robots and drones are to play a vital role. Further on, all kinds of physical assets will have their own AI-enabled digital twins. With the better and deeper connectivity being guaranteed through 5G communication, we will have fully connected and intelligent oil rigs, pipelines, and fields. Downstream, midstream, and upstream processes and procedures are all set to be competitive and cognitive. As the nations around the world are batting for decarbonization, the oil and gas sector is being compelled to be disruptive through AI-like technologies and tools. This chapter has focused on AI applications exclusively for the oil and gas sector. So, many use cases are being explored, experimented, and explained by oil and gas companies and academic institutions.

Bibliography

1 Choubey, S. and Karmakar, G.P. (2021). Artificial intelligence techniques and their application in oil and gas industry. *Artificial Intelligence Review* 54 (5): 3665–3683.

2 Koroteev, D. and Tekic, Z. (2021). Artificial intelligence in oil and gas upstream: trends, challenges, and scenarios for the future. *Energy and AI* 3: 100041.

3 Sircar, A., Yadav, K., Rayavarapu, K. et al. (2021). Application of machine learning and artificial intelligence in the oil and gas industry. *Petroleum Research* 6: 379–391.

4 Nguyen, P.H.M. (2013). Contract lifecycle management on the sell-side: a case study in upstream oil and gas industry

5 Baaziz, A. and Quoniam, L. (2014). How to use big data technologies to optimize operations in upstream petroleum industry. arXiv preprint arXiv:1412.0755.

6 Ali, S.S., Nizamuddin, S., Abdulraheem, A. et al. (2013). Hydraulic unit prediction using support vector machine. *Journal of Petroleum Science and Engineering* 110: 243–252.

7 Agwu, O.E., Akpabio, J.U., Alabi, S.B., and Dosunmu, A. (2018). Artificial intelligence techniques and their applications in drilling fluid engineering: a review. *Journal of Petroleum Science and Engineering* 167: 300–315.

8 Anifowose, F., Labadin, J., and Abdulraheem, A. (2015). Improving the prediction of petroleum reservoir characterization with a stacked generalization ensemble model of support vector machines. *Applied Soft Computing Journal* 26: 483–496.

9 Aissani, N., Beldjilali, B., and Trentesaux, D. (2009). Dynamic scheduling of maintenance tasks in the petroleum industry: a reinforcement approach. *Engineering Applications of Artificial Intelligence* 22 (7): 1089–1103. https://doi.org/10.1016/j.engappai.2009.01.014.

10 Wanasinghe, T.R., Wroblewski, L., Petersen, B.K. et al. (2020). Digital twin for the oil and gas industry: overview, research trends, opportunities, and challenges. *IEEE Access 8*: 104175–104197.

11 Cooper, J., Dubey, L., and Hawkes, A. (2022). Methane detection and quantification in the upstream oil and gas sector: the role of satellites in emissions detection, reconciling and reporting. *Environmental Science: Atmospheres* 2 (1): 9–23.

3

Artificial Intelligence for Sophisticated Applications in the Oil and Gas Industry

3.1 Introduction

Artificial intelligence (AI) [1] can be defined as the ability of a machine to mimic the intelligence possessed by humans to categorize and identify things based on which processes are carried out. AI is the future of technology. AI working in mysterious ways is one perception for the non-technical people. But it all boils down to the training data. The training data should be large and perfectly labeled, which is the scope of covering all scenarios, including the exception cases as well. This leads to a perceptive AI.

The oil and gas industry is one of the profitable energy sectors in the world. They are a major influential player in the global economy. Since it also falls under the non-renewable energy resource, care must be taken in maximizing the resources as well as care for it. Implementing AI in the oil and gas industry can deliver both with minimal wastage. The oil and gas industry has so many facets. AI can aid the industry right from exploration of oil and gas to the delivery of the refined resources to the consumers. Even the delivering method of the resources can be taken care of by AI.

When an infrastructure as big as such an industry is enhanced with the help of AI, there are so many advantages that can come out of this model. This could seriously reduce the risks involved in this energy industry. The governance and risk management of the oil and gas industry will also be enhanced. This would lead to economic growth as well as management of the resources.

The Power of Artificial Intelligence for the Next-Generation Oil and Gas Industry: Envisaging AI-Inspired Intelligent Energy Systems and Environments, First Edition. Pethuru Raj Chelliah, Venkatraman Jayasankar, Mats Agerstam, B. Sundaravadivazhagan, and Robin Cyriac.

3.2 Oil and Gas Industry

The oil and gas industry [2] is one of the most eminent sectors in the world. Usage of oil and gas over the years has left its toll on the planet. Hence, it is imperative to take care of the remaining resources and to utilize it properly with maintenance and governance. The processes of the oil and gas industry is divided into three main segments, as given below:

- **Upstream**: Production and Exploration
- **Midstream**: Transportation
- **Downstream**: Refining and Marketing

These segments take care of the whole process of exploring oil and gas raw resources to delivering it to the consumers. The segments are classified based on certain processes involved in the oil and gas industry. This is the standard that is followed in the industry to take care of the phases involved. Understanding the processes that are involved in this segment can help automate the processes with AI. A clear picture of all the processes and actors involved would be beneficial for the smooth transition in the scenario of AI being replaced in one of these three or all segments.

3.2.1 Upstream: Production and Exploration

This phase is the quest in identifying the crude oil and natural gas fields. Once the identification is completed, then comes a series of activities such as obtaining legal permissions and production activities including onshore and offshore drilling. This phase deals with a lot of surveys and research for the hunt of availability of resources. Once the identification is done, then the production stage starts. The production phase is where the companies begin drilling. Smart monitoring of the drilling data such as duration and depth can be useful in providing efficiency and performance metrics. Based on these metrics, the production stage might be adjusted to achieve efficiency.

3.2.2 Midstream: Transportation

The next phase is the transportation, which entails storage as well. In this phase, there are three processes that are taken care are of, as given below

- Transportation
- Storage
- Trading

Transportation in this stage is usually done by tankers or pipelines if the resource is unrefined. Trading also happens between companies for this crude oil.

Table 1 Industry codes and its description for oil and gas.

Code	Description
2212	Natural gas distribution
221210	Natural gas distribution
237120	Oil and gas pipeline and related structures construction
484220	Specialized freight (except used goods) trucking, local
484230	Specialized freight (except used goods) trucking, long-distance
486	Pipeline transportation
4861	Pipeline transportation of crude oil
486110	Pipeline transportation of crude oil
4862	Pipeline transportation of natural gas
486210	Pipeline transportation of natural gas
4869	Other pipeline transportation
486910	Pipeline transportation of refined petroleum products
486990	All other pipeline transportation
493190	Other warehousing and storage (bulk petroleum storage)

So, the midstream phase involves the processes of these three stages. There are industry codes that are associated with this segment. It is known as the North American Industry Classification System. This is used for the purpose of collecting, analyzing, and publishing statistical data related to the U.S. business economy. There are specific codes that are contributing to all the sectors in the world. The oil and gas industry has some codes that are often used, which are listed in the Table 1.

These codes can help with smart tracking and when AI is integrated within the system. The outcome of this automation would be smooth and flexible.

3.2.3 Downstream: Refining and Marketing

The final segment of this industry is the downstream phase. This segment deals with refining and marketing of the resources. Though the goal is very simple as to purifying the oil, but the process involved with refining is complex. There are a number of processes that are involved in refining depending on the end product. The marketing strategy is also important in this segment. Though the world runs on fuel, there are so many political as well as economical agendas that can be carried out with the help of this energy resources. The more efficient the industry is in exploring and delivering the resources, the profits would see an increase with keeping the overheads to a minimum. AI can assist a lot in this segment, thereby keeping efficient track of data, and based on this, data strategies can be drawn up on how to proceed for the sales of the resources.

3.3 Artificial Intelligence

The perception of a machine to synthesize and infer information is known as AI. AI has its roots in machine learning. The IT industry grows and innovates every day, thereby boosting AI in almost all the fields. AI can be used as an aid in almost every field in the world. The oil and gas industry is no exception to it. If implemented properly, AI can take care of the oil and gas industry as well as the nonrenewable resources, which has to be left alone to replenish itself.

Artificial intelligence mainly has four types that are listed as below:

- Purely Reactive
- Limited Memory
- Theory of Mind
- Self-Aware

Each type of AI deals with a certain type of perception of data. There are two ways to achieve the implementation of AI, which are listed as below:

- Machine Learning
- Deep Learning

Both of these ways help the machines in determining the output based on the inputs that are given for evaluation [3]. The models that are chosen to train the AI in these ways are also important. There are various models that are involved in training based on what type of AI is required for the project. But once the machine learns and understands the concepts, the evolution that would follow on its own would be massive.

3.4 Lifecycle of Oil and Gas Industry

There is a lifecycle involved in the oil and gas industry. In order to survey the best adaptations of AI in the oil and gas industry, one must understand the lifecycle of the oil and gas industry. The lifecycle of the oil and gas industry [4] is described in Figure 1. It describes the phases and time taken for each phase to be conducted.

3.4.1 Exploration

Exploration phase in the oil and gas industry takes care of the land acquisition and procuring license necessary for the following procedures. Substantial data are monitored and used to do exploratory drilling as well as to identify which would be the best location for mining the resources.

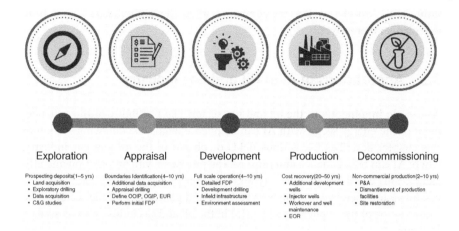

Exploration	Appraisal	Development	Production	Decommissioning
Prospecting deposits(1–5 yrs)	Boundaries Identification(4–10 yrs)	Full scale operation(4–10 yrs)	Cost recovery(20–50 yrs)	Non-commercial production(2–10 yrs)
• Land acquisition	• Additional data acquisition	• Detailed FDP	• Additional development	• P&A
• Exploratory drilling	• Appraisal drilling	• Development drilling	wells	• Dismantlement of production
• Data acquisition	• Define OOIP, OGIP, EUR	• Infield infrastructure	• Injector wells	facilities
• C&G studies	• Perform initial FDP	• Environment assessment	• Workover and well	• Site restoration
			maintenance	
			• EOR	

Figure 1 Lifecycle of the oil and gas industry.

3.4.2 Appraisal

The appraisal phase deals with boundary identification. This phase also deals with the viability of the land and acquisitions of data, which goes with the drilling of oil and procurement of the necessary materials and licenses.

3.4.3 Development

The development phase is a full-scale operation which ranges from drilling and building infrastructure to sustain the mining of oil and gas. Infrastructure planning and building is taken care for long-term support.

3.4.4 Production

The production phase involved the long span of drilling and extracting the resources. The production phase deals with adding additional wells. Enhanced oil recovery is taken care of in this phase. Maintenance of the oil wells is pertinent in this phase as well.

3.4.5 Decommissioning

This phase is a noncommercial phase, which deals with post production of oil and gas mining. Restoration is done after the extraction in this phase. The infrastructure that was built for the extraction of oil and gas will be de-structured in this phase. After this, the restoration process of the site would be started.

The oil and gas industry has a moral obligation to take care of the land resources. Hence, the restoration is a vital part of the entire process. This safeguards the future of the world as well as the companies allow the natural resources apt time to replenish themselves.

3.5 Applications of AI in Oil and Gas industry

AI can aid imminently in the oil and gas sector. The possibility of the enhancement of the energy sector with AI can be phenomenal. The following section of this chapter describes the various fields or phases in the oil and gas industry, where AI can help enhance and improve the processes with automation and efficiency. This can boost productivity and turnover of the most economically successful sector, thereby boosting both revenue and governance of the sector in the future. Some of the applications of AI[5] in the oil and gas industry are described in Figure 2.

For ease of understanding, the applications of AI in the oil and gas industry are categorized according to the lifecycle of the industry and the categorization as furnished in Table 2. This may guide in implementation of the applications in the respective departments to enhance the processes.

3.6 Chatbots

Bots are automated processes that carry out a specific task at a scheduled time or after a trigger process. Chatbots are bots that are tailored to a specific list of queries and answers. Based on the selection of answers by the users, one is taken further down a stream to fulfil the requirement of the user or provide the

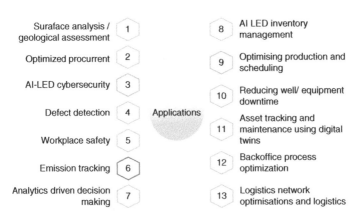

Figure 2 AI based applications in the oil and gas industry.

Table 2 Lifecycle phase-based AI applications.

S. no	Lifecycle stage	Possible AI applications
1	Exploration	Chatbots
2	Appraisal	Optimized procurement
		Digital twin
3	Development	Drilling
		Production
		Reservoir management
		Emission tracking
4	Production	Process excellence and Automation
		Inventory management
		Well monitoring
		Optimizing production and scheduling
		Logistics network optimizations
5	Decommissioning	Asset tracking and maintenance
		Reducing well and downtime

necessary support in order to resolve the issues. Chatbots are the new wave in the present technology. As time progresses, the chatbot is also getting enhanced to automate almost all the queries. On the off chance that it does not get resolved with the set of queries, an agent can be connected to assist further. Previously, the chatbot was able to only follow a set of questions and answers. As time progressed, real live chatbot were created, which can perceive and infer accordingly, as described in Figure 3. One such example is the Google's AI assistant that can make real phone calls to place order and enquire about anything on behalf of the user.

Exploration and production firms, sometimes referred to as upstream oil and gas corporations, locate deposits, drill wells, and collect raw materials from the earth. AI chatbots assist in addressing the unique requirements of upstream oil and gas enterprises. A few of the use cases are as follows: offer workers on-demand support both in the office and on the job. AI chatbots enable workers to ask inquiries, access real-time information, and call for assistance in an emergency quickly and without having to interrupt what they were doing previously. To swiftly obtain information such as lubricant specs, average reservoir porosity, and specifics regarding defects, the chatbot establishes connections to both internal and external sources of the business.

Additionally, as operators move from one location to another, voice-enabled bots can respond to queries and give status updates. It can also provide onsite

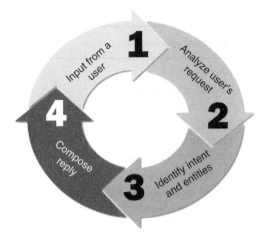

Figure 3 Input of the user to AI Chatbot.

employees with real-time solutions workers who are currently performing maintenance on the field can provide unstructured input to an AI chatbot. It may then examine these problems and provide immediate fixes based on a vast knowledge and insight repository. Reducing new hires' learning curves, AI chatbots retain a large database of process knowledge, which can be helpful to new hires. Additionally, they can aid in shortening the time needed for data collection, interpretation, and simulation. The bot may use the business' software applications to assist staff in building type curves, finding the best well spacing, and prioritizing infill drilling areas.

AI-based chatbots [5] can avoid possibly expensive problems. The oil and gas industry are vulnerable to a number of upstream challenges, including pipeline and drilling concerns, oil spills, and health and safety risks. For instance, a production line flaw might result in significant losses and the revenue loss of millions. When deployed effectively, AI chatbots can alert people to situations that could have disastrous results and offer suitable answers. Activities that take place after crude oil and natural gas are produced are referred to as downstream oil and gas production. O&G players further upstream are located closer to the customer. AI chatbots can provide product information, respond to client inquiries, and provide specialized product recommendations, thanks to their proficiency with natural language. Most questions about customer service are straightforward and frequently asked. AI chatbots assist in deflecting questions and are made to handle an overabundance of customer service agent demand. This lightens their workload and gives them more time to concentrate on activities that offer value to the company. AI chatbots can open up new sales channels. They are able to comprehend customer wants and present them with the ideal products. Additionally, chatbots with RPA integration can automate the entire procedure, from taking down customer information in the CRM system to concluding the sale.

3.7 Optimized Procurement

Building interconnected digital supply networks (DSNs) with the use of AI-driven customized procurement processes can help Oil and Gas companies be more dynamic, adaptable, and effective in their planning and execution. AI can offer new insights [6] from data processing and analysis of exceedingly complicated and big volumes of data to procurement of specialists' decision-making powers to address classic issues.

By assisting businesses by understanding key procurement categories, automating purchasing, identifying critical supply chain and non-critical supply chain bottlenecks, and gaining traction in the planned and actual figures by the supplier, material, geography, and other company-specific dimensions have been carried out. Leveraging an AI-based solution can help alleviate some of the current challenges in Oil and Gas procurement, as described in Figure 4.

The application of AI in the oil and gas industry's supply chain is a topic that various oil and gas corporations are investigating. These areas include prediction of crude oil and finished goods market prices: This aids in making pricing contract decisions and optimization of the crude basket, logistics, warehouse,

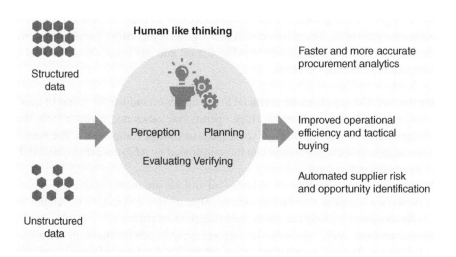

Figure 4 AI based optimized procurement.

inventory, and shipping processes. This assists in ensuring that the best type of crude oil is chosen, supplied, and handled properly.

Risk management investments are appropriate to mitigate the risks associated with changing supply, demand, and prices, boat tracking the deliveries to manage the demand from end consumers and the warehouses, and planning and time management: Planning and scheduling enable the business to make better use of its assets and resources and ensure that stocks are available to complete the orders quickly.

Robotic process automation [7] is being implemented, which will have a significant impact on corporate processes in terms of productivity, efficiency, and accuracy. Robots will replace repetitive tasks that need a lot of human labor.

The oil and gas industry has a lot of potential for using AI in its supply chain processes. Demand forecasting is made easier by predictive capabilities: When the inventory does not stock in-demand materials, businesses lose money. Network planning and predicting demand are becoming more autonomous thanks to AI, that allows merchandisers to be more proactive rather than a passive participant. Knowing what to anticipate allows the merchandisers to modify the number of products required and guide them to areas where the most demand is anticipated. Lower operational costs are the result of this. Customer service is being redefined by chatbots. A bot can manage 80% of all consumer interactions. AI has the power to customize interactions between clients and logistics companies. Intelligent warehouses are more effective: a completely automated facility called a "smart warehouse" completes the majority of work using software or automation. Logics can be used to schedule deliveries such that no deadlines are missed.

Delivery times and costs are getting better thanks to algorithms: In the perspective of logistics, every mile and every minute count. In order to plan the shortest routes for deliveries, businesses can utilize a generic algorithm for planning the routes. Each function supported by AI in the oil and gas value chain involves a component of the supply chain.

Upstream: The supply chain is crucial to the upstream industry in terms of moving materials and equipment. Huge operational losses might result from the slight delay or unavailability of spare parts or critical components. The operations crew and the warehouse can be coordinated by AI to ensure the availability of vital components.

Midstream: In the midstream industry, oil and gas are mostly stored and transported via a variety of vehicles, such as ships, trucks, and rail. Here, AI aids in effective planning and execution, selecting the best path, etc.

Downstream: With the downstream sector, AI aids refiners in blending strategies, demand forecasting, price estimation, and customer relationship management.

3.8 Drilling, Production, and Reservoir Management

AI can be used by the oil and gas industry to mine and drill for raw hydrocarbons and other items that are used to create fuel. AI will assist by creating algorithms that provide precise and accurate intelligence that directs both on- and off-shore drilling operations [8]. Precision drilling will increase the penetration rate while lowering the possibility of oil leaks, mishaps, and fires. By locating inefficient locations, AI aids oil and gas businesses in increasing production [9]. Large amounts of unprocessed production data can be used to train AI systems, allowing them to automatically recognize patterns and refine the data to produce analytics.

Another crucial aspect of the oil and gas sector is reservoir management [10]. To increase the longevity of the production cycles, oil and gas reservoirs should be continuously maintained and controlled using geology, reservoir engineering, seismic interpretation, and specific production technology. ML is being trained with short-term data to predict long-term production as per Figure 5.

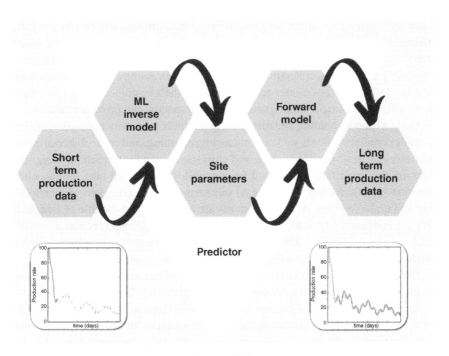

Figure 5   ML-based long-term production prediction.

The industry previously required a wide range of professionals, but with the most recent technology, prices can be decreased and a single solution can be used. AI collects historical information that can be utilized to model, categorize, and help locate new and efficient drilling sites.

3.9 Inventory Management

The oil and gas business benefits from the addition of enterprise resource planning (ERP) and the optimization of inventory [11], logistics, and warehouse management through the use of AI, machine learning, smart track-and-trace technology, and cloud networks. They also make it possible for transparent shipments, digital category management, and smart procurement. To schedule maintenance and prevent equipment failure, IoT-connected sensors and intelligent devices provide fleet data such as vehicle performance, fuel consumption, and inventory. By digitizing and automating the material master request (MMR) authorization process, low-code AI-based ERP solutions can speed up document approval, do away with manual intervention, eliminate approvals of paper-based material requests, and provide 100% accuracy and traceability throughout the MMR process. AI has significant advantages for data mining. AI systems have the capacity to gather, evaluate, and interpret TG into prompt actions. As a result, adding AI to the inventory control system enables the business to advance more quickly and produce solutions that are more effective for a particular circumstance. By monitoring, collecting, recording, and processing each customer's data and interests, businesses can better understand their customers' needs to create more effective strategies, anticipate customer demands, and provide products.

Along with only holding and distributing products, inventory management also includes crucial elements such as forecasting, planning, and control. Because the technology can accurately analyze and correlate demand information, recognize and manage variations in demand for a certain product, and take into account location-specific demand, adopting AI solutions can lessen the risks of overstocking and understocking. AI-based systems are adaptable and capable of assessing any circumstance, which is crucial for efficient planning, inventory management, and delivery scheduling. The business can increase customer satisfaction while reducing expenses by decreasing errors and inventory management issues.

The product distribution in commercial and retail consumers at the conclusion of the downstream oil and gas value chain, which extends from the oil fields to the refineries, is a complicated process. A Business 4.0 framework and the application of digital technology can facilitate the efficient operation of the supply chain. For organizations to safeguard profits and maintain competitiveness, they must be able to optimize the entire value chain. Effective planning of supply chain and processing of the steps are two of the essential elements for improving the overall organization efficiency, as in any industry. However, due to the compartmentalized nature of the operations, inventory management of downstream processes poses a significant issue. The management of the inventory by several organizations using competing KPIs frequently leads to less-than-optimal decisions that reduce enterprise value. The refinery inventory planning throughout the supply chain is different than depots form of inventory planning. This is caused by the refinery's boundary between push and pull, which is present since the facility is asset-intensive and runs as efficiently as possible to lower unit costs and downtime expenses. Products are continuously produced as a result of refinery activities, and it has to be separated from the refinery at specific intervals to avoid issues in the tank top. The pull demand from the lower part of the supply chain and the type of transportation in a bulk way such as ships, pipelines, and trains to supply the products to depots or to send it to customers determine the rate at which goods are removed from the refinery. The inventory approach also has an impact on planning for manpower and infrastructure sizing. The nature of the commodity of the firm, where the cost of the product is volatile, further complicates the entire planning process. A quick rise or fall in market prices may has an effect on the value of the stock held and has a big financial impact on the business. Therefore, accurately coming up with a plan for inventory levels in the downstream requires consideration of demand forecasts, preserving refinery, transportation, and the prognosis for cost on the market. In order to efficiently manage and reduce the variations in the results, various businesses decide to operate on a push model in supply as opposed to a demand-driven approach because of the numerous variables and moving parts. This frequently involves paying for transportation, which results in expensive logistical costs and activities that are counterproductive in terms of controlling inventory and replenishment. Businesses has to include demand estimates into inventory rather than depending on a push approach. Instead of using fixed targets throughout the year, rolling targets should be used to exchange demand, inventory costs, and transportation. This will simplify logistics processes and cut down on transportation expenses, which will be reflected in the price of the final product. Additionally, any extra production can be purposefully disposed of outside the system. Additionally, having access to the system's anticipated inventory can assist in reducing some price exposure concerns through the use of trading instruments. The "new oil," or data and decision-making supported by data, is what Oil and Gas businesses must focus on if they are to successfully implement this strategy. Downstream divisions of the energy sector companies have historically been slower to adopt digital technology, but they are now starting to invest in building the platforms required to gain a competitive edge. Businesses must integrate a number of technological pillars to support this journey in order to utilize Business 4.0 behaviors and pursue the desired results.

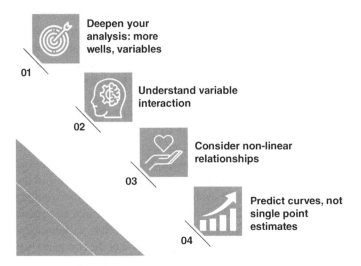

Figure 6 Well monitoring POV of AI.

3.10 Well Monitoring

The safety of the workforce as well as the environment as a whole will be guaranteed by monitoring of gas stations, oil fields, plants, mines, and machinery. AI-powered cameras, robots, drones, and other devices can monitor accidents to lessen the possible damage. Regular equipment inspections and risk assessments will assist businesses in implementing preventative steps to steer clear of unforeseen events [12]. As described in Figure 6, AI can be implemented in well drilling and monitoring by following the steps.

3.11 Process Excellence and Automation

The study conducted by McKinsey revealed some crucial facts. It discovered that, on average, rigs in the North Sea were operating at 82% of capacity when examining production there. This is a very negative news for operators because the desired result should be close to 95%. Unexpected downtime and maintenance were the cause of this subpar performance. Too frequently, operators connected Internet of Things (IoT) sensors to the manufacturing machinery but only used them to observe and influence the machinery's existing operations. The data were not being sent into an analytics tool that would enable the identification of patterns and trends that could notify the operations personnel before a problem developed.

Predictive maintenance [13] is made possible by AI-powered analytics tools like OpenText Magellan. To apply predictive models that can immediately pinpoint potential issue areas, data from IoT sensors from one rig can be combined with data from sensors used in all operations, from other operations, and from production systems. The solution can be fully automated so that if physical maintenance is required, it orders the necessary parts and schedules the work right away.

This drastically cuts downtime. According to Forbes, 13% of offshore equipment has downtime and is at least 15 years old. The effects of predictive maintenance fueled by analytics powered by AI can be astounding. The experts cite one instance when 500 rigs were used to maximize productivity. The analysts cite one instance in which production was optimized across 500 rigs a million times more quickly than was previously feasible. Although the use of AI-powered analytics solutions is still in its infancy, it will have a huge impact on operational excellence. One will be able to see exactly what is occurring across all of the operations in real-time and have the knowledge one needs to make the best decisions on how to maximize productivity as well as recognize new opportunities and working methods.

3.12 Asset Tracking and Maintenance/Digital Twins

Data can be used to predict a variety of events. By evaluating the data, together with other technologies like AI, machine learning, Big Data, and analytics, it is possible to gain insightful information and make the necessary corrections as described in Figure 7. The digital twin [14] needs precise, high-quality, comprehensive, and dependable data for successful results. The data are gathered from a variety of sources, including internal sensors built into the assets and external ones like drone surveillance and visual or manual inspections. The technology should ideally rely on a wide range of data, such as part numbers, specifications, operational data, environment, and others.

In general, businesses can gather and use data from many sources as the input to the models to get and anticipate the operating health of the assets. Inbuilt sensors, SCADA, data from surveillance or inspection drones, remote sensing, visual inspections, and other sources are examples of typical sources. The operators can ascertain how the physical assets in the oil and gas facilities function in real-time by using the data as input to the models. Additionally, the modeling enables the technical teams or operators to run what-if analyses to determine how real assets and processes might function or react under various conceivable circumstances. Once a digital twin has been created, it is crucial to make sure that it contains up-to-date information that accurately reflects the asset's present status, including any changes. In theory, real-time monitoring and simulations aid in providing the asset's correct state. Oil and gas businesses may regularly evaluate their on-site

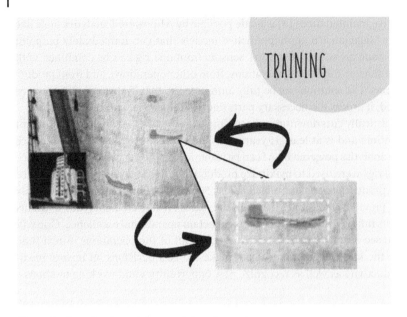

Figure 7 Data for the training model to detect damage.

and remote field assets using drones, ensuring that the digital models are built using the most recent data. The enterprises should be able to better model and plan their drilling and operations with the aid of digital technologies, which should also enhance asset management and preventative maintenance. The digital twins are helpful in supplying insightful information, that subsequently paves the way for more data-driven decisions and actions, which is important given the complexity of the oil and gas exploration and extraction activities. As a result, they enable the businesses to optimize every aspect of their operations across the ecosystem, from where to dig and how to extract resources to processing and shipping the finished goods.

The use of technology can automate repairs, detect anomalies in assets or processes, and replicate procedures. Additionally, it gives oil and gas firms the ability to foresee and address possible issues, avoiding failures and downtime. The digital twins may be able to spot deterioration and early indications of approaching equipment breakdown using sophisticated inspection data. Therefore, it benefits the business to be proactive and depart from the conventional approach of acting or reacting to problems after they have occurred. The technical teams are ideally able to plan and carry out remedial maintenance prior to problems. Companies can gather a wide range of data from the assets by continually monitoring them with a variety of built-in sensors and conducting frequent drone inspections. The digital twins can gather insights, recognize, and fix numerous performance and maintenance issues before failures by being fed data. In addition to being less expensive, doing this also avoids downtime, lost productivity, increased damage from faulty components, etc. The businesses can examine assets and obtain real-time data using a variety of sensors and technologies, including IoT, drones, automatic and manual methods, and automated ways. This gives them the chance to change and optimize the systems and procedures to boost performance while also understanding how the rigs and wells are performing in real-time. The digital twins give the technical teams visibility, real-time data, and an understanding of the status and condition of their assets through the use of sensors and other technology. Fortunately, remote asset monitoring is now possible thanks to drones, robots, sensors, and a variety of communication technologies that span public and private networks. For instance, they might keep an eye on their oil wells and make adjustments or improvements to their operations based on geological data and productivity patterns to maintain consistency in the output. Remotely inspecting the assets also enhances efficiency while lowering costs. Teams may keep an eye on assets, processes, the environment, and other factors that affect productivity remotely. As a result, the energy firms may conduct frequent, economical virtual inspections that cut down on site visits, lower their need for labor, and save them time and money.

3.13 Optimizing Production and Scheduling

Artificial intelligence (AI) at an enterprise can revolutionize mainly the upstream operations. Data integration, the creation, deployment, and upkeep of numerous machine learning (ML) models, as well as combining data science and oil and gas domain experience, are all necessary for a successful implementation. Since the availability of high accurate data from different point of systems, including various types of data such as financial, production well, field infrastructure, reservoir modeling and economic planning for the business of production, is the first requirement for a production optimization solution's successful implementation [15]. For AI metering, for instance, the data must consist of frequently occurring readings that can be utilized as training data and also for testing data that can be used as target data for training the models. This information must be accessible for numerous pads in a typical oil and gas field, each of which has data about the production wells.

For optimizing the production application, the number of wells can easily range from hundreds to thousands. A productive set of training data and the chosen data model which can enable fast and accurate processing of all data that are required for the organization and unification of all data are required for

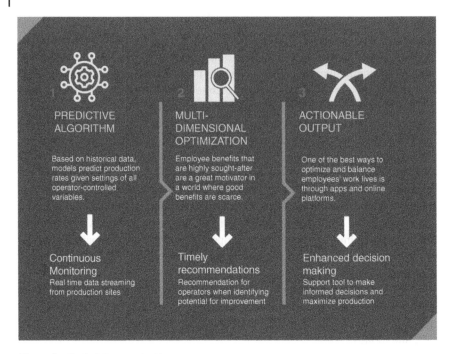

Figure 8 Optimizing production.

optimization of production application. The process of creating an effective machine learning model is extremely iterative and often calls for domain expertise, extensive investigation, and testing of various features, strategies, and models as described in Figure 8.

It might be difficult to manage hundreds or thousands of machine learning models. Every model must first be trained and tweaked using hyperparameters. Once trained, these models must be "productionized" in order to produce predictions, and the outcomes must be stored and given access via the application. In case of performance degradation, the models must be retrained or updated, and the quality of the forecasts must be regularly evaluated. The size of these operations needed for a field-level reading solution presents numerous challenges.

3.14 Emission Tracking

The most efficient method for monitoring carbon emissions [15] at a corporation right now is AI. There are numerous strategies to effectively assist businesses in lowering their carbon footprint. To stop emissions from accumulating to dangerous levels in the atmosphere, we must first monitor emissions. Currently, several academics are working to develop a brand-new technique for monitoring leaks from oil and gas corporations. In pipelines that carry gases from oil and gas production sites to refineries, cameras are installed. The cameras [16] capture pictures of the locations where methane leaks occur. Then, by comparing them to their databases, AI machines are utilized to find the leaks. Finally, AI systems provide precise forecasts regarding the environment's catastrophic effects that those emissions will have so that technicians can properly fix them before gas leaks spread too far or cause serious harm. Systems with AI are in the development phases. Even so, there has been a promise of better outcomes, and past performance demonstrates that these systems will be a reliable source. AI combined with video cameras can be used by oil and gas businesses to measure cost savings, site security, and regulatory compliance automatically. Customers save money, time, and effort thanks to continuous monitoring, vision AI-powered insights, and rapid alerting, which also enables quick and precise prevention and resolution.

When the spread of methane gas is discovered in gas companies, researchers devised a deep learning model to train the system to detect the shape of a breach. This made it easier for researchers to pinpoint the source of methane emissions, such as a landfill or a gas pipeline, and to distinguish between methane and hydrocarbons in a single image. The technology for finding methane leaks is said to have an 87 percent success rate. Multiple user sources, including previously unknown leaks, can detect more leaks, which is a result of biomethane leaks from buildings, businesses, and urban infrastructure caused by poorly supervised activities. A tried-and-tested system of fixed and pan-tilt-zoom thermal imagers is used by maintenance employees to evaluate gas tank levels, find probable gas leaks, and track facilities. Additionally, AI software may analyze thermal images to discover growths, calculate heat loss, and carry out other jobs, enhancing a variety of operations.

3.15 Logistics Network Optimizations

Logistics and transportation are difficult in the oil and gas sector. To consider everything from beginning to conclusion, there are too many nodes or decision-making points. By planning and managing transportation services, AI helps midstream enterprises. Teams in the upstream sector of the economy can better align and coordinate their efforts to reduce delivery times with the use of AI. Supply chain managers occasionally find it difficult to develop a thorough strategy to get ready for successful supply network accounting when confronted daily with expanding globalization, extended product portfolios, increased complexity, and changeable consumer demand. The absence of complete insight into current product portfolios brought on by unforeseen events, manufacturing closures, or transportation problems makes this attempt more difficult. Various items, replaceable parts, and a smart supply chain structure are typical components. A typical smart supply structure is made up of numerous products, spare parts, and other components that provide precise results. In many supply chain businesses, these products or parts can be described using a number of attributes with a wide range of values. This could lead to a wide number of product variations and uses. Additionally, products and components are constantly phased in and out, which can have a bullwhip effect on the supply chain at all points and cause proliferation and uncertainty.

By incorporating AI in supply chain and logistics, supply chain managers can enhance their decision-making by anticipating emerging bottlenecks, unforeseen irregularities, and remedies. Production scheduling will become more consistent as a result, as it would otherwise be highly variable owing to reliance on factory operation management. Furthermore, supply chain AI has created accurate projections and quantified anticipated results across numerous timetable phases, enabling the scheduling of more optimal alternatives as and when such execution bottlenecks occur. Additionally, products and components are constantly phased in and out, which can have a bullwhip effect on the supply chain at all points and cause proliferation and uncertainty. By incorporating AI in supply chain and logistics, supply chain managers can enhance their decision-making by anticipating emerging bottlenecks, unforeseen irregularities, and remedies. Production scheduling will become more consistent as a result, as it would otherwise be highly variable owing to reliance on factory operation management. Furthermore, supply chain AI has created accurate projections and quantified anticipated results across numerous timetable phases, enabling the scheduling of more optimal alternatives as and when such execution bottlenecks occur.

3.16 Conclusion

Artificial intelligence (AI) is one of several disruptive themes that will have a big impact on oil and gas corporations as they build the future of the energy sector. The lack of readily accessible hydrocarbon deposits, which forces businesses to explore remote reserves that are difficult to find, expensive to explore and produce, and risky, is just few of the many issues the energy sector is currently facing. Sustainability considerations are another factor that may cause demand for oil and gas to shift away from dirty sources and toward cleaner ones. In addition, the industry's asset-heavy structure and high operational costs make it challenging and slow to adapt to fresh consumption trends.

Cost issues continue, despite the fact that pandemic-related worries have largely subsided. In this situation, businesses must forgo traditional methods of cost-cutting and invest in technological advancement. Market conditions have altered, necessitating evolution. Businesses that can improve their effectiveness, efficiency, and alignment with the market's direction will endure and prosper. Employee transfer in the energy industry is the biggest problem for the business when compared to other sectors. By maintaining intelligent chatbots, a central repository for past and present data is created, which can speed up the induction of new employees. By implementing AI-based digital transformation tools and processes, businesses may get more direction and real-time insights based on critical data in each of the aforementioned scenarios. Utility industry digital transformation solutions that incorporate intelligent analytics offer insights in the form of both visual visuals and debriefing reports. One of the main technologies that will drive enterprises through the digital transformation is AI. Although AI benefits all businesses to differing degrees, the oil & gas sector may stand to benefit the most. One of the most lucrative, albeit dangerous, industries to work in is oil and gas production. AI is used to boost equipment and resource availability, increase efficiency and security, monitor controllable losses in real-time, and optimize business processes. Hence, there are a lot of scenarios where AI could aid and grow the oil and gas industry into the future with innovations.

References

1 Choubey, S. and Karmakar, G.P. (2021). Artificial intelligence techniques and their application in oil and gas industry. *Artificial Intelligence Review* 54 (5): 3665–3683.

2 Koroteev, D. and Tekic, Z. (2021). Artificial intelligence in oil and gas upstream: trends, challenges, and scenarios for the future. *Energy and AI* 3: 100041.

3 Sircar, A., Yadav, K., Rayavarapu, K. et al. (2021). Application of machine learning and artificial intelligence in the oil and gas industry. *Petroleum Research* 6: 379–391.

4 Nguyen, P.H.M. (2013). Contract lifecycle management on the sell-side: a case study in upstream oil and gas industry

5 Hojageldiyev, D. (2018). Artificial intelligence in HSE. *Abu Dhabi International Petroleum Exhibition & Conference*, November 2018, Abu Dhabi. OnePetro.

6 Baaziz, A. and Quoniam, L. (2014). How to use big data technologies to optimize operations in upstream petroleum industry. arXiv preprint arXiv:1412.0755.

7 Ali, S.S., Nizamuddin, S., Abdulraheem, A. et al. (2013). Hydraulic unit prediction using support vector machine. *Journal of Petroleum Science and Engineering* 110: 243–252.

8 Agwu, O.E., Akpabio, J.U., Alabi, S.B., and Dosunmu, A. (2018). Artificial intelligence techniques and their applications in drilling fluid engineering: a review. *Journal of Petroleum Science and Engineering* 167: 300–315.

9 Al-Yami A.S., Al-Shaarawi A., Al-Bahrani H., Wagle V.B., Al-Gharbi S., and Al-Khudiri MB (2016). Using Bayesian network to develop drilling expert systems, *SPE Heavy Oil Conference and Exhibition*, December 2016, Kuwait. Society of Petroleum Engineers.

10 Anifowose, F., Labadin, J., and Abdulraheem, A. (2015). Improving the prediction of petroleum reservoir characterization with a stacked generalization ensemble model of support vector machines. *Applied Soft Computing Journal* 26: 483–496.

11 Newstyle, D. and Opuene, E.G. (2022). Inventory management techniques and financial performance of listed oil and gas companies in Nigeria. *BW Academic Journal* 11: 87–109.

12 Aulia, A., Rahman, A., and Quijano Velasco, J.J. (2014). Strategic well test planning using random forest. *SPE intelligent energy conference & exhibition*, April 2014, Utrecht, The Netherlands. Society of Petroleum Engineers.

13 Aissani, N., Beldjilali, B., and Trentesaux, D. (2009). Dynamic scheduling of maintenance tasks in the petroleum industry: a reinforcement approach. *Engineering Applications of Artificial Intelligence* 22 (7): 1089–1103. https://doi.org/10.1016/j.engappai.2009.01.014.

14 Wanasinghe, T.R., Wroblewski, L., Petersen, B.K. et al. (2020). Digital twin for the oil and gas industry: overview, research trends, opportunities, and challenges. *IEEE Access* 8: 104175–104197.

15 Kandziora, C. (2019). Applying artificial intelligence to optimize oil and gas production. *Offshore Technology Conference,* May 2019, Houston, Texas. OnePetro.

16 Cooper, J., Dubey, L.. and Hawkes, A. (2022). Methane detection and quantification in the upstream oil and gas sector: the role of satellites in emissions detection, reconciling and reporting. *Environmental Science: Atmospheres* 2 (1): 9–23.

4

Demystifying the Oil and Gas Exploration and Extraction Process

4.1 Process of Crude Oil Formation

Crude oil is a fossil fuel that is formed from the remains of dead plants and animals that lived millions of years ago. The process of crude oil formation, also known as petroleum genesis, is complex and lengthy, occurring over a period of millions of years. The process begins with the accumulation of organic matter, such as plants and animals, in a marine or lacustrine environment. As these organisms die, their remains sink to the bottom of the water body and are buried by sediment. Over time, heat and pressure from the overlying sediment cause the organic matter to undergo chemical and physical changes.

The first stage of crude oil formation is the transformation of the organic matter into a substance called kerogen. Kerogen is a waxy, insoluble substance that is found in sedimentary rocks and is rich in carbon. As the temperature and pressure increase, the kerogen begins to break down and releases hydrocarbons, which are molecules composed of hydrogen and carbon. The hydrocarbons are then subjected to further chemical reactions, resulting in the formation of various types of crude oil. The specific type of crude oil that is formed depends on the characteristics of the organic matter, the temperature and pressure conditions, and the length of the time for which the process has been occurring.

There are several factors that can affect the process of crude oil formation. One of the most important factors is the presence of oxygen. The absence of oxygen, or anerobic conditions, is essential for the formation of crude oil. If oxygen is present, the organic matter is likely to be converted into natural gas instead of crude oil. The process of crude oil formation is not a straightforward or linear process. It is a complex process that occurs over millions of years and is influenced by a

The Power of Artificial Intelligence for the Next-Generation Oil and Gas Industry: Envisaging AI-Inspired Intelligent Energy Systems and Environments, First Edition. Pethuru Raj Chelliah, Venkatraman Jayasankar, Mats Agerstam, B. Sundaravadivazhagan, and Robin Cyriac.
© 2024 The Institute of Electrical and Electronics Engineers, Inc.
Published 2024 by John Wiley & Sons, Inc.

variety of factors. Despite this complexity, the process of crude oil formation is essential to our modern way of life, as crude oil is the primary source of energy for transportation, heating, and electricity generation.

4.2 Composition of Crude Oil

One of the main components of crude oil is hydrocarbons. Hydrocarbons are molecules composed of hydrogen and carbon atoms. They are the primary source of energy in crude oil and are responsible for its flammability [1]. The most common types of hydrocarbons found in crude oil are alkanes, cycloalkanes, and aromatic hydrocarbons.

Alkanes, also known as paraffins, are linear or branched-chain hydrocarbons that are characterized by their high energy content and low reactivity. They are the primary components of natural gas and make up a significant portion of crude oil. Cycloalkanes, also known as naphthenes, are hydrocarbons that contain a ring structure. They have a lower energy content and higher reactivity than alkanes. Aromatic hydrocarbons are a class of hydrocarbons that contain a ring structure and are characterized by their strong, pleasant odors. They are less common in crude oil but are important because they are used to produce a wide range of products, including plastics, rubber, and synthetic fibers.

In addition to hydrocarbons, crude oil also contains a variety of other organic compounds, including sulfur, nitrogen, and oxygen compounds. These compounds can have a significant impact on the properties of crude oil, including its viscosity, density, and corrosion resistance. The composition of crude oil also varies based on its geographical location and the geological conditions under which it was formed. For example, crude oil from the Middle East tends to be more viscous and has a higher sulfur content than crude oil from the United States.

4.3 Crude Oil Classification

Crude oil is classified based on its chemical composition, physical properties, and location of origin [1]. There are several different methods for classifying crude oil, including

1) **American Petroleum Institute (API) gravity**: This is a measure of the density of crude oil relative to water. Crude oils with a high API gravity are considered light, while those with a low API gravity are considered heavy.

2) **Sulfur content**: Crude oils are also classified based on their sulfur content. Those with a high sulfur content are referred to as "sour," while those with a low sulfur content are referred to as "sweet."

3) **Location of origin**: Crude oils are often named after their geographic location of origin. For example, "West Texas Intermediate" (WTI) is a type of crude oil that is produced in the United States, while "Brent" is a type of crude oil that is produced in the North Sea.

4) **Chemical composition**: Crude oils can also be classified based on their chemical composition, including the types and proportions of hydrocarbons they contain.

5) **Physical properties**: The physical properties of crude oil, such as viscosity and pour point, can also be used to classify it.

Light crude oils are more valuable than heavy crude oils because they are easier to refine and produce higher yields of gasoline and other valuable products. Sour crude oils are also typically less valuable than sweet crude oils because they require more extensive refining to remove the sulfur.

4.3.1 Other Types of Crude Oil

In addition to the various methods of classifying crude oil based on its chemical composition, physical properties, and location of origin, there are also several different types of crude oil that are produced around the world. Some examples of these types of crude oil include the following:

1) **Conventional crude oil**: This is the most common type of crude oil and is produced by drilling into reservoirs of oil in the Earth's crust. It is a mixture of hydrocarbons and other organic compounds that can be refined into a range of products, including gasoline, diesel, and aviation fuel.

2) **Unconventional crude oil**: This includes oil sands, also known as tar sands, and shale oil. Oil sands are a type of heavy, viscous oil that is found in deposits in countries such as Canada and Venezuela. Shale oil is a type of oil that is extracted from shale rock formations using hydraulic fracturing, or fracking.

3) **Bio-crude oil**: This type of crude oil is produced from biomass, such as plant matter or animal fat, using a process called pyrolysis. It can be used as a substitute for fossil-based crude oil in the production of fuels and chemicals.

4) **Natural bitumen**: This is a type of heavy, thick oil that is found in deposits around the world, including Canada and Venezuela. It is a more viscous form of crude oil and requires more extensive processing to refine it into usable products.

5) **Extra-heavy crude oil**: This is a type of very heavy, thick oil that is found in countries such as Venezuela and Canada. It is more challenging to extract and refine than conventional crude oil and requires specialized processing techniques.

4.4 Crude Oil Production Process

The production of crude oil involves several steps, beginning with the exploration and identification of potential oil deposits and ending with the refinement of the crude oil into useful petroleum products. The first step in the production process is oil exploration, which involves searching for and identifying potential oil deposits. This can be done through a variety of methods, including studying geological and geophysical data, conducting aerial and satellite surveys, and drilling exploratory wells, to test the composition and productivity of the rock formations. Once a potential oil deposit has been identified, the next step is to drill a production well to extract the oil. There are several methods of drilling, including traditional rotary drilling, hydraulic fracturing (also known as fracking), horizontal drilling, and offshore drilling.

After the oil has been extracted from the ground, it is brought to the surface and transported to a refinery, where it can be processed and refined into useful products. The refining process involves separating the various hydrocarbons in the crude oil through distillation and then further refining them through processes such as cracking and reforming. The final step in the crude oil production process is the marketing and distribution of the refined petroleum products, which include products such as gasoline, diesel, and heating oil. These products are then transported to storage facilities or distribution centers and sold to consumers through a variety of channels.

The operations in the oil and gas industry can be broadly classified into three, namely, upstream, midstream, and downstream, as shown in Figure 1. These stages are regulated by the central and state governments, and there are different laws governing their production and distribution. International policies control their movement across countries and continents [2].

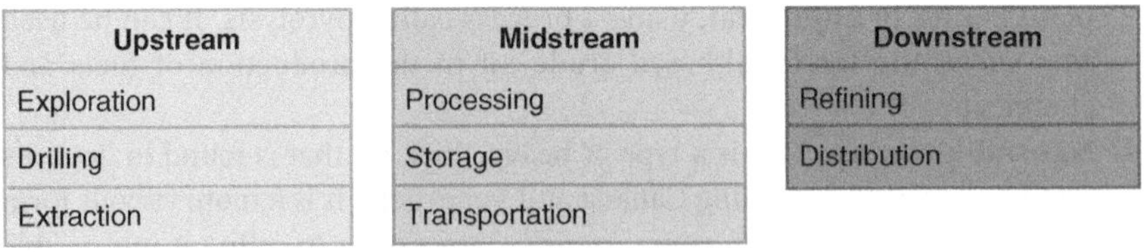

Upstream	Midstream	Downstream
Exploration	Processing	Refining
Drilling	Storage	Distribution
Extraction	Transportation	

Figure 1 Operations phases in the oil and gas industry.

4.5   Oil Exploration

Oil exploration is the process of searching for and identifying potential reserves of crude oil and natural gas. There are several techniques that are commonly used in oil exploration to identify and confirm the presence of oil and gas deposits. Some of the most common techniques include the following:

1) **Geologic analysis**: This involves studying the geology of an area to identify the types of rock formations and sediments that are present. This can help determine the likelihood of oil and gas deposits being present in a particular area.
2) **Geophysical surveys**: Geophysical surveys are a key tool used in oil exploration to gather data about the subsurface of an area. They involve the use of specialized equipment to measure and record physical properties of the Earth, such as its gravitational field, magnetic field, and electrical conductivity. These data can be used to identify the types of rocks and sediments that are present, as well as potential oil and gas deposits. There are several different types of geophysical surveys that are used in oil exploration, including the following:

    Seismic surveys: These surveys use seismic waves to create a map of the subsurface of an area. Seismic waves are generated by a small explosion or by striking the ground with a heavy weight, and the waves that are reflected back to the surface are recorded by sensors called geophones. The data that are collected can be used to identify underground rock formations and potential oil and gas reservoirs.

    Gravity surveys: These surveys measure the gravitational field of an area to identify denser rock formations, which may indicate the presence of oil and gas deposits. The measurements are typically taken using a device called a gravimeter, which is a highly sensitive instrument that is able to detect small variations in the gravitational field.

    Magnetic surveys: These surveys measure the magnetic field of an area to identify variations in the magnetic properties of the rocks and sediments. This can be used to identify underground rock formations and potential oil and gas deposits.

    Electromagnetic surveys: These surveys measure the electrical conductivity of the Earth to identify variations in the conductive properties of the rocks and sediments. This can be used to identify underground rock formations and potential oil and gas deposits.
3) **Drilling**: Once potential oil and gas reserves have been identified, the next step is to drill exploratory wells to confirm the presence of the deposits and to determine their size and quality. This can be done using a variety of drilling methods, including rotary drilling and directional drilling.

4) **Well logging**: Once an exploratory well has been drilled, well logging techniques can be used to gather more detailed information about the geology of the area and the characteristics of the oil and gas deposits. This can include methods such as resistivity logging, which measures the resistance of rocks to the flow of electricity, and sonic logging, which measures the velocity of sound waves through the rock.

5) **Reservoir modeling**: Once the size and characteristics of the oil and gas deposits have been determined, reservoir modeling techniques can be used to develop a plan for extracting the resources in an efficient and economically viable manner. This can include methods such as computer simulations and mathematical models that help predict the flow of oil and gas through the reservoir.

4.6   Oil Extraction

Oil extraction involves the drilling of production wells to extract oil. There are several different types of drilling techniques that are used in the oil and gas industry, including the following:

1) **Rotary drilling**: This is the most commonly used drilling method in the oil and gas industry. It involves the use of a rotary drill bit that is attached to a drill string, which is rotated to create a hole in the ground. The drill bit is made of hard, wear-resistant materials, such as tungsten carbide or diamond, and is able to cut through a variety of rock formations.

2) **Directional drilling**: This is a specialized form of drilling that is used to create wells that deviate from a vertical path. It is often used to access oil and gas deposits that are located in challenging or hard-to-reach areas, such as offshore or in urban areas. Directional drilling involves the use of specialized equipment and software to control the direction and inclination of the well as it is being drilled.

3) **Hydraulic fracturing**: This technique, also known as fracking, is used to extract oil and gas from shale and other low-permeability rock formations. It involves the injection of a mixture of water, sand, and chemicals into the well at high pressure. It will lead to the formation of fractures in the rock, which causes the oil and gas to flow more freely, making it easier to extract. The oil is then brought to the surface through the well.

4) **Horizontal drilling**: This is a specialized form of drilling that involves the creation of a horizontal wellbore that is parallel to the surface of the Earth. It is often used in combination with hydraulic fracturing to extract oil and gas from shale and other low-permeability rock formations. The oil is brought to the surface through the well using a series of pipes.

5) **Offshore drilling**: In offshore drilling, the oil is extracted from underwater oil deposits using specialized drilling platforms. The oil is brought to the surface through the well and then transported to the shore via pipelines or tanker ships.

4.6.1 Bringing Extracted Crude Oil to the Surface

There are several techniques that can be used to move crude oil up the well through a pipe if the pressure is low. These techniques are known as artificial lift methods, and they involve using mechanical means to lift the oil to the surface [3]. Some common artificial lift methods include the following:

Rod pumping: In this method, a rod is used to lift the oil to the surface. The rod is connected to a pump at the surface, which helps lift the oil to the surface.

Electric submersible pumps (ESPs): In this method, an electric motor and pump are placed in the wellbore, and the pump is used to lift the oil to the surface.

Gas lift: In this method, a small amount of gas is injected into the wellbore, which helps lift the oil to the surface.

Plunger lift: In this method, a plunger is used to lift the oil to the surface. The plunger is typically made of metal or rubber and is lowered into the wellbore. When it reaches the bottom of the well, it creates a pressure wave that helps lift the oil to the surface.

4.6.2 Enhanced Oil Recovery

Enhanced oil recovery (EOR) is the process of using various techniques to increase the amount of oil that can be extracted from an oil reservoir. Some common techniques used in enhanced oil recovery include the following:

Water flooding: Water flooding is used in the oil and gas industry to increase the recovery of hydrocarbons from an oil reservoir. It involves injecting water into the reservoir through injection wells, with the goal of increasing the pressure within the reservoir and forcing the oil toward production wells. The water that is injected into the reservoir can be freshwater, seawater, or a combination of the two. There are several benefits to using water flooding as a method of enhanced oil recovery. It can be an effective way to increase the overall recovery of oil from a reservoir, and it is relatively inexpensive compared to other methods of enhanced oil recovery. Additionally, water flooding can help reduce the viscosity of the oil, which results in easier production.

Thermal recovery: It involves injecting heat into the reservoir to reduce the viscosity of the oil and make it easier to produce. There are several techniques that can be used to deliver the heat to the reservoir, including steam injection, in situ combustion, and cyclic steam stimulation. One of the main benefits of thermal recovery is that it can be an effective way to increase the overall recovery of oil from a reservoir. It can be particularly useful in heavy oil reservoirs, where the oil is more viscous and more difficult to produce using traditional methods. Thermal recovery can also help reduce the amount of water that is required for enhanced oil recovery, which can be beneficial in areas where water is scarce.

Chemical injection: It involves injecting chemicals into the reservoir in order to alter the properties of the oil or the rock formations, with the goal of making it easier to produce oil. There are several different types of chemicals that can be used for this purpose, including surfactants, polymers, and alkaline chemicals. Chemical injection can be an effective way to increase the overall recovery of oil from a reservoir, particularly in reservoirs where the oil is more viscous or where the rock formations are poorly permeable. It can also be a relatively inexpensive method of enhanced oil recovery compared to other techniques. However, there are also some potential drawbacks to chemical injection. The injected chemicals can be expensive, and there is a risk of damaging the reservoir rock if the chemicals are not properly formulated or if they are injected at the wrong concentration. Additionally, there are environmental concerns associated with chemical injection, including the potential for chemical spills and the risk of contaminating groundwater. As a result, it is important to carefully design and implement a chemical injection program to ensure that it is effective and does not cause negative impacts on the reservoir or the surrounding environment.

Gas injection: It involves injecting a gas, such as natural gas, nitrogen, or carbon dioxide, into the reservoir in order to increase the pressure within the reservoir and force the oil toward the production wells. Gas injection can be an effective way to increase the overall recovery of oil from a reservoir, particularly in reservoirs where the oil is more viscous or where the rock formations are poorly permeable.

Microbial enhanced oil recovery (MEOR): It involves injecting microorganisms into the reservoir, with the goal of altering the properties of the oil or the rock formations in order to make it easier to produce the oil. The microorganisms can be either naturally occurring or genetically modified, and they can produce a variety of substances that can affect the oil or the rock formations, including enzymes, surfactants, and gases. It can also be a relatively inexpensive method of enhanced oil recovery compared to other techniques. However, there are also some potential drawbacks to MEOR. It can be difficult to predict the behavior of the microorganisms in the reservoir, and there is a risk of damaging the reservoir rock if the microorganisms are not properly selected or if they are injected at the wrong concentration. Additionally, there are environmental concerns associated with MEOR, including the potential for contamination of the reservoir and the risk of releasing genetically modified microorganisms into the environment.

These techniques are typically used when the natural pressure in the reservoir is low or when the flow of oil has slowed and can be an effective way to extend the life of an oil well.

4.7 Processing of Crude Oil

Usually, a land-based oil field has many wells. This oil is transported from the wells to a point in or near the oil field for preliminary processing through a network of collection lines. In addition to water content in the liquid and gaseous form, oil generated from practically every reservoir contains various gases, including natural gas. As a result, from the well, a separator is used to extract and separate natural gas and the water content from crude oil. The raw crude is cleaned of any solid impurities, such as sand, as well as some dissolved salts. Following this first treatment, the crude oil, as well as any condensate (light hydrocarbon liquids produced during the process of separation of onsite natural gas), is put into leased containers beside the well for short-term storage. Before loading oil into the pipeline from storage, an automated system or a technician verifies the amount of oil to be transported as well as the quality of the oil to make sure that the water and sediment particles are small and under acceptable limits. Skid-mounted automated custody transfer (ACT) systems measure and test the oil at various plants. A technician examines the ACT device on a regular basis and validates the BS&W measurement [1].

4.7.1 Oil and Natural Gas Storage

Aboveground tanks or subterranean chambers are often used to store oil. It is also sometimes kept on tanker ships, depending on market circumstances and supply/demand dynamics. Crude oil storage tanks are generally massive cylindrical carbon steel containers. Storage tanks are painted with bright colors to reflect the Sun's energy and therefore reduce evaporation. Some tanks have a large lid that floats on the upper surface of the oil stored in it. The floating lid, whether exposed to the elements or housed beneath a separate peaked roof, inhibits evaporation and decreases oil interaction with moisture and oxygen. A tank farm is a facility having numerous storage tanks [4].

Crude oil, natural gas, and propane are also stored in salt caves. The caves are formed by pumping water into subterranean salt domes or beds to dissolve the salt lining. The saline water is subsequently pumped out, leaving a hollow cavern behind. It produces an impermeable barrier to hydrocarbon liquids and gases.

Natural gas can also be stored in liquefied form as liquified natural gas (LNG) and compressed form as compressed natural gas (CNG). LNG storage allows for easy transportation, but it also requires high energy expenditure to convert natural gas to liquid form and back to gas when it is being used, while CNG is stored in high-pressure tanks, and this storage is mainly for transportation purposes, mainly for vehicles.

4.7.1.1 The Liquefaction Process

The liquefaction process is a method to convert natural gas, which is primarily composed of methane, into a liquid form called LNG. The process involves cooling natural gas to −260°F, at which point it becomes a liquid and takes up about 1/600th the volume of its gaseous state. Natural gas obtained from different regions will differ in their composition. Initially, the natural gas is passed through filters and separators to remove impurities such as water vapor, nitrogen, and carbon dioxide. These impurities can cause problems during the liquefaction process and could damage the equipment. Then, the pre-treated natural gas is cooled to the liquefaction temperature of −260°F. The most common method of cooling is to use propane or ethane refrigerants, although other methods such as the use of helium or nitrogen can also be used. The cooled natural gas is then passed through an expansion turbine, which reduces the pressure and causes the natural gas to expand and cool even further. This further cooling causes the natural gas to liquefy. After the expansion process, the LNG may still contain some natural gas liquids (NGLs) like ethane, propane, and butane. These NGLs can be removed through fractionation to obtain pure LNG.

The liquefaction process allows natural gas to be transported and stored more easily and efficiently than in its gaseous form. LNG enables different countries to buy natural gas with less cost from producing nations. LNG can be shipped by tankers or trucks and then re-gasified at the point of use through regasification terminals or by being vaporized back into a gas and sent through pipelines

4.7.1.2 The Regasification Process

Regasification is the process of converting LNG back into its gaseous form so that it can be transported through pipelines and used as fuel. The process involves heating the LNG, which is stored at extremely low temperatures (−260°F), to a temperature at or above its boiling point, which causes it to turn back into a gas. The process is explained as follows:

The LNG is unloaded from the tanker or truck at a regasification terminal and then transferred to a storage tank. It is then heated to its boiling point by running warm water or steam through the storage tank. This causes LNG to vaporize into gas. The gas is then passed through filters to remove any remaining liquid droplets. The gas is compressed in order to reach the pressure required for transportation through pipelines. It can also be metered to measure the flow rate and energy content of the gas, allowing for accurate billing and control of the gas supply. The regasified natural gas is then distributed through pipelines to various customers such as power plants, industrial facilities, and homes [5].

The regasification process is relatively simpler than the liquefaction process, but the terminal and equipment are expensive to build and maintain.

4.7.1.3 LNG Storage

Single containment, double containment, and total containment are the three types that are used more often while developing onshore storage sites for LNG. The first option is the most straightforward and inexpensive one. It has several feet of thermal insulation in between the inner tank made of 9% nickel steel and the outer tank made of carbon steel. The inner tank is freestanding and has an open top. The external tank is constructed of carbon steel. The top of the inner tank is insulated by a suspended ceiling, which is supported by a steel roof that also gathers any vapor that boils off. The term "single containment" refers to the fact that the external tank does not hold any LNG that escapes from the interior tank. Rather, the released LNG would flow into another system, which is nothing but a ditched area around the tank. In a double-containment system, it is possible for the outer tank, which is made of fortified concrete and reinforced by an earth or rock wall, to contain any leaking LNG from the interior tank. This system is comparable to that of the single containment in this respect. Even though the inner tank is covered by a steel roof, the roof is unable to trap any of the vapor that is released from the inner tank. The third design for the tank is referred to as the full-containment design. The double-containment system gets an additional concrete roof in addition to its existing concrete external walls. This roof has the capacity to retain any vapor that may be produced as a result of a break in the interior nickel tank. This is the most expensive of the three alternatives. Comparatively, the price of the double-containment system is around fifty percent more. It makes it feasible to have the least amount of space possible between the tanks and the equipment.

4.7.2 Oil and Gas Transportation

The movement of crude oil and natural gas from the point of production to the point of consumption is referred to as oil and gas transportation. Oil fields are generally situated in isolated areas and must be transported across great distances to their destination. As a consequence, oil and gas transportation is a vital stage in the production and distribution of these commodities. Pipelines, tanker boats, railcars, and tank trucks are among the several modes of oil and gas transportation.

4.7.2.1 Pipeline

Long-haul pipelines, often known as mainlines or trunk lines, transmit crude oil, natural gas, or other petroleum products across long distances of hundreds or thousands of miles. Short-haul pipelines have smaller diameter pipes with fewer delivery stations. The consumer usually gets the same grade of oil that was placed into the delivery line. The flow rate of oil via a pipeline is determined by the diameter of the pipe, condition of interior wall, pressure of the pumping system, terrain, and pipeline orientation. A particular oil batch travels at around 3 to 8 miles per hour via a mainline, driven by centrifugal pumps spaced every 150 km based on the geography. They may be made of a number of materials, such as steel, plastic, or concrete, and can be buried below or supported aboveground.

Short-haul lines, also known as spur lines, stub lines, or delivery lines, transmit crude oil, natural gas, or other petroleum products across short distances. These lines are typically batch-operated, meaning that the client gets the identical grade oil that was poured into the line. Pipeline junctions are known as hubs, or marine terminals if they are near a port. Such facilities often have a large amount of storage space. There are no pipeline entrance points nearby at certain manufacturing locations. In such circumstances, barges, trucks, or railcars transport the pipe to a separate tank facility in a pipeline-served region. Pipeline transport, on the other hand, is the most popular method of transferring crude oil to refineries for conversion into a variety of products [1].

4.7.2.2 Transporting Natural Gas by Pipeline

The natural gas that has been collected from an underground deposit is transported via pipe systems to one or more separation units, where undesirable components such as water, sand particles, and crude oil are removed. The raw gas will often include traces of NGLs and water vapor. Examples of NGLs include butanes, propane, and ethane. Sometimes, the raw gas may also contain pentane. There is also the possibility of the presence of non-hydrocarbon gases such as hydrogen sulfide, carbon dioxide, helium, and nitrogen. The raw gas is then passed through a collecting system in which a collection of pipes delivers the gas to a gas plant where production levels may be aggregated in order to facilitate more cost-effective later processing.

When the processing of the natural gas is complete, the gas is then pumped into a transmission pipeline with a large diameter so that it may be then delivered to the customer. Compressors are used to pump gas into pipes at entrance locations and to maintain sufficient pressure along the length of pipelines. Local distribution companies (LDCs), which are also known as gas utilities, take control of the gas at significant facilities. After adjusting the gas pressure and adding an odorant, which aids in leak detection, the LDCs inject the gas into their distribution. The gas passes via pipes with an even smaller diameter on the last segment of the path, which are known as service lines.

4.7.2.3 Finished Product Pipelines

The petroleum pipeline, which transports end products such as aviation fuel, gasoline, and diesel fuel from trading centers, refineries, or marine terminals to markets, plays a significant role in the delivery of petroleum products to clients. The delivery sites might be a product distributor's facilities, such as a fuel wholesaler, a transportation business, another trade firm, or an export company. In many situations, one of the products is often given in enormous quantities. Products, on the other hand, may be transferred by a pipeline in lower quantities, with numerous products lined up in a particular order. The behavior of pipeline flow and the molecular characteristics of each product type require no physical barrier between batches, resulting in very minimal mixing. If zero percent mixing or contamination is only acceptable, then a sphere or other mechanical device may be used to keep them apart.

4.7.2.4 Pipeline Inspection and Maintenance

An oil and gas pipeline needs to be carefully inspected and managed so that the products can be delivered as per the quality standards demanded in the contractual agreement. It is also required to maximize the economic life of the pipeline so that it functions securely and effectively to meet the demands and concerns of all stakeholders. Leakage of oil or gas is a significant indicator of possible issues. Third-party damage to pipelines and their equipment from excavation and agricultural activity, corrosion, and mechanical failures, including manufacturing flaws and construction flaws, natural hazards such as ground motion, weather, and erosion due to water flow, are some of the causes of pipeline damage. In order to deal with these problems, pipeline operators implement a wide range of solutions, some of which include the following: compliance with all applicable standards; careful mapping of pipeline location; quality control of all pipeline- and construction-related operations; public education regarding pipeline routes and operation; regularly monitoring the pipeline route for signs of intrusion or leakage; use of telephone call centers to assist excavators; and installation of corrosion control systems such as cathodic protection devices.

Pipeline inspection is the process of examining pipelines for defects or damage in order to ensure their safe and reliable operation. There are several different methods of pipeline inspection, including visual inspections, ultrasonic testing, and inline inspection (ILI) [1].

1) Visual inspections involve physically inspecting the pipeline for signs of damage or wear, such as corrosion, dents, or cracks. This can be done by trained personnel who walk or drive along the length of the pipeline or by using aerial or remote-sensing technologies to inspect the pipeline from above.
2) Ultrasonic testing involves using sound waves to detect defects or anomalies within the pipeline. This can be done by sending sound waves through the

pipeline and measuring the time it takes for them to be reflected back. Any deviations from the expected time can indicate the presence of a defect or anomaly within the pipeline.

3) Inline inspection (ILI) involves using specialized tools that are inserted into the pipeline to inspect the interior of the pipe. These tools can use a variety of technologies to examine the pipeline, including magnetic field sensors, ultrasonic sensors, or video cameras. ILI tools can be used to inspect pipelines for a variety of defects, including corrosion, cracks, and dents.

Pipeline maintenance includes activities such as cleaning the pipeline, repairing any defects or damage that has been identified, and replacing worn or damaged components. Regular maintenance can help extend the lifespan of the pipeline and ensure that it is operating safely and efficiently. Some of the maintenance techniques are as follows:

1) Using smart pigs, which are advanced internal line inspection (ILI) equipment that go and search through a pipeline for any metal loss, wall deformation, and fractures. To complete their task, smart pigs use modern technologies such as ultrasonography and magnetic field readings.
2) Another method is an electrical survey, which is based on the idea that corrosion is linked to electron flow. Technicians monitor the voltage differential between the pipeline and the neighboring soil, as well as the amount of current flow between two sites on the pipeline.
3) A third strategy, known as direct assessment (DA), is applied in situations when ILI equipment cannot be employed, which uses statistical analysis to identify the most probable issue areas along a pipeline. Based on the assessment, the laborers dig and inspect the coating and other conditions of the pipe at specific locations, including measuring the thickness of pipeline walls.
4) Finally, pipeline operators must conduct an official risk assessment. Simply put, risk is assessed as the product of the chance of an event occurring and the quantitative assessment of the negative effects of that occurrence. As an example, the likelihood of a catastrophic oil pipeline crack might be as low as 0.5%. If a pipeline ruptures in a densely populated region, the consequences might include millions of dollars in terms of property damage, economic interruption, and loss of life.

Pipeline operators also organize and carry out a variety of pipeline repair and maintenance programs. Pipeline maintenance includes things like cleaning the area around pipelines so that signs and markings can be seen and so that evidence of tiny breaches or third-party incursions may be more easily found in the event that they occur [1]. Other pipeline maintenance tasks include repairing or replacing damaged pipes, reducing stress concentration points on pipes, burying pipelines deeper for increased protection, maintaining cathodic protection systems, and repairing or replacing damaged cathodic protection systems.

4.7.3 Tanker Ships

Crude oil is also carried by maritime tanker ships across the globe. It is the most cost-effective means of distribution, but second only to a pipeline. The first economically viable crude oil tankers were built in the 1870s and 1880s, but World War I was the driving force behind the creation of the vessels that are currently utilized all over the world. The T2 was the most prevalent tanker design decades later, during World War II. It was just over 500 ft long and held around 16,500 tons of oil.

Today's crude tankers must have two steel hulls so that if the exterior hull is broken, the oil will still be kept inside the interior hull. For the same reason, some containers feature twin bottoms. Tankers are classified based on size. The term supertanker refers to very large crude carriers (VLCCs) and ultra large crude carriers (ULCCs) that can transport more than 2 million barrels of cargo. Oil can be pumped in and out of small- or medium-sized oil tankers at piers. But supertankers are often too huge to do so. Instead, they must load and unload their goods at offshore facilities or even shift their cargos to smaller tankers or barges at their destination port.

4.7.3.1 LNG Tanker Ships

The storage tank or containment system of an LNG tanker has a gas-tight seal for many reasons, such as to prevent gas vapor from mixing with air, to insulate the LNG from heat, to restrict boil-off, and to prevent the ship base from becoming too cold so that it does not become brittle and eventually break. The process in which natural gas evaporates from LNG causing its temperature to reduce and thereby maintaining LNG at a very low temperature without any artificial refrigeration is called boil-off. Thermal insulation is also provided to maintain low temperature for LNG as the boil-off gas is removed from the storage tank. There are a few stable designs that have been widely accepted, and these include the self-supporting independent tank and the membrane tank. The shape of the independent tank may be either a sphere (a Moss or Kvaerner Moss) or a box and is both sturdy and massive. Large metal tanks that extend halfway above the deck are used to store the LNG in the Moss layout. Both kinds of storage tanks can handle the pressure of LNG. The membrane tank is a compact metal box with a thin, flexible membrane. A strong insulation is maintained between this metal membrane and the tanker's hull. With the insulation in place, the membrane surface may contract and expand in reaction to changes in temperature, transferring pressures from the LNG tank to the hull of the vessel.

Care must be taken while loading a cargo so as not to shock the LNG tank with sudden heating. During the trading period, it is common practice to keep a little quantity of LNG, known as LNG heel, in the tank. By not needing to cool down the tanks before each loading, time and money can be saved, and the containment system is saved from the potential strain of several thermal cycles (warm-to-cold-to-warm).

4.7.4 Railcars

Railcars are used to transport oil and oil products within the oil industry. They are a cost-effective and efficient mode of transportation, particularly for long distances or for areas without pipelines. Railcars come in various sizes and designs, depending on the type of oil or oil product being transported. Some railcars are designed to transport crude oil, while others are used to transport refined products such as gasoline, diesel, and jet fuel. In recent years, there has been increased scrutiny of the safety of transporting oil by rail, particularly after several high-profile derailments and spills. As a result, there have been efforts to improve the safety of rail transport in the oil industry, including the use of stronger and more advanced railcars.

4.7.5 Tank Trucks

Tank trucks are used to transport oil and gas products by road. They are typically used for short-distance travel or for transporting small quantities of these products. Tank trucks are equipped with a cylindrical container, called a tank, which is used to hold the oil or gas. The tank is usually made of metal, such as steel or aluminum, and is designed to withstand the pressure and temperature of the product being transported. Tank trucks can be used to transport a variety of oil and gas products, including crude oil, gasoline, diesel, and propane. They may be used to transport oil and gas from a production facility to a storage facility or to a refinery for processing. Tank trucks may also be used to deliver oil and gas products to retail outlets, such as gas stations or industrial users.

There are various safety considerations when using tank trucks to transport oil and gas products. Tank trucks must be properly maintained and inspected to ensure that they are in good working order. Drivers must also be trained in the proper handling and transportation of these products, as well as in emergency response procedures. In addition, there are regulations in place to ensure the safe transport of oil and gas products by tank trucks.

4.8 Overview of Refining

The refining of crude oil is a process that involves separating the various hydrocarbons and other chemical compounds contained in crude oil into more useful products. These products include gasoline, diesel fuel, heating oil, and jet fuel. A refinery is a complicated industrial facility that performs a number of processes to convert raw crude oil into consumable products. Refinery operators often take a variety of crude oil types known as crude slate from diverse sources as feedstock and employs a sequence of processing units to turn them into final products. These final items are often categorized in the marketplace as follows:

- Liquefied petroleum gas (LPG), gasoline, and naphtha are examples of light distillates.
- Middle distillates include kerosene and associated jet engine fuels, as well as diesel fuel.
- Heavy distillates and residues include heavy fuel oil, lubricating oils, paraffin wax, asphalt/tar, and petroleum coke.

The refinery normally transforms crude oil into final products in four stages: separation/distillation, conversion, enhancement, and blending.

4.8.1 Separation/Distillation

Distillation is the process of separating crude oil into different compounds which are again processed during the various stages in the refinery. As the first step, leftover water and salts are removed from the crude. It is important to remove water and salt from crude oil prior to distillation, as these impurities can interfere with the refining process and reduce the efficiency and yield of the distillation. There are several methods that can be used to remove water and salts from crude oil prior to distillation. One such method is gravity separation in which the crude oil is allowed to settle in a tank, causing the heavier water and salt to sink to the bottom, while the lighter oil floats to the top. The water and salt can then be removed from the bottom of the tank. Another method is centrifugation in which the crude oil is spun at high speeds in a centrifuge, which separates the water and salt from the oil based on their densities. Heat treatment is another method in which the crude oil is heated to a high temperature, which causes the water and salt to evaporate. The vapors can then be condensed and removed from the oil. Chemical treatment is also done where chemicals are added to the crude oil to react with and remove the water and salt.

After removing the water and salt impurities, the process of distillation starts. Crude oil is a chemically heterogeneous liquid which includes several different hydrocarbon compounds, each with its own boiling point. The initial boiling point of a molecule decreases with decreasing size. For example, at about 85°F, lighter petroleum gases and butanes will start to boil. Between 85°F and 185°F, the gasoline component vaporizes. Even significant residual hydrocarbons will begin to boil about 800°F. Before entering into the distillation tower, crude oil is passed through a furnace where superheated steam increases its temperature to about 750°F.

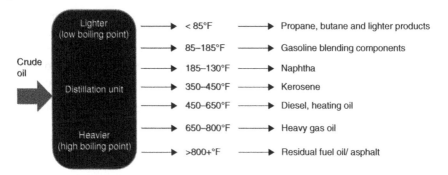

Figure 2 Crude oil distillation/separation process.

The vapor at 750°F or above is introduced at the bottom of the distillation tower or column. As the vapor rises, it cools and condenses, which is then collected in trays kept at different intervals. The distillation tower consists of a stack of horizontal trays at a distance of 1 or 2 ft. As more liquid accumulates in each tray, it is drawn out at different heights in the tower [6]. Lighter items are collected from the top of the column, while heavier liquids are removed from the bottom trays, as shown in Figure 2.

4.8.2 Conversion

The most in-demand and lucrative by-products of the refining business are gasoline and the other chemicals together referred to as middle distillates. In order to maximize profits, it is desirable to transform as many molecules from crude oil into these two product classes as feasible. This process, known as conversion, encompasses a wide range of activities that take hydrocarbons in other forms and transform them into gasoline and intermediate distillates. Conversion, which transforms lower-value products into higher-value ones, is also known as upgrading. Conversion may also be used to produce petrochemicals and lower the viscosity of residual fuel oil.

4.8.3 Enhancement

Enhancement is the application of specific techniques to eliminate undesirable components or compounds from a hydrocarbon during the refining process. Hydro-processing, amine treatment, solvent extraction, and sweetening are the most typical enhancements applied in industry. Here is a brief overview of each technique:

Hydro-processing: This process involves using hydrogen to remove impurities from refined products, such as sulfur, nitrogen, and other contaminants. Hydro-processing can help improve the quality and performance of the products and is commonly used in the production of gasoline, diesel, and other fuels.

Amine treating: This process involves using an amine solution to remove impurities, such as hydrogen sulfide and carbon dioxide, from natural gas and other gases. Amine treating can help improve the quality and purity of the gas and is often used in the production of clean burning natural gas for power generation and other applications.

Solvent extraction: This process involves using a solvent to extract impurities from a product, such as heavy metals or other contaminants. Solvent extraction is commonly used in the production of edible oils, as well as in the refining of crude oil to produce high-quality lubricating oils.

Sweetening: This process involves removing impurities, such as hydrogen sulfide and mercaptans, from natural gas and other gases to improve their quality and make them more suitable for use. Sweetening is often used to make natural gas suitable for transportation and use in homes and businesses.

4.8.4 Blending/Finishing

Most finished petroleum products on the market are the result of a refinery merging a variety of hydrocarbon molecules from various operations. The most obvious example is gasoline. Straight-run gasoline has an octane value that is too low for direct usage in current engines as it leaves the distillation tower. To make the gasoline usable to customers, blendstocks are mixed into it. One key gasoline blendstock is ethanol, which is added downstream, often at a wholesale gasoline distributor's storage depot, rather than at the refinery. This provides the refiner more options when it comes to selling gasoline. Some other types of blendstocks that are commonly used in the oil and gas industry are diesel blendstocks and jet fuel blend stocks. They are formulated to meet specific specifications for a particular application. They are also used to manage the price and quality of the final products and to optimize the use of resources.

4.8.5 Types of Refineries

Following are the four classifications of refineries based on their complexity.

4.8.5.1 Basic or Topping Refineries

These refineries are primarily used in China. They use distillation to separate the components of crude oil into various products. The term "topping" refers to the process of converting heavy, high-boiling point hydrocarbons into lighter, lower-boiling point products, such as gasoline. Topping refineries are larger in scale and are designed to process a wide range of crude oils into a wide range of products. They are typically used to produce fuels for global markets and are often located in areas with access to large crude oil deposits.

4.8.5.2 Hydroskimming Refineries

They are types of oil refineries that are designed to process crude oil into a range of products. The term "hydroskimming" refers to the process of separating the various components of crude oil based on their specific gravity, rather than their boiling points, as is done in conventional refining processes. In a hydroskimming refinery, the crude oil is heated and passed through a series of separation stages, where it is subjected to various physical and chemical processes to separate the different components of the oil. The separated components are then further processed to produce a range of end products, such as gasoline and diesel. The main drawback of this type of refinery is that it produces a lot of residual oil. They are typically used to produce fuels for local markets and are often located in areas with access to small- or medium-sized crude oil deposits.

4.8.5.3 Conversion Refineries

It is also called the cracking oil refinery. It has all the features of topping and hydroskimming refineries. It also incorporates catalytic cracking and hydrocracking units. A cracking refinery uses catalysts and high temperature to reduce the amount of residual oil and produces more valuable products. Compared to the other refineries, a cracking refinery also contains a vacuum distillation unit, fluid catalytic cracking (FCC) unit, and an alkylation unit. The main advantage of these type of refineries is that they are capable of producing high-value products. But the disadvantage is that their operating cost is higher.

4.8.5.4 Coking or Complex Oil Refineries

They have all the features of a conversion refinery plus an additional coking unit. This unit makes it possible to convert very heavy crude oil to lighter fractions, thereby producing quality and high-value products.

Refineries are also classified based on their processing capacity. They can be small refineries having a capacity of less than 50,000 barrels per day (bpd), mid-sized refineries with a capacity of 50,000–200,000 bpd, large refineries having a capacity of more than 200,000 bpd, and super-large refineries capable of producing more than 500,000 bpd. The size and capacity of an oil refinery depend on a variety of factors, including the type of crude oil being processed, the products being produced, and the location of the refinery.

4.9 Marketing and Distribution of Oil and Gas

This is the final stage in the production of oil and gas. Marketing of oil and gas products involves determining the demand for various products, as well as the price at which they can be sold. Companies will often enter into contracts with large customers, such as airlines and trucking companies, to secure a steady source of demand for their products. They may also invest in advertising and promotional campaigns to increase awareness of their products and build their brand. Distribution of oil and gas products can take many forms, depending on the product and the location of the market. For example, gasoline is typically transported by trucks or pipelines to retail gas stations, where it is sold to consumers. Jet fuel, on the other hand, is often transported by pipelines or trucks to airports, where it is delivered directly to airlines. The end-users of oil and gas are varied, and they include industrials, power generation, residential, commercial, and transportation sectors. The distribution to these end-users will be based on the service offered by the supplier and the kind of usage the end-users want. The final marketing and distribution of oil and gas products is a complex and highly regulated process, which requires significant infrastructure and expertise.

4.10 End of Production

Oil and gas wells must be capped and abandoned when their reserves have been exhausted to the point that further extraction will result in economic loss. To avoid poisoning the aquifer with saltwater, this is mandated by law. Well casing removal and salvaging is the first phase in the process. After that, cement or mechanical plugs are installed. These effectively block off the well's high-pressure and high-permeability production sections. A cement cap is sometimes installed at the top of a well to prevent groundwater seepage into the supply of fresh water. Similarly, onshore wells in the ocean are sealed off and left abandoned, and all underwater equipment is retrieved. Only the deck equipment is normally retrieved from an abandoned offshore rig. The legs of the deck may be cut off at the bottom of the water, and the rig can then be pulled up onto dry ground for repair or disposal. The construction may also be tipped over so that it rests flat on the bottom, creating a fantastic artificial reef for fish and other aquatic creatures.

4.11 Factors Influencing the Timing of Oil and Gas Exploration and Production

There are several factors that can influence the timing of oil and gas exploration and production process, including the following:

4.11.1 Physical and Technical Factors

The location of the prospective or discovered oil or gas, together with the projected volumes and ease of extraction, is a crucial factor in determining the speed and complexity of the operation. The location and depth of the underlying geology can have a role. For example, working in deep water offshore presents greater challenges than in shallow water. Hard infrastructure, accessibility and transportation, water, and electricity are also taken into account. If they are not already there, International Oil Companies (IOCs) will need to construct and provide them before the operations begin.

4.11.2 Social and Political Factors

Government capability and community responsiveness are two aspects that impact investment choices and the rate with which operations move.

Capacity of the government: The government is in charge of establishing and executing the regulatory environment. If the regulatory environment is not stable and predictable, or if it is not favorable to exploration or production, it might delay or prevent development activities. The degree of coordination between layers of the government, such as national or local, and across government departments, such as those responsible for natural resources, the environment, labor, and the financial plan, among others, is also important in determining how quickly the government can establish a conducive regulatory system, as are accountability, transparency, and civil service capacity.

Community reaction: The local population's mindset may determine whether or not to invest initially, as well as whether or not to pursue further exploration and production. A community may petition the government or take direct action to influence or prevent oil or gas production. Community reactions to exploration and production are influenced by a variety of factors, including regulatory and governance factors (such as government accountability and income transparency), legal considerations (such as land ownership), and company participation in the areas in which they operate [7].

4.11.3 Business Coordination Factors

IOCs may share responsibilities via joint venture frameworks, in which the IOC or the government provides access to develop and produce assets based on agreed-upon rights to reserves. Several corporations have different proportionate holdings in joint venture activities, and operators may have the ultimate responsibility for the technical and broader socioeconomic consequences of operations. This might have an effect on their capacity to interact with communities as well as the visibility of their community engagement initiatives. The level of government participation in manufacturing, as well as the speed with which government decisions are made and the kind of engagement by other enterprises, may all influence how quickly contracts are signed and how readily technical and political difficulties may be resolved.

4.12   Non-revenue Benefits of the Oil and Gas Industry

Job creation and local content: In the oil and gas life cycle, many types of jobs are created: direct skilled and less skilled employment, indirect jobs, and induced jobs, which are jobs produced as a consequence of wages and salaries spent by people engaged in direct and indirect occupations. Many of the IOCs' operational countries in the developing world provide few chances for locals to get unskilled or semi-skilled labor in indirect services. As a result, firms may occasionally assist local enterprises in entering the supply chain by providing capacity development in order to deliver competitive local products, services, and skills. Other corporate efforts include increasing the ability of local businesses to participate in the supply chain, such as via education and skill development.

Infrastructure development: Oil fields have to be accessible by road, sea, or air for the most inaccessible locations, and they must have access to essential utilities such as water and power. Companies may spend in developing these if they do not already exist at an exploratory location. Companies will invest in infrastructure to achieve two goals: the business case for the company's operations and compliance with local laws and infrastructure investment needs particular to the location. Companies and governments may coordinate their interests and goals in the development of infrastructure and investment planning. The more this happens, the better the long-term value to communities.

Social investment: Many IOCs expressly establish community involvement and investment initiatives in the regions where they operate. Social baseline and impact assessments may be carried out in response to regulatory obligations, in assisting corporations in assessing social risk, identifying priorities for social investment, and monitoring. It is frequently more beneficial for both the firm and the community when this is part of the basic business strategy. Social investment may refer to infrastructure and local content development, philanthropic contributions, and expenditure on non-business activities. To be most successful, social investment should begin early in the life cycle and become an integrated component of the business strategy and development, with the goal of mitigating the effects of oil and gas firm presence. Social investment often consists of programs that provide education and skill development, as well as environmental, health, and safety services, with the goal of benefiting the local community.

4.13   Conclusion

The oil and gas business sector is one of the world's biggest and a significant contributor to the global economy. Oil and gas production is the broad process of extracting oil and natural gas from oil fields and converting them into usable petroleum products. It involves a series of phases, beginning with site discovery and ending with product distribution to enterprises and the general public. In this chapter, we have discussed the different processes involved in the production of oil and natural gas starting from the formation of crude oil. An overview of all the three stages of production in the oil and gas industry, namely, upstream, midstream, and downstream, has been provided. It explains how crude oil is turned into useful products through exploration, drilling, transportation, storage, refining and final marketing and distribution. In addition, the readers are also provided with information about the various factors influencing the timing of oil and gas production and some of the non-revenue benefits of this industry.

References

1 Hilyard, J.F. (2012). *The Oil and Gas Industry – A Nontechnical Guide*. PennWell Corporation.

2 Reyes, J. (2022). Content Specialist, SafetyCulture. An article on "A Simple Guide to Oil and Gas Production". https://safetyculture.com/topics/oil-and-gas-production/ (accessed 03 January 2023).

3 Devold, H. (2013). *Oil and Gas Production Handbook - An Introduction to Oil and Gas Production, Transport, Refining and Petrochemical Industry*. ABB Oil and Gas.

4 Solken, W. https://www.wermac.org/others/oil_and_gas_transportation.html (accessed 24 December 2022).

5 AZO Materials (2017). Article on – "Liquefied Natural Gas – Simpler Natural Gas Transportation". https://www.azom.com/article.aspx?ArticleID=14373 (accessed 11 January 2023).

6 U.S. Energy Information Administration – EIA (2012). Crude oil distillation and the definition of refinery capacity. https://www.eia.gov/todayinenergy/detail.php?id=6970#:~:text=Crude%20oil%20is%20made%20up,products%20boil%20off%20and%20are (accessed 23 January 2023).

7 Darko, E. (2014). *Short Guide Summarising the Oil and Gas Industry Lifecycle for a Non-technical Audience*. Overseas Development Institute.

5

Explaining the Midstream Activities in the Oil and Gas Domain

5.1 Introduction

Since producers required a mechanism to process and separate the various products from oil and gas, storage and transport of these products to a refinery, and eventually to customers, lead to the establishment for the midstream industry. As production is often remote from refining plants and demand centers, storage and transportation are essential components of midstream oil and gas activities [1]. Upstream, midstream, and downstream oil and gas operations can be used to categorize the complete oil and gas energy value chain. Companies in the midstream industry offer connections between the upstream and downstream market divisions. While some businesses specialize in the processing and storage of crude oil and natural gas, others offer logistics, pipeline construction, railcar supply, transportation, transloading, and other services [2]. In the midstream industry, digitization replaced outdated data processing techniques with automated ones, increasing productivity and effectiveness in all sectors. The organization's operating costs have decreased, and the output has improved as a result of the adoption of cloud computing, IoT, robotics and automation, 3D technologies, etc. The midstream industry also has to deal with social and environmental issues, cyberattacks, and product manufacturing, storage, and transportation obstacle issues.

This chapter may be divided roughly into two parts. The first part of the article discusses the numerous activities carried out in the midstream sector, while the second part provides information on the industry's digital transition and the constraints it faces.

The Power of Artificial Intelligence for the Next-Generation Oil and Gas Industry: Envisaging AI-Inspired Intelligent Energy Systems and Environments, First Edition. Pethuru Raj Chelliah, Venkatraman Jayasankar, Mats Agerstam, B. Sundaravadivazhagan, and Robin Cyriac.
© 2024 The Institute of Electrical and Electronics Engineers, Inc.
Published 2024 by John Wiley & Sons, Inc.

5.2 Role of Midstream Sector in Oil and Gas Industry

One of the three phases of oil and gas energy operations is the midstream sector. The upstream and downstream sectors are two of the other phases.

Upstream operations, which comprise activities involved in drilling and producing hydrocarbons in significant quantities, represent the initial stage of oil and gas operations. The second step, known as midstream operation, comprises the preparation, storage, and delivery of crude oil, natural gas, and NGLs to refineries. Prior to being delivered to the final customers, the items must also be compressed and treated to remove waste. Downstream operations, which include the processing, marketing, and delivery of crude products to the end customers, are the final stage of oil and gas activities. The crude oil is converted into diesel, jet fuel, gasoline, and other fuels in the downstream phase for use in automobiles, aircraft, and residences, among other things [3].

The upstream stage of the production of oil and natural gas is where drilling and oil extraction take place. The midstream phase then begins, during which the workers move the unprocessed hydrocarbons from the well into a central facility using a pipeline. After storing the hydrocarbons, the facility processes and separates them into their various components. The finished goods are transported farther downstream via a pipeline, truck, or train, where they are stored and further processed into consumable commodities.

Midstream assets include all resources utilized in the midstream oil and gas sector for the processing, storage, and transportation of oil, natural gas, natural gas liquids (NGLs), and other hydrocarbons. Assets include gas storage facilities, fractionating and dehydration tanks, fluid compressors, oil pipelines, and LPG and LNG storage plants. Midstream assets represent significant financial investments in several economies, amounting to billions of dollars. [3]. Integrated and independent midstream firms are the two primary business categories involved in the midstream oil operations. To optimize the amount of oil and gas they produce, integrated oil firms incorporate upstream, midstream, and downstream activities in their operations. Businesses engaged in oil and gas activities that require access to midstream assets might get services from independent midstream companies.

Midstream operation is a highly regulated industry in practically every nation. Numerous federal and state rules control the pricing, interstate pipeline transportation of oil and gas throughout the nation, and domestic state-level distribution operations of oil and gas businesses [2]. For instance, in the United States, the Federal Energy Regulatory Commission is the entity in charge of midstream regulation (FERC). The Petroleum and Natural Gas Regulatory Board (PNGRB) in India is in charge of overseeing the midstream and downstream industries. Similarly, every other country has its own regulatory body for controlling and regulating the oil and gas industry.

5.3 Midstream Oil and Gas Operations

The following are the main operations in the midstream sector:

5.3.1 Field Processing

A gas–oil separation unit separates the combination of oil, natural gas, and NGLs that are released when oil is retrieved from an oil field. It also separates the water from oil and gas. While the oil, gas, and NGLs flow into their appropriate storage units, the water is recycled or pumped into a disposal well [3]. Measurement of the product, the elimination of trash, and temporary storage of the product prior to transportation to the refineries are additional operations that take place during field processing. The process of calculating the mass or volume of generated hydrocarbons is known as oil and gas measurement.

Oil measurement: Oil is sold in barrels of crude oil (BBL), which are 42 US gallons in volume [2]. One of two methodologies can be used to determine the volume after removing water atoms and sediment from crude:
1) Oil is measured directly from storage tanks in runs for modest amounts (1–100 barrels per day). A run is when oil is extracted from a lease area and transported somewhere for processing. In order to separate the water molecules and sediments from the mixture, we first take a sample of crude and put it in a centrifuge for a test known as a "shake-out." After that, a measuring strap is lowered into the storage tank to determine the run's volume. This tool has a weighted end and an outlet valve that opens to release liquids into a neighboring pipeline or vehicle. To determine the precise volume of oil pulled, a second measuring strap is obtained (and the quantities are compared) when the storage tank is virtually empty.
2) Operators often utilize lease automatic custody transfer (LACT) devices, an automated measuring method, for high volumes (100–100,000 barrels of oil per day). They are designed to precisely quantify and sample different hydrocarbon liquids.

Measurement of natural gas: Orifice meters are commonly used to measure natural gas. Orifice meters help in determining the volumetric flow of natural gas by detecting the differential pressure between the upstream and downstream sections of a partially blocked conduit (orifice). Natural gas is sold in British thermal units (MMBTu) or thousands of cubic feet (MCF) [3].

In order for natural gas to be of adequate quality to be transported via facilities and consumed by end users, contaminants must be removed from the gas stream. Some deposits produce natural gas that is so pure when it leaves the well that it may be delivered straight to the pipeline without any further treatment. Natural gas may include contaminants that might pose risks or cause other issues as the liquid components are removed from it. H_2S, which natural gas may or may not include, is the major contaminant. H_2S is corrosive to pipelines and hazardous to people (perhaps lethal at certain quantities). Therefore, it is preferable to get rid of H_2S as quickly as possible throughout the conditioning process. The threat caused by water vapor is another issue. Gas hydrates, particles that can block pipelines, valves, and gauges, are ceated when water reacts at high pressures with the elements in the gas. This is especially true at low temperatures. In the subsurface, natural gas may be combined with nitrogen and other gases. Prior to sale, these additional gases must be separated. Benzene, toluene, ethylbenzene, and xylene, sometimes known as BTEX, are high-vapor pressure hydrocarbons that are liquid at surface temperature and pressure and are extracted and treated separately. Water is removed from the gas stream via dehydration. Refrigeration, absorption by liquid desiccants, and adsorption by solid desiccants are the three basic methods of dehydration. Glycol that absorbs water is exposed to the gas when a liquid desiccant is used. Heat regeneration is a technique that may be used to evaporate the water from the glycol. Afterward glycol can be used again. Crystals having large surface areas that draw water molecules are solid desiccants, also known as molecular sieves. Simply heating the solids over the boiling point of water will regenerate them. Finally, just chilling the gas to a temperature below the water condensation threshold can remove enough water to transport the gas, especially for gas collected from deep, hot wells. The most used method among the three afore-stated methods is glycol dehydration. The process of sweetening involves removing H_2S and, occasionally, CO_2 from the gas stream. Amine treatment is the most used technique. The H_2S will react with the amine solution in the presence of the gas stream in this procedure, separating it from the natural gas. After that, heat is applied to the contaminated gas solution to separate the gases and regenerate the amine. When there is a market for it, the sulfur gas can be sent to a sulfur-recovery factory to produce elemental sulfur, which can be sold [4]. The sulfur gas can also be disposed of by flaring or incineration.

5.3.2 Storage

Aboveground and underground atmospheric (non-pressure) and pressure tanks hold crude oil, natural gas, LNG, LPG, and other petroleum products. Refineries, terminals, bulk plants, marine loading/unloading facilities, and pipeline storage sites have storage tanks [5].

Tank farms are clusters of storage tanks at manufacturing locations, refineries, distribution hubs, and bulk factories. Crude oil, intermediate stocks, petroleum feedstocks, and finished petroleum products are kept in vertical and horizontal aboveground atmospheric and pressure tanks. Storage tank size, form, configuration, and design and operation are determined by the quantity and kind of items stored and by governmental or organizational laws. Aboveground vertical tanks can have double bottoms and cathodic protection to prevent leaks and corrosion. Horizontal tanks can be double-walled or vaulted to prevent leakage. Large underground storage reservoirs are typically used to store natural gas. Depleted gas reservoirs, aquifers, and salt caverns are the three primary categories of subsurface storage. Natural gas can be stored as LNG in aboveground tanks in addition to subsurface storage. Depleted gas reservoirs are the most prevalent and typical type of subsurface storage. Formations known as depleted reservoirs have previously had all of their recoverable natural gas extracted from them. Aquifers are porous, permeable rock formations under the earth's surface that might be repaired and utilized as natural gas storage spaces. Salt caverns are underground salt deposits that are ideal for storing natural gas.

Pipelines and ships are the primary modes of transport for the bulk of crude oil and petroleum products on their way to terminals for storage. Transporting crude oil and other petroleum products from ports to refineries, other terminals, bulk facilities, gas stations, and eventually to customers requires the use of pipelines, maritime tankers, railcars, and tank trucks, among other modes of transportation. Terminals are able to be operated and owned by a wide variety of entities, including big oil companies, pipeline companies, autonomous terminal operators, significant industrial or commercial customers, and distributors of petroleum products. The majority of the time, bulk plants obtain petroleum products through the use of railroads or tank trucks, often from terminals rather than directly from refineries. They are typically smaller than terminals. Tank trucks transport goods from bulk facilities to gas stations and customers for redistribution. Oil firms, wholesalers, or independent proprietors may run bulk facilities.

Storage is crucial in ensuring that the supply and demand of oil in the global market are balanced. Large amounts of oil and gas are stored in the storage tanks, which offers market participants a competitive edge in setting oil prices [3]. When it comes to setting market prices for petroleum goods, the petroleum industry will have the least amount of power if storage tanks run out of oil and gas. The price of crude oil and its refined products will thereafter be significantly under the producers' control in the global market.

5.3.3 Transportation

Since crude oil and gas must be transported to other places for additional refining and delivery to the end users, transportation is indeed a crucial part of the midstream industry [3]. Crude oils, compressed and liquefied hydrocarbon gases, liquid petroleum products, and other chemicals are transported from their place of origin to pipeline terminals, refineries, distributors, and customers using pipelines, marine vessels, tank trucks, rail tank cars, and other means. The natural liquid form of crude oils and other liquid petroleum products is maintained for transportation, handling, and storage efficiency. Both in their gaseous and liquid forms, hydrocarbons must be entirely contained in cylinders, pipes, tanks, or other containers before usage. The most crucial feature of liquefied hydrocarbon gases (LHGs) is that they take up a little amount of space when stored and transported as liquids before expanding into a gas when needed [6].

5.3.3.1 Pipelines

Since they are safer, quicker, and more cost-effective than alternative methods, pipelines are the most popular method for moving crude oil and gas across distant locations. Domestic, commercial, and industrial transportation use pipelines. Aboveground, underwater, and underground pipes with diameters ranging from a few centimeters to a meter or more are used to carry substantial volumes of crude oil, natural gas, and liquid petroleum products. These pipelines are also used to transport greenhouse gases. Pipelines transport crude oil, compressed natural gas, and liquid petroleum products over long distances at speeds ranging from 5.5 to 9 km per hour. These materials are moved through the pipelines by large pumps or compressors that are placed along the length of the pipeline at intervals ranging from 90 to over 270 km. The distance between pumping or compressor stations must be determined, in part, by the size of the pipeline, the product viscosity, the pump capacity, and the kind of terrain that must be crossed. Pipeline pumping pressures and flow rates are regulated throughout the system to ensure that the product is constantly transported through the pipeline, in spite of the factors that can affect the operation.

Pipelines may be roughly classified into two groups [7]. The first category includes petroleum pipelines that carry crude oil or NGLs. The crude oil or NGL from the producing wells is collected by the collecting pipeline systems. The crude oil pipeline system is then used to deliver it to a refinery. Petroleum is delivered via pipeline networks for refined products to storage or distribution facilities after it has been converted into goods like gasoline or kerosene. The second category consists of natural gas pipelines, which are used to carry gas wells' raw materials to distribution systems in various towns and cities. The finished product is subsequently distributed to specific houses through the distribution system. Additionally, three separate pipeline types – gathering systems, transmission systems, and distribution systems – are used in this operation. The raw materials are gathered from producing wells using the natural gas collection pipeline system. Large transmission pipeline networks are then used to transfer it from the plants to ports, refineries, and cities across the nation. Finally, the product is distributed to households and businesses through a network as part of the distribution systems.

The main benefit of pipelines is that they do not emit carbon dioxide into the environment, unlike other modes of transportation like trains, trucks, and tankers, which might cause acid rain, ozone depletion, and poor air quality. Most of them are sealed, which lessens the impact on animals. Additionally, the sealed design prevents more highly volatile substances from spilling, thus lowering the chance of an explosion while in transit. Both wildlife and people are vastly safer as a result of this. Compared to other modes of transportation, pipelines need very little space above ground. This is because, with the exception of locations with natural rock formations, the majority of pipelines are buried beneath the earth's surface. Additionally, because pipelines pass through less heavily inhabited regions, the effects of accidents there are significantly less severe.

Despite the fact that pipelines are practical and seem like a better alternative than other forms of transportation, there are still things to be concerned about, such as construction, spills, and leakage. For initial construction, pipelines need a substantial investment and several municipal and federal approvals. People who live near pipelines, wildlife, and plants all suffer when there is an oil leak. Gas leaks via pipes that have lost their structural integrity run the risk of igniting or disseminating poisonous gases.

5.3.3.2 Trucks

From producing facilities to refineries or from refineries to final customers, trucks are utilized to carry oil and gas. Frequently, a single tanker truck may transport up to 34,000 liters of product or around 200 barrels of oil [7]. Trucks transport the least amount of petroleum products, but they make up for it with their accessibility and adaptability. Wherever there is a little distance to be traveled, they are still the best choice. Tanker trucks are an essential part of the business since fueling stations sometimes lack railway tracks and have limited storage space.

The least energy-efficient way to transport oil and gas across land is by tanker trucks since it requires roughly three times as much energy to transfer the same quantity of oil by tanker as it does by rail. Of all the ways to move oil and gas, trucks are perhaps the most hazardous. It is rated first in terms of monetary damage, past environmental spills, and threat to human life of all transportation options.

5.3.3.3 Rail Transportation

Since the goods may be loaded onto several locomotives for further transit to a refinery or storage facility, railcars are regarded as one of the preferred and affordable modes of transporting oil and gas. Rail transportation continues to be quite perfect in terms of cost and efficiency. When oil has to be transported fast over shorter distances, rail often has low fixed costs and quick construction timeframes. Rail is a preferred mode of transportation if there is considerable demand for oil in a region where trucks, tankers, and pipelines are restricted in various ways.

Transporting oil by train has a significant risk of accidents, and oil spills are very harmful because of the damage they inflict to the environment. The dynamics of the local wildlife are changed, the land can be degraded but almost never entirely repaired, and pollution from spills can have an adverse effect on drinking water, which in turn affects human health [7].

5.3.3.4 Ship or Marine Vessels

Crude oil and natural gas stored in storage containers may travel significant distances aboard ships. Due to the enormous amount of merchandise they can transport at once, large ships may drastically lower midstream transit costs. A tank barge is a fantastic example because it can travel hundreds of miles over the sea and carry up to 27,000 barrels of petrochemicals and natural gas to chemical refineries. Large ships called crude oil tankers are used to deliver crude oil from various nations to refineries all over the world. Although oil tankers come in various sizes, refined petroleum products are often transported in smaller boats, while crude oil is typically transported in much bigger tankers. Transporting cryogenic gases over sea requires specially designed vessels with reservoirs or compartments that are extremely well insulated. The most common example of cryogenic gas is LNG. After being unloaded at the delivery port, the LNG is transported to either storage facilities or regasification plants for further processing. It is possible to carry LPG in the form of a liquid or in a cryogenic state in marine vessels and barges that are either insulated or not insulated. LPG may also be carried as cargo in containers aboard boats and barges used for maritime transportation.

5.4 Technological Advancements in Midstream Sector

Leading midstream oil and gas companies employ digital transformation as one of their key advancements to boost asset utilization through increased recovery, quicker production, higher efficiency, decreased downtime, capex, and regulatory events. Finding and utilizing strategies to transform operations more quickly than the competition is a major problem for larger midstream oil and gas firms [8]. Whether oil sector businesses can make daring decisions required to shift to the new energy future will determine their existence. The oil and gas business will be greatly aided by technological advancements. Additionally, the cost of effective technological utilization will be oil operations that are safer, more efficient, and ecologically responsible. Let us examine how cutting-edge technology and the Internet of Things (IoT) are assisting oil and gas industries in accelerating digital transformation and navigating the present economic landscape [9].

5.4.1 Cloud Computing

Businesses can grow their data management and storage using cloud computing, which increases the flexibility of their infrastructure budgets. Higher operational efficiency is made possible by the high computational power provided by cloud platforms, which also encourages the use of additional technologies including intelligent asset management of physical assets. Companies can remotely monitor and operate wells, fields, and production facilities with the help of cloud platforms. When assets require repair, we will have unplanned disruptions, which can be minimized, and better decisions can be made about when to act, with the use of data analysis in the cloud. Several Industrial Internet of Things (IIoT) modules are the basic building blocks of a digitized oilfield. These IIoT devices are responsible for gathering data on the oilfield's functioning. The information might be stored in the cloud, which provides an alternative for storage that can scale to accommodate extremely large amounts of data collected. Cloud technology is often more cost-effective than owning servers for businesses due to the fact that companies are only charged for the amount of storage space that they actually utilize in their day-to-day operations. This can cut spending [10].

In the midstream industry, refineries use sensors with predetermined replacement and maintenance schedules to track pressures inside storage tanks. These data may be analyzed in the cloud to provide condition-based maintenance, which can cut costs and labor hours. Oil and gas firms have made significant investments in a selected few cloud-based systems, including Microsoft Azure and Amazon Web Services [11].

Let us go through some major cloud computing benefits in the oil and gas industry:

- **Improved agility**: Cloud computing increases a company's agility, allowing it to make quick choices. Because of effective data mining, new models may be introduced in a relatively short time frame.
- **Cost savings:** By storing their data on the cloud, businesses may avoid the significant one-time costs of purchasing big servers and the challenges of dealing with space constraints. Data centers provide all of the expertise that is required to keep remote servers up and running, with the added benefit of reducing the likelihood of a power outage and including cost mitigation provisions in the contract in the event that servers fail. This results in a significant increase in productivity.
- **Better security and compliance:** In contrast to the early days of cloud computing, when data security was a major concern, data centers have now made significant investments in their security services by effectively deploying cutting-edge protection against physical and online threats. They have also made it a priority to recruit more security staff, which helps the client manage

security compliance and eliminates the need for clients to have qualified security specialists working in-house.

- **Increased efficiency:** Supply chain management, customer data, papers, and other resources are all in one place, making them easily accessible to anybody who requires them. Cloud-based applications, such as Google Docs, allow numerous individuals in various places to collaborate on the same document at the same time, which can boost efficiency.

- **Lower environmental impact:** Oil and gas companies have long been the target of criticism for the harm they cause to the environment. They now need different techniques in order to lessen their impact on nature. Cloud computing is quite successful in this context. In order to accomplish this objective, data centers place a high priority on energy efficiency and green technology developments. Cloud computing and improvements in artificial intelligence are enabling oil and gas corporations to find and prevent natural gas flare-ups and leaks, both of which emit massive volumes of carbon dioxide. Companies detect and shut down methane leaks using sensors, artificial intelligence, and cloud computing as soon as they occur.

5.4.2 Internet of Things

The IoT is rapidly transforming the oil and gas sector by linking assets, people, goods, and services and simplifying data flow. It allows real-time decision-making; enhances asset performance, process, and product quality; and opens up new possibilities. IoT technologies enable oil and gas companies to effectively monitor environmental conditions and ensure compliance with emissions and waste requirements. Real-time monitoring reduces reaction time and expenses. Such response aids in preventing non-compliance and thereby lowering sanctions. This is critical since sanctions for leaks and emissions are nearly equivalent to the amount of material spilled, thereby doubling the company's loss.

5.4.2.1 Benefits of IoT in Midstream

- **Real-time visibility**: IoT solutions enable oil and gas organizations to continually monitor their facilities and track ships or delivery trucks by providing a real-time insight into equipment performance and safety status, environmental conditions, and fleet activities.

- **Predictive maintenance**: The data collected by IoT sensors deployed in oil and gas plants allow for predictive analytics and root cause investigation. They assist in identifying pre-failure problems and performing preventative maintenance to minimize catastrophic downtimes and ensure smooth operation and wiser asset usage. This capability is particularly useful in the maintenance of remote offshore installations, which sometimes lack visibility and rely on human checks. Using IoT optimizes maintenance plans to eliminate unnecessary service appointments while assuring maximum equipment life.

- **Hazard management**: IoT technologies detect the presence of combustible gases and poisonous vapors in the surroundings and aid in the prevention of gas leaks and oil spills.
- **Lower environmental impact**: The oil and gas sector is one of the most significant contributors to climate change, accounting for more than 40% of worldwide greenhouse gas emissions from direct operations as well as fuels produced. To minimize their carbon footprint, upstream operators may use IoT-enabled leak detection and predictive maintenance to reduce airborne contaminants and flaring, while downstream operators can use IoT to increase energy efficiency.
- **Regulatory compliance:** IoT solutions give comprehensive data on facility operations, paving the door for improved compliance with industry norms and regulations.
- **Fleet administration**: To maximize carrier usage and promote preventative maintenance, IoT systems provide real-time monitoring of tanker and vessel position and condition, idle time, cargo status, and so forth. This capability may be enhanced further with IoT-enabled geofencing. Oil and gas firms can designate routes and boundaries for each carrier in order to save fuel usage and ensure that their tankers follow the selected route and arrive on time.
- **End-to-end pipeline connectivity:** Ensure safe and cost-effective oil and gas transportation by monitoring pipeline metrics such as temperature, flow, and pressure using IoT. The information gathered is utilized to detect anomalies, optimize human efforts, and avert mishaps caused by equipment failure. SCADA-based applications use sensors to monitor and enhance environmental protection by assessing air or water quality at stations and important sites, leveraging the IoT.
- **Digital twin**: IoT data are also used to construct digital twins, which are virtual clones of real assets that replicate its operations, states, and lifetime. If a problem occurs, a digital twin is utilized to investigate it, determine the underlying cause, and devise a repair strategy. It improves asset management and performance and allows for comparative asset performance monitoring.

5.4.3 Robotics and Automation

Robotics and automation can assist organizations in the oil and gas sector in identifying and mitigating risk before it becomes a problem. Robotics and process automation not only speed up operations but also eliminate the need for manpower, enhancing efficiency and decreasing human-induced mistakes. Oil in bodies of water is created by naturally occurring seepage or by spilling from a production platform. Natural seepage moves at a slower pace with time. Identifying the source of oil existing outside a drilling site is crucial for upstream oil and gas firms to ensure regulatory compliance. Because pinpointing the specific reason can be challenging when there are several sites of naturally occurring seepage, artificial intelligence (AI) robots are utilized to do detailed on-site studies [9].

Working conditions in the midstream industry, while somewhat better than in the upstream sector, can nevertheless be hazardous. In this case, robots and automation technologies address dangers while improving workplace safety. There are several possibilities for robots and automation in preserving pipeline integrity, which is a primary emphasis of the midstream business. Unmanned aerial vehicles (UAVs), often known as drones, are rapidly becoming an important inspection facilitator by combining high-resolution sensors and cameras to capture visual data on the status of assets such as rigs, platforms, tanks, columns, and elevated structures. This can reduce costs by eliminating the need to install and dismantle scaffolding and improve safety by eliminating the need for inspectors to operate at heights or enter dangerous confined places. The tank conditions may be examined for deformation, cracks, corrosion, or coating degradation utilizing UAV technology. Furthermore, UAV-based methane detection technology is utilized to detect and quantify methane and leakage of other hydrocarbon gases [12]. Unmanned, untethered autonomous surface vehicles (ASV) and autonomous underwater vehicles (AUV) employ geofencing data to monitor and report real-time pipeline conditions in the water. These data, when applied to machine vision algorithms, may swiftly analyze and identify if a pipeline infrastructure repair is required, without the need for human involvement. Smart pigging systems employ magnetic flux leakage testing and ultrasonic testing within pipes to identify break or leak possibility such as cracking, erosion, or thickness irregularities. Other pipe analysis technologies generate a three-dimensional image of the complete pipe segment, highlighting blockages, out-of-alignment portions of the pipeline, pipeline surface deformations, and more. All this is highly useful for pipeline maintenance planning.

The advantages of automation and robots in the oil and gas industry are obvious. Fixed cameras, drones, and subsea robotics could considerably decrease or even eliminate the labor required to undertake remote asset surveillance and inspection and scaffolding work onboard platforms. Such technology could reduce inspection costs, enhance worker health and safety, and reduce emissions.

5.4.4   3D Technology

The 3D technology suite, which includes 3D visualization, animation, and printing, is also helping oil and gas companies reach key performance indicators across sectors.

5.4.4.1   Safety and Training

The advantages of adopting 3D animation and virtual reality for oil and gas safety education and training are enormous. Workers must be able to respond appropriately and rapidly in life-threatening circumstances. Operators may use 3D technology to mimic situations and teach staff safely without putting themselves in danger. Trainees, for example, can respond to a virtual oil spill scenario and repeat exercises until competency is acquired. Because no physical equipment or precise position is required, virtual reality technology can result in huge cost cutting. This technology may also be utilized to improve corporate safety rules and processes. Corporations can match safety processes with industry regulatory standards by using training metrics and equipment operational specifications.

5.4.4.2   Exploration and Planning

3D modeling offers value by lowering the expenditures of iterative procedures inherent in exploratory drilling. Geologists and engineers, for example, may utilize 3D models built from real-time data to see realistic depictions of oil and gas deposits hundreds of feet deep or underwater. This improves planning and enhances prospect development accuracy. Engineers can also employ 3D animation to complete several design iterations prior to final submission. They may tweak their ideas while saving money for the final product this way.

5.4.4.3   Replacement Parts

Oil industry equipment is often complicated, extremely sensitive, and expensive to replicate. The most major advantage of 3D printing in the midstream segment is the ability to produce replacement components on-site. Oil and gas pipelines with lengthy and complicated geometries can be challenging to repair and maintain, especially in isolated places or areas with a volatile sociopolitical climate. Furthermore, sending components to faraway places might prolong the time it takes to recover from an accident. Process disruptions may be avoided by 3D printing replacement components. There are thousands of kilometers of oil and gas pipelines throughout the world, and each line's structural integrity is vital. Pigs have been employed for generations to check that these lines are operating properly. If there is a problem with the pig, reverse engineering technologies may quickly 3D print a replacement one on-site. Smart pigs have been introduced as a result of 3D scanning technology, which scans and then 3D prints legacy pieces to detect the pig's displacement as it passes through the pipeline [13]. In some cases, 3D printing may create single-part fabrications that outperform multi-component alternatives.

5.4.5 Manufacturing and Execution Systems

Manufacturing facilities, various operational technologies such as supervisory control and data acquisition (SCADA), and various computing systems are all brought together in what is known as production and execution systems (MES). Because the operations in oil and gas equipment manufacturing are sophisticated, engineers seek strategies to monitor and manage the continuous working processes. MES provides information for optimizing operations by delivering data that are recent, precise, and easily accessible. Customers have the capability to have a sizable effect, if they are able to make timely and well-informed decisions on productivity improvement and have a better grasp of how their facilities are functioning in real-time. The oil and gas sector may benefit from the intelligent design that MES can give for their industrial systems through integrated control.

In terms of the benefits of using MES, it is crucial to remember that the software assists its users in effectively monitoring expenses. MES software can track all critical data on the factory floor. This includes, among other things, labor, materials, downtime, and tools. All of these factors are constantly updated, providing for precise cost records and projections. Second, MES software aids in increasing uptime. To elaborate, the application will analyze the schedule to discover gaps and will arrange repairs without requiring human involvement. Maintenance can be scheduled before the machinery shuts down, and in this way downtime can be reduced. Furthermore, not being able to properly categorize inventories and other items may result in a lot of needless labor. Finally, quality control is essential. Without an MES, obtaining quality control input may take hours. There is a possibility that during that time period, an arguably larger number of products were manufactured, all of which would now need to be abandoned since they are considered waste. When using an MES, quality control information is supplied in real-time. This allows production to be halted as soon as an issue emerges so that technicians may rectify the problem and get the machines back up and running as quickly as feasible, with the least amount of waste as possible [14].

From the original contract through the final product, digital checks are performed on the authenticity of all operational procedures. Virtual reality and even simulators can be used to mimic in-production stages throughout the idea conception and system-building phases. The new, optimized digital process instructions are evaluated during the system development phase. Finally, the refined output combines optimal manufacturing with precise methods and quality data. A functional MES is a data hierarchy that is synergistic and integrated from the manufacturing production floor to the enterprise level of an organization. The advantages of the MES include seamless integration between these layers, which connect higher and lower levels with real-time data. Companies can transfer assets more conveniently, handle communication more effectively, and monitor resources and equipment to determine which ones are delaying the output. The MES technology improves data management, production execution, and operational effectiveness, allowing manufacturing companies to swiftly turn data into profit.

5.5 Midstream Sector Challenges

5.5.1 Cyber-Attacks

The hydrocarbon supply chain is geographically extensive and incorporates several information technology (IT) and operational technology (OT) systems from the reservoir to the refinery and everywhere in between. To guarantee that the product flows effectively and profitably, these systems require ongoing cybersecurity protection. IT and OT systems are interconnected and intertwined [15].

The Colonial Pipeline incident in May 2021 is a perfect example of a cyberattack. The hack did not damage the company's OT systems. The successful penetration of the pipeline's IT network, on the other hand, compelled the firm to shut down its pipes, resulting in significant downtime, costly repair operations, and loss of revenue. Within a 2-hour period, the attackers took 100 terabytes of data. Following the data theft, the attackers attacked the Colonial Pipeline IT network with ransomware, resulting in the shutdown of 5500 miles of the pipeline that provided 45% of the petroleum to the East Coast of the United States. Consumers in the eastern United States flocked to petrol stations, triggering short-term shortages. To avoid additional delays in fuel deliveries, the Colonial Pipeline's CEO approved a $4.4 million ransom payment to the hackers.

The growing monetary demands and frequency of ransomware attacks and data breaches have an influence on more than just a company's operations and operational expenditures. They also raise cyber insurance prices and compel corporate policyholders to reduce coverage.

5.5.2 Environmental Considerations

5.5.2.1 Greenhouse Effect

As the midstream business model relies largely on fossil fuels to transport oil and gas, the air quality deteriorates and GHG emissions increase due to its activities [16]. These GHG emissions include methane (CH_4) and carbon dioxide (CO_2), both of which are present in variable amounts in natural gas. As a result, natural gas leaks during transportation can emit a considerable amount of CH_4 and CO_2 into the atmosphere, exacerbating the greenhouse impact. Climate change caused by human-made GHG emissions has now received nearly unanimous support from the scientific community. The transportation industry accounts for 14% of annual worldwide GHG emissions, while the petrochemical and refining industries contribute around 4% only.

5.5.2.2 Ecological Risk

There is also a greater chance of unintentional leaks, spills, and explosions from pipelines, trains, and ship tankers transporting petroleum products, which can affect the environment. Once the pipeline's structure is damaged, gas will leak and spread swiftly in massive amounts in a short period of time, producing severe mishaps such as explosions and poisoning, resulting in heavy losses of life and property as well as substantial environmental contamination. The rolling of machinery and vehicles during pipeline construction disturbs the soil; destroys natural vegetation around the pipeline; discharges more wastewater, waste gas, and waste residue; produces more waste soil; and disrupts wildlife habitats in general [17].

Oil and gas firms are being pushed to acknowledge their part in climate change and are finding it increasingly difficult to get operating licenses owing to societal acceptability issues, among other things. Previously, if an oil and gas project was demonstrated to be technically feasible, it would have been approved. This is no longer true since public engagement in environmental decision-making has gained importance.

5.5.3 Social Concerns

5.5.3.1 Increased Use of Renewables

Countries all across the globe have implemented regulations to increase the use of renewable energy, which has impacted the oil and gas business. The European Commission has established requirements that push all European Union nations to boost the use of renewable energy in three sectors: electricity, heating and cooling, and transportation, with the purpose of lowering GHG emissions. Renewable Fuel Standard (RFS), Energy Independence and Security Act of 2007 (EISA), and other initiatives in the United States require increasing use of total renewable fuels, advanced biofuels, biomass-based diesel, and cellulosic biofuels. Asian countries are likewise boosting their renewable energy industry by establishing goals and putting policies in place. By 2025, the ASEAN (Association of Southeast Asian Nations) has set a collective goal of obtaining 23% of its main energy from renewable sources.

Because of the continuous interest in renewables, the demand for oil and gas products will decrease in future.

5.5.3.2 New Mobility Approaches

Mobility trends are quickly shifting. Important indicators to watch in specific locations include the transition from private car ownership to public transportation and vehicle sharing. Furthermore, electric car penetration is beginning to increase, bringing competition to traditional gasoline powered vehicles.

5.5.3.3 Energy Conservation

Nations all over the world are committing to sustainability by boosting energy efficiency. One significant trend here is the transition to a more sustainable world, which is generally accomplished through increased electrification of economic sectors.

5.5.3.4 Maximum Oil Demand

Projects will become more technically and financially challenging as easy-to-recover oil sources continue to dry up. Efficiency increases and the possibility of substituting other energy sources for oil are putting doubt on the ongoing development of oil consumption in industrialized countries.

5.5.3.5 Supply Security and Safeguarding Supply in the Event of an Interruption

On a worldwide scale, oil is the most important energy source. The breakdown of supply jeopardizes countries' economic progress and citizens' wellbeing. In a broad sense, supply security is maintaining supply availability and cost.

5.5.3.6 International Markets

Because of the changing nature of the global demand structure, it is critical to diversify the portfolio of petroleum products, including petrochemicals, in order to compete with a bigger number of players using various marketing and sales techniques. It is also necessary to develop plans to deal with the complexities of trading and future protection activities.

5.5.4 Regulations

There are several rules imposed on the oil and gas industry by various nations' state and central governments. Existing rules can hinder the progress and deployment of new technology by impeding an operator's capacity to solve possible difficulties cost-effectively through the use of the most innovative technology. Global environmental and safety laws are becoming more stringent, leaving compliance for the oil and gas industry more difficult. Energy corporations are still under heavy scrutiny from the government and customers, who want stronger regulation to minimize environmental harm and save lives. Even though these restrictions restrict and tighten product extraction, manufacture, and distribution, it is vital to ensure health and environmental safety while maintaining lucrative sale and trade. It covers practically every topic, including the transfer of rights, pipeline transportation, environmental impact assessments, decommissioning, waste disposal policies, and reform recommendations.

5.6 Conclusion

The midstream oil and gas industry, which focuses on the processing, storage of crude oil and natural gas, and its transportation is seen as a low-risk and highly regulated sector of the oil and gas industry. Flow assurance, which requires the correct transportation and storage of petroleum products as dictated by market movements, is a vital obligation of the midstream industry. The capacity and flexibility of the midstream supplier are critical in keeping the entire production process running. Advances in technology have the potential to make the midstream industry safer, more operationally efficient, and environmentally conscious. The industry, regulators, and communities will all profit from improved scientific innovation. Understanding and tackling the midstream sector's basic concerns is also crucial to generating safety, environmental, and operational improvements.

References

1 George-Sharpe, L., Patel, R., Pérez-del-Rosario, K., and Ramanan, A. (2019). Technology commercialization challenges in the midstream oil & natural gas sector. Prepared for the National Petroleum Council Study on Oil and Natural Gas Transportation Infrastructure. https://www.energy.gov/sites/default/files/2022-10/Infra_Topic_Paper_4-1_FINAL.pdf (accessed 10 December 2022).

2 DXP/IFS (2019). Blog on 'What Is Midstream Oil & Gas?'. https://ifsolutions.com/what-is-midstream-oil-and-gas-industry/ (accessed 15 December 2022).

3 CFI Team (2022). Resources on midstream oil operations. https://corporatefinance institute.com/resources/valuation/midstream-oil-operations/ (02 January 2023).

4 Staff, Office of Air and Radiation, U.S. Environmental Protection Agency (2015). Document on 'Regulatory Impact Analysis of the Proposed Emission Standards for New and Modified Sources in the Oil and Natural Gas Sector'. https://www3.epa.

gov/ttn/ecas/docs/ria/oilgas_ria_proposed-nsps_2015-08.pdf (accessed 18 December 2023).

5 Sölken, W. https://www.wermac.org/others/oil_and_gas_transportation. html#gsc.tab=0 (accessed 19 November 2022).

6 Kraus, S.R. (2011). Storage and transportation of crude oil, natural gases, liquid petroleum products and other chemicals. https://www.iloencyclopaedia.org/part-xvii-65263/transport-industry-and-warehousing/item/946-storage-and-transportation-of-crude-oil-natural-gases-liquid-petroleum-products-and-other-chemicals (accessed 10 December 2022).

7 Leibel, A., Seagram, C., Carly McMann, C. et al. Transportation of oil and gas. https://web.uvic.ca/~djberg/Chem300A/GroupLM_OilGasMovement_Proj1.pdf (accessed 10 December 2022).

8 AVEVA Team (2019). Digital acceleration in the midstream oil & gas industry. IndSol_Digital-Acceleration-Midstream-OandG_EN.pdf.coredownload.inline.pdf (accessed 13 January 2022).

9 Neiman, E. (2022). Oil and gas: digital prosperity during a downturn?. Whitepaper. https://www.actsoft.com/wp-ontent/uploads/2022/01/TeamWherx_Oil_and_Gas_WhitePaper.pdf (accessed 12 November 2022).

10 Offshore Technology (2020). Cloud computing helps to optimize data management in the oil and gas industry. https://www.offshore-technology.com/comment/cloud-computing-oil-gas-companies/ (accessed 05 January 2023).

11 Dogra, M. (2021). Impact of cloud adoption in oil & gas industry. Blog. https://www.hcltech.com/blogs/impact-cloud-adoption-oil-gas-industry#:~:text=Cloud%20computing%20in%20the%20oil,data%20on%20on%2Dpremise%20servers (accessed 23 January 2023).

12 Al-Walaie, S.A., Bahwal, O.B., Alduayj, S.S. et al. (2021). Emerging robotic technologies for oil and gas operations. *Journal of Petroleum Technology*. https://jpt.spe.org/emerging-robotic-technologies-for-and-gas-operations (accessed 11 January 2023).

13 Team Imaginarium (2022). 3D printing – revolutionizing the oil & gas industry. https://imaginarium.io/3d-printing-revolutionising-the-oil-gas-industry/ (accessed 13 Febraury 2023).

14 Alkady, R. (2022). The impact of manufacturing execution systems (MES) in oil & gas operations. Egypt Oil & Gas. https://egyptoil-gas.com/features/the-impact-of-manufacturing-execution-systems-mes-in-oil-gas-operations/ (accessed 10 December 2022).

15 Arctic Wolf Networks (2021). Cybersecurity in oil & gas: how to strengthen your cyber resilience in an evolving threat landscape. Whitepaper. https://arcticwolf.com/resources/category/white-paper/ (accessed 22 December 2022).

16 Pelegry, E.A. and López, M.B. (2018). The oil industry: challenges and strategic responses. https://www.orkestra.deusto.es/images/investigacion/publicaciones/informes/cuadernos-orkestra/the-oil-industry-challenges-strategic-responses.pdf (accessed 22 December 2022).

17 Donev, J. Midstream oil and gas industry. https://energyeducation.ca/encyclopedia/Midstream_oil_and_gas_industry#:~:text=Environmental%20Impacts&text=Since%20the%20midstream%20industry%20relies,natural%20gas)%20and%20carbon%20dioxide (accessed 30 December 2022).

6

The Significance of the Industrial Internet of Things (IIoT) for the Oil and Gas Space

6.1 Overview of IIoT

The evolution and application of Internet of Things (IoT) in business processes and other areas are recognized by Industrial Internet of Things (IIoT). It appears that the incorporation of Industries 4.0 is paying further focus on the effectiveness of the advanced manufacturing evolution. IIoT is a sort of network which manages complex scientific sensors attached to detectors used in automation technology and its associated software as shown in Figure 1, whereas IoT is a type of system that connects materials or things via a wireless or wired network.

Integrated surveillance devices help firms discover shortfalls as well as prior issues, resulting in cost savings while trying to assist with business information endeavours. IIoT must have tremendous opportunities in business for quality assurance, environment friendly and defendable procedures, supply chain tracking, and entire supply competence. IIoT seems to be crucial in industrial processes which include predictive maintenance (PdM), extensive product support, resource management, and location tracking.

6.1.1 Functioning of Internet of Things

IoT is an interconnection of acute gadgets that collaborate to build systems that track, gather, transmit, transfer, as well as evaluate information.

The logical ports, communications, and applications which facilitate the use of detectors in the production process is termed to be the industrial internet of things (IIoT) infrastructure as shown in Figure 2. Owing to the execution of 5G

The Power of Artificial Intelligence for the Next-Generation Oil and Gas Industry: Envisaging AI-Inspired Intelligent Energy Systems and Environments, First Edition. Pethuru Raj Chelliah, Venkatraman Jayasankar, Mats Agerstam, B. Sundaravadivazhagan, and Robin Cyriac.

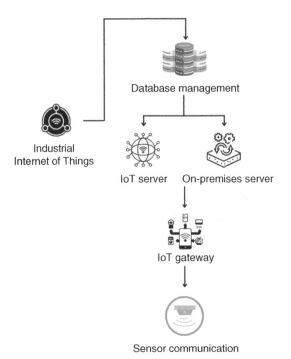

Figure 1 Overview of IIoT.

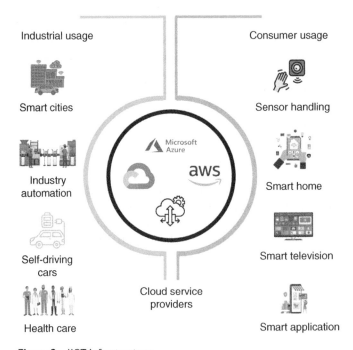

Figure 2 IIOT infrastructure.

technology, that guarantees higher bandwidth update intervals, edge computational modeling and computer communication techniques are becoming more crucial.

Each business IoT ecosystem consists of the following elements:

- Connected equipment is capable of sensing, communicating, and recording transactions regarding itself
- Data communication framework for community or individual use
- Metrics and implementations that deduce data analytics from source data
- A repository for storing the information obtained from IIoT equipment and
- Individuals.

6.1.1.1 Conceptual Architecture of IIOT

The notion of the IoT is carried to the business sector through IIoT. Each organization utilizes its own collection of gadgets with limited integrations. Identified the complexity, there is not a universal solution which will rectify all the problems.

The key elements of the conceptual architecture of IIoT as shown in Figure 3 are explained below:

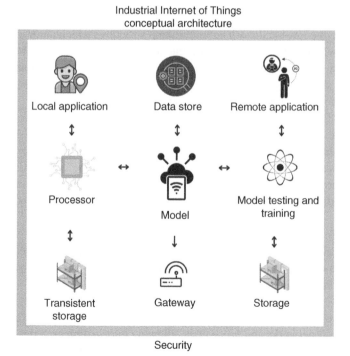

Figure 3 Conceptual architecture of IIOT.

6.1.1.2 Industrial Control System (ICS)

The phrase "Industrial Control System" refers to a combination of both hardware and software used for regulating a vital framework. They are typically constructed with distributed control systems (DCS), programmable logic control (PLC), supervisory control and data acquisition (SCADA) devices, remote terminal units (RTU), control servers, human–machine interface (HMI), intelligent electronic devices (IED), and a wide range of other business processes.

6.1.1.3 Devices

Detectors, actuators, and translation services are some manufacturing systems that interrelate with ICS, Dynamic Data Warehouses, Streams, and Processing units to provide information to the application's end-user level.

6.1.1.4 Transient Data Store

Transient Data Store would be the captive of expert design that transiently stores the instantaneous description of object classes to ensure longevity throughout procedure and framework inadequacies, along with network problems.

6.1.1.5 Local Processors

Local processors are the low-bandwidth computational structures which analyze information rapidly. Thus, they can be incorporated only along the gadgets to handle data. Such a microcontroller is divided into various categories, such as information filters, site managers, system administrators, regulation propellers, frequency sensors, computational methods, and gateways.

6.1.1.6 Application

These implementations focus on providing genuine insight into site activities; they assist employees in handling equipment, communicating along other processes, as well as processing information. Warning message, notifications, and image processing help them make appropriate decisions and determined options.

6.1.1.7 Channels

Information is transferred between the framework and the implementation process through the channels. This enables centralized procedures, satellite links, API, routers, and other components.

6.1.1.8 Gateways

Gateways connect sub-networks and procedures, allowing information exchange among distinctive IIOT networks. Smart signal access points, information transmission procedures, and other components are included.

6.1.1.9 Collectors

Collectors use networking technologies to gather information from gateways. Such types of gadgets differ from sector to sector depending on the requirements.

6.1.1.10 Processors

Processors serve as the center of all IIoT solutions. The primary responsibilities include advanced analytics, channel estimation, numerical simulations, and process-level handling.

6.1.1.11 Permanent Data Store

They are long-term backup and recovery systems which are connected to the IIoT structure. They function as a chronicler for equipment, supplying various types of information to the processing units for powerful analysis handling and framework prepping. It comprises a huge number of concurrent processing data warehouses, cloud services, database systems, RDBMS, transparent data frame, and so on.

6.1.1.12 Models

In any IIoT solution, there are two frameworks. The first is the analytical model and the other is the data model. The data models framework the information, whereas the analytical models are developed specifically for the sector needs. Models are essential in any IIoT method and are usually constructed by combining information in the perpetual storage arrays, individual knowledge, and business norms. The systematic representations are developed on statistical information using huge statistics sets or progressive appliance mechanism.

6.1.1.13 Security

Security is the significant characteristic of the IIoT-based framework. It extends across the transmission lines from the origin to the client. It includes information consent, encoding and verification, authorization, user access, access controls, and other security features.

6.1.1.14 Fog Computing

Fog is a modern framework that incorporates unique demands to protocol stack with regard to efficiency and configurability. It provides green technology, legitimate information exchange, high bandwidth, secondary storage on endpoints, and streamlined distribution of resources, which are open issues that need to be addressed for prospective automated processes.

6.1.2 IIOT Viewpoints

The Industrial Internet Reference Architecture (IIRA) points of view are formed through scrutinizing the numerous IIoT practice belongings established by the

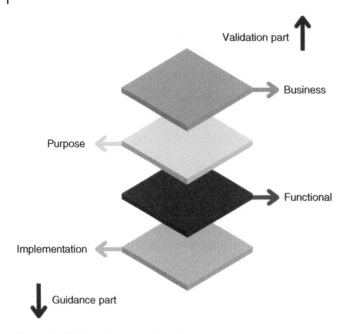

Figure 4 IIoT architecture viewpoints.

IIC elsewhere, determining the appropriate framing and recognizing the related investors of IIoT schemes of concern.

Conception of the IoT is carried to the business level through IIOT. Each organization utilizes its own collection of gadgets with limited integrations. With the complexity identified, there is not a universal solution which will rectify all the problems.

The four viewpoints of IIoT as shown in Figure 4 are as follows:

- Business
- Usage
- Functionality
- Implementation

6.1.2.1 Business Viewpoint

By constructing an advanced manufacturing effective algorithm in its commercial and legal context, the business viewpoint focuses on the challenges of recognizing shareholders and their business – vision, principles, and goals. It also explains the IIoT data analysis method, which yields the original objectives by sticking to essential functionalities.

6.1.2.2   Usage Viewpoint

This viewpoint subscribes the alarms of predictable gadget usage and typically signified as a series of events concerning individual or rational users that transport the proposed mechanism in order to achieve the system's fundamental capabilities.

These concerned stakeholders typically include system engineers, product managers, and other stakeholders such as individuals involved in the specification of the IIoT system under consideration and who represent the users in its ultimate usage.

6.1.2.3   Functional Viewpoint

This viewpoint emphasizes on the efficient mechanisms of a business analytics scheme, their construction and interrelations, as well as the system's relationship and interactions with peripheral fundamentals to provision the overall system's conventions and actions.

6.1.2.4   Implementation Viewpoint

The execution perspective is concerned with the knowledge required to implement efficient gadgets, as well as their communication schemes and lifecycle procedures.

These fundamentals are associated by events and are valuable to the organization's competences. Such issues are particularly vital to structure and constituent designers, inventors and integrators, and system machinists.

6.1.3   Benefits of IIoT

Remote monitoring, which eliminates the need for workers to go on the spot and perform tedious labor-intensive checks, is not the only gain by means of IoT for oil and gas construction and dispersal.

Now look at some of the other reimbursements of IoT in the oil and gas sector:

- Permit for instantaneous tracking of gadgets, fleet, and ecological circumstances, as well as amended exposure and control progressions.
- Authorization for desired equipment maintenance and diminished associated costs and exertion.
- Progress operative protection through the transportation of risky on-the-spot operations to automated machines like robot systems and Unmanned Aerial Vehicle (UAV).
- Present computerization, like as mechanized outflow and splintering resistor.
- Reduce non-productive time and downtime by optimizing manpower.
- Decrease the destructive ecological impression of oil and gas manufacture and circulation.

Prognostic preservation is one of the greatest widespread reimbursements of IIoT gadgets used in the production sector. Businesses can use actual information produced by IIoT structures to expect when an appliance will need maintenance.

It permits essential preservation work that can be carried out before a fault arises. This is particularly useful on a construction line where mechanism failure can result in work stoppages and high costs. An organization can improve its operational efficiency through carefully addressing prolongation issues.

An additional advantage of IIoT is resource tracking. Resource management systems can be used by dealers, producers, and clients for tracking the site, position, and alignment of products throughout the procurement. If goods are spoiled or in danger of being damaged, the arrangement sends instant alerts to those involved people giving the opportunity to them to take instant or protective act to rectify the condition.

The IIoT also progresses capability supervision. Industrial equipment is liable to attire and slit, which can be intensified by factory circumstances. Devices can screen vibrations, temperature, and other aspects that might lead to optimal working circumstances.

6.1.4 Security in IIoT

IIoT security concerns trunk from an enlarged occurrence surface and the requirement for far-flung access. Supplementary communication channels, data stores, ports, and endpoints are created as more devices and sensors come online. If left unprotected, this increased interconnectivity represents more vulnerabilities.

IIOT is divided into three groups:

- Local area networks (LAN)
- Endpoint management
- Information processing

6.1.4.1 Ensuring IIoT LAN

Producers and further IoT manufacturing manipulators must prioritize safety inside their LAN. Though a minor business may solitarily need to organize the same security procedures across processers and servers, an IIoT facility will present exclusive trials due to the variety of equipment in usage across multiple sites.

Numerous IIoT systems also were not constructed to focus on safety. Emphasizing safety inside an IIoT local network involves defending all your gadgets from unauthorized users, irrespective of their own advanced automation mechanism.

6.1.4.2 Endpoint Security for Users

Despite the fact that certain elements of IIoT demonstrate automation and technology, maintenance workers, management staff, as well as technicians must always converse only with the control unit. Malicious hackers perceive such access points as elevated objectives.

Unless the absence of access points strictly outlined user privileges and non-linear authorization, one's IIoT systems are susceptible to suspicious activities as well as expensive distractions. Application server management is necessary for industry players desiring optimized protection, as it is for retail chains.

6.1.4.3 Data Transmission Security

A further risk factor for IIoT production facilities is the enhanced information communicating throughout the channels of IIoT nodes as well as infrastructure. The number of information retail locations and portals that need to be secured increases in combination with a number of smart machineries. Violation of confidential documents might pose a security risk, which ends up causing electrical malfunction or starts causing identity of service outages.

6.2 Technical Innovators of Industrial Internet

Detector manufacturers and innovators can already start taking an entirely new method like integrating intellectual capacity for self-tracking and regulation to enhance the efficiency of their parts and accessories. Correspondingly, progressions in sensor industrial automation enables the creation of reduced elements and systems. It would thus include significant advances in micro-system innovations including such exterior and huge quantities of fabrication methods, which the manufacturers are gradually working to implement.

Microelectronics is seeking innovative online services for the objects in IoT implementations including wearable technology, automobiles and mass transit, and expendable tracking systems for medical products and generating even more smart city and intelligent home implementations more than what we can constitute.

Researchers and engineers face several challenges in this digital era as we assign technologically additional gadgets to the cyberspace, such as the way to service a radio signal into their established equipment property investment, which not only ends up making extremely tiny gadgets but also minimizes the spatial coverage for escalating transistors. In addition, people are continuing to work to meet the client's demands for ergonomic design in IoT commodities.

The following stands some typical IIoT scenarios besides business contexts:

- Applications for advanced manufacturing industries and automated depository
- Virtual and prescriptive management

- Monitoring of consignment, products, and transference
- Logistics that are linked
- Smart grid and smart metering
- Environment-friendly solutions
- Applications for smart cities
- Smart farming and livestock tracking
- System of industrial security
- Optimization of energy consumption
- Heating, ventilation, and air conditioning in the workplace
- Monitoring of manufacturing equipment
- Asset tracking and intelligent logistics
- Monitoring of ozone, gas, and temperature in industrial settings
- Workers' safety and health (conditions) are monitored
- Asset performance monitoring
- Use cases for remote service, field service, remote maintenance, and control

6.2.1 Industrial Control Systems (ICS)

Control mechanisms and connected systems including equipment, processes, channels, and control systems often used to perform automated manufacturing processes are referred to as Industrial Control Systems (ICSs). Every ICS feature is diversely based on sectors and is constructed to effectively perform various tasks digitally. ICS systems and procedures are now utilized in almost every manufacturing industry and essential infrastructure, including production, power generation, and water systems.

The most popular types of ICS are

- Supervisory Control and Data Acquisition (SCADA) systems and
- Distributed Control Systems (DCS).

Control system products are mainly dependent on predominant embedded framework platforms used in a variety of gadgets such as interfaces and Ethernet. These are frequently deployed profitable applications and command lines depicted to support industry processes and control networks and devices. SCADA is a major subset of the ICS.

6.2.2 Supervisory Control and Data Acquisition (SCADA)

SCADA functionalities are aimed at providing supervision and control. SCADA systems are constructed of gadgets dispersed across different venues. SCADA controllers could procure as well as transmit information and they are incorporated with such a Human–Machine Interface (HMI) that it becomes an illustration of an interface that provides centrally managed visibility and management of a broadening range of operational interfaces.

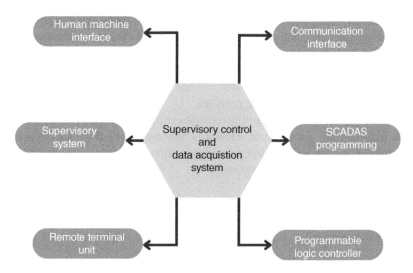

Figure 5 Supervisory control and data acquisition system.

SCADA is mainly utilized for long-distance management and surveillance of field sites via a central console. Rather than employees who have to drive long distances to complete tasks or gather information, a SCADA system can streamline this procedure. Real-time gadgets control operations including nozzle, breaker opening and finishing, information gathering from sensing devices, and tracking the local ecosystem for alarm conditions.

It consists of the following elements as shown in Figure 5:

- Communication Interface
- Programmable Logic Interface (PLC)
- SCADA Programming
- Supervisory systems
- Remote Terminal Units (RTU)

Although such thread processes have been distributed over huge regions and made accessible via Network Communications Server farms, a Decentralized Control Scheme has been used. The DCS would be a category of controller design that controls and monitors set points located across a whole factory.

6.3 IoT for Oil and Gas Sector

The oil and gas sector was one of the first to endorse about the IoT. Lower oil costs, reduced production prices, variability of both demand and supply, increasing upstream working capital, economic recession, challenge for resource management, increasing prices, environmental legislation, rapidly growing population, tax rates, safety regulations, and an increased workload are all hurdles for the oil and gas industry. Changes in climate, as well as other health and sustainability issues, pose significant barriers to accessing mineral wealth such as oil products, petroleum products, and gas. Intense competition and complicated market dynamics add to the intense pressure.

Implementing IoT-based smart utility remedies enhances field communication, reduces maintenance expenditure, enhances the digital oil field framework, reduces energy consumption, mines robotization, increases equity safety and protection, and thus improves the productivity.

6.3.1 Utilizing IIoT in Oil and Gas

The following elements suggest IOT applications throughout the oil and gas sector:

- Excavating management
- Stream tracking
- Oil refinery tracking
- Subsea surveillance
- Logistics transportation
- Robust and protection
- Pollution control
- Increased revenue

6.3.1.1 Excavating Management
Drilling is a significant factor of the oil and gas sector's operations. The IoT demonstrates to be a huge benefit for enhanced drilling performance. Even as rig got deeper, it produces highly unsafe conditions. To extract oil by drilling, rig operators should carry accurate readings. Errors occur when deep-water drilling is done improperly. IoT devices are useful for minimizing hazard and trying to effectively carry out complicated functions. Utilizing sensor data, intelligent technologies can also warn relevant personnel well ahead in time of any drilling mistakes.

6.3.1.2 Stream Tracking
The most severe problems confronting the oil and gas industry is pipeline leakage. It causes significant economic, ecologic, and infrastructural damage to the firm. IoT facilitates in the tracking of the pipeline components and elements such as pipes, pumps, and filters. Without IoT, industries should depend on human personnel to carry out regular inspection and servicing. IoT diminishes the requirement for manual inspections by instantaneously tracking transmission lines. Actual data can substantially reduce huge dangers connected with pipeline leaks

and other inappropriate circumstances. Personnel can be deployed immediately to tackle any problems that may pose a major risk.

6.3.1.3 Oil Refinery Tracking

IoT screens metrics such as temperature and flow rate. Every component of the processing plant requires substantial standard measures and data analysis. If accomplished, this is a lengthy process as well as costs the company a substantial amount of money. Some implementations necessarily require accurate measurements in real-time.

The IoT enables quite precise information collection in areas where human resources cannot reach. Sensors can be installed at numerous locations that are challenging for workers to obtain and gather more information. This facilitates in the refinery's 24-hours surveillance.

Pertaining with one survey, oil and gas firms can ramp up production by 6–8% by effectively utilizing information.

6.3.1.4 Subsea Surveillance

The significant proportion of offshore oil and gas extraction occurs in harsh conditions. At some of these offshore platforms, there are only a few elements to interact. Monitoring temperatures, pressures, and other equipment has become a difficult and costly task. IoT helps in overcoming these hurdles to create a reliable tracking system.

Considerably more tracking locations could be linked utilizing Low-Powe Wide Area Network (LPWAN). One such application offers an affordable remedy for tracking oil and gas offshore rigs. Numerous leak sensors can be connected to oil wells distributed along a greater territory. Every one of these detection systems can transfer real-time information to a centralized location. The data can then be used to monitor various drilling and oil extraction mechanisms.

6.3.1.5 Logistics Transportation

Authentic ship and fleet surveillance is critical for fleet operators. The ability to monitor oil and gas container ships is equivalent to offshore tracking. There is no connectivity at sea, so the workers should depend on satellite internet. If the workers need information from the ship's environment, they have few possibilities. The information from such ships is enormous in its own. IoT offers simple data analysis from such locations. LPWANs enable simple monitoring of regions of a ship that are not regularly visited by ship personnel. Sensory tracking devices focus on providing protection in addition to comfortability in collecting information from restricted environments.

Some cargo ship components should be interconnected owing to real-time prerequisites, although certain non-operational components are not required to be interconnected in real-time. As a result, small IoT networks prove to be a great substitute to wired sensors.

6.3.1.6 Robustness and Protection

Gas and oil locations are generally situated in hazardous and rural locations. The labor environment at all these sites is hazardous to the workers who work there. IoT systems allow remote management of machinery and operational processes, eradicating the requirement for people to visit a location without previous understanding of the situation. Devices and sensors and image aspiration could provide precise information on the issue and help determine the right plan of action. IoT in oil and gas fields can substantially reduce worker accidents and deaths. Deaths and injuries among gas and oil laborers are diminishing, and IoT can assist in reducing them even further.

Collisions can be exorbitant to industries both economically and in terms of their reputation. Oil producers can ensure a secure workplace for their staff by implementing IoT-enabled safeguards.

6.3.1.7 Pollution Control

Oil and gas production is enhanced with the integration of IoT outcomes in this segment by functioning more productively. IoT solutions are advantageous both environmentally and economically. The carbon pollution produced by the plant could be drastically decreased with effective operations and management. It makes a significant contribution to reducing the environmental impacts of oil drilling and extraction processes. Oil and gas producers can therefore hold their ethical value of not adversely affecting the environment while deploying troops.

6.3.1.8 Increased Revenue

The oil and gas industry has been intensely popular in past few years. Pioneering companies are looking for ways to improve profitability and efficiency. Quick premium metrics can endanger long-term success in business. Investing in IoT will guarantee that, companies are much more able to reduce expenditures. Technology devices can drastically enhance efficiency improvements, prediction supply in the market, and optimize overpriced mechanisms. Companies have increasingly lowered their operating costs by removing excess investments and human resource expenses.

6.3.2 Excellence in Operations

- Intelligence on location
- Monitoring and control of emissions, as well as release management
- Monitoring of pipelines and equipment
- Predictive maintenance

An example architecture for the oil and gas industry using Microsoft Azure is shown in Figure 6:

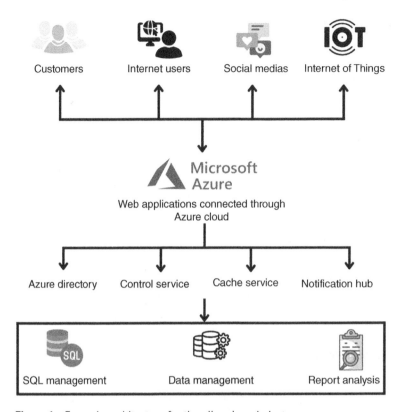

Figure 6 Example architecture for the oil and gas industry.

While constructing an intelligent system, adaptability and stability seem to be the two vital aspects to consider. The IoT environment ought to be customizable to accommodate uncertainty in traffic, and equipment protection has been essential to guarantee its safety. The Azure IoT Control systems are consistently needed to safeguard the works by enhancing, and yet parallelization should be incorporated just at the edge of the system.

6.3.3 Device Management

Detectors can transmit and receive information from and to the gateways but upon having to register only with virtualization access point. This must have a remote support functionality, which permits to increase, stimulate, disable, eliminate, and upgrade the same features of a machine.

6.3.4 Device Connectivity

Huge number of messages will be obtained within a moment from a massive number of systems, culminating in numerous thousands rather than potentially millions of notifications a day though. This same framework must be able to handle large quantities of signal consumption through a solitary inevitable conclusion.

6.3.5 Transformation and Storage

While receiving signals, a framework must include a methodology for identifying, converting, as well as tracking them toward multiple storage media for digital preservation as well as constructing for subsequent analysis.

6.3.6 Presentation and Action

Utility infrastructure could perhaps support local signal position visibility through the use of data in a tabular functionality. Moreover, because certain communications could comprise warning updates, an efficient system such as IoT must include a methodology for offering real notices to admissible processes.

6.3.7 Microsoft Azure

The IoT package would be a corporate technology that enables you to even get up and run rapidly by offering a collection of expandable, preloaded workarounds for prevalent IoT environments including the Global Positioning System (GPS) tracking and preventative analysis. Such options were indeed configurations of the Microsoft Azure IoT infrastructure, which was originally mentioned.

6.4 Rebellion of IoT in the Oil and Gas Sector

IoT is a sophisticated approach that allows for digitization, predictive analysis, forecasting, advanced analytics, and so forth. One of the key driving influences is characterizing the technological growth by offering businesses with improved operational processes and computational efficiency for higher management level.

The technique is both simple and evolved, allowing for the effective evolution of businesses via:

- Real-Time Monitoring
- Cutting-edge Analytics
- Expensive Hardware and Software
- Gateway Connectivity with Multiple Channels

- Workflow Automation
- Intelligent Inventory Management
- IoT Logistics and Transportation

Enterprises searching for remedies to overcome obstacles consider IoT as one of their effective approaches. Such enterprises are energized by IoT's capability to implement innovation in a novel way for minimizing manufacturing difficulties such as device inspections, unexpected idle time, controlling temperature and humidity, and virtual handling.

6.4.1   Improved Operational Efficiency

IoT-based systems are more innovative variants of technology adoption in the oil and gas sector. High-quality sensor systems capture reliable information from resources and transfer it instantaneously to the customer's centre console. This obviously benefits clients by recognizing trends of oil demand and analyzing oil usage in all facets.

With vastly increased demand to increase economic viability and decrease pollution, the IoT sector provides the opportunity to perform in a cost-effective way. Numerous oil and gas firms are working to improve business performance by creating extraordinary marketing strategies in order to remain viable in a less unsure market condition.

Subsistence in this challenging situation requires a proactive solution that analyzes operational processes analytically and decisively. This enhances the sector's innovation effect through including improved sensing devices and supplying a one-stop shop for oil and gas difficulties. Below are a few instances as to how IoT can aid in production excellence:

- Lower maintenance expenses
- Better oil and gas lifecycle management
- Appropriately timed site visits

6.4.2   Optimize Inventory Levels Based on Actual Usage

The application of IoT enhances the way oil and gas organizations run their stock levels. It offers a more efficient, precise, and streamlined procedure for dealing with massive amounts of oil and gas. Moreover, with the incorporation of smart IoT devices, leaders need to set their responsibilities and baselines, beyond where an incredibly quick alert alerts them, informing them of a specific action to be taken.

This can be analyzed even in the early stages of the process, making an IoT solution moving ahead with complete factory automation absolutely awesome.

6.4.3 Improve Stockroom Management

Using detectors and other equipment to run efficient industrial operations allows businesses to obtain meaningful information and truly comprehend how to ensure maximum commercial and operational advantages. Due to a rise in the number of devices with sensors, devices that can evaluate crucial data are now available on the website, enabling oil and gas firms to cut investment expenditures by up to 20%.

Information gathered from sensors and smart resources aids oil and gas companies to become more accurate and effective, as well as to generate additional value by resolving unexpected favorable interruptions.

6.4.4 Return on Investment (ROI)/Revenue

A lot has been engaged inside the oil and gas industry in the recent times, which has led to increased outcomes and efficient profitability. Tectonic intelligence information in source rock gas drilling can assist companies in modifying their oil and gas wells and hydraulic fracturing procedures based on an improved knowledge of the anatomy of shale ladles.

Detectors deep within the drilling operation or during drilling send a flow of information to producers that describes how the rig actually does work. Pipeline devices detect the force and can provide the controller with consistent constraints. All these point to a huge expansion of IIoT in the oil and gas sector, which will lead to greater ROI and revenue-based outcomes.

6.4.5 Real-Time Monitoring

The production of oil and gas (O&G) near the coast is much more complicated and riskier. Because of the location and isolation of offshore oil and gas, O&G confronts issues in obtaining an up-to-date as well as true depiction of its production process.

As a result, technicians are often unsure of non-specific equipment and process variables, raising the risk of investment malfunction and complex and expensive breakdowns. In a worser situation, inadequate supervisory ability can cause catastrophic blasts that impact the environment, laborer's lifestyles, and the legitimacy of industries. Today, the IoT enable trustworthy and effective offshore surveillance.

6.4.6 Removing Manual Measuring Processes

IoT-enabled procedures almost completely eradicate the requirement for physical work and reduce the risk of inconsistencies in outcomes. It involves digging, surveying, and protective supplies throughout elevated offshore or off-site operational processes, in addition to digging and industrial production and utilizing robotic and semi-automatic machineries instead of humans. Such devices can regulate and share information in a control panel in a region where the majority of their operations are handled at remote locations, despite the fact that the bandwidth and potentially edge computer technology are needed to restore the most crucial details to the fundamentals of transfer.

6.4.7 Reduction of Safety Risk

It is a vital phase to keep employees secure in the oil and gas business. With a collaborative effort from the company, the integration of sensory technologies offers extensive safety regulations.

The integration of IoT-based products and services enhances the operation of the oil and gas industry and surveillance systems in massive failure regions. Also, it helps in handling errors well in advance of their occurrence. Thereby, it seeks to minimize potential dangers and integrate advanced components to identify major issues that may affect workers' health.

6.4.8 Hurdles in the Oil and Gas Sector

During the past centuries, this segment has indeed been characterized by huge turnaround, tremendous commercial reverts, and considerable mass within the financial systems of a provided nation and the entire world. One such depiction is still valid today, and yet the expedition inside this sector is not effortlessly conceivable.

Administrators in the oil and gas worth chain face significant encounters daily like

- Aging equipment and legacy systems
- Hazardous environments

6.4.8.1 Aging Equipment and Legacy Systems

In being completely obvious, we imply influential super-machines, massive drills, tankers, and complex surveillance systems which accomplish vital estimations to reinforcement and matrix while preserving the safety of employees. They seem to be tough, require diligent study, and more necessarily involve continuous supervision and a rapid turnaround to carry out these service requirements. Currently, numerous wells depend on primitive tools and tracking tools. Improving them is highly expensive in terms of both cash and workforce; however, the associated shutdown is even more expensive.

6.4.8.2 Hazardous Environments

The ecosystem and ease of access are two additional factors that make it challenging to preserve wells as well as other portions of the oil and gas distribution chain. Accumulations are quite often discovered in distant installations. Numerous oilfields are constructed in the risky northern oceans, and gas fields pass though such extreme conditions including such as deserts and snowy mountains. Each and every breaking or overflow is much more challenging to prevent and fix due to the challenging atmosphere, availability, and unsafe working conditions.

As a consequence, the functionalities of an IIoT are incredibly beneficial in the oil and gas sector.

6.5 Oil and Gas Remote Monitoring Systems

Utilization of IoT frameworks within the oil and gas field represents the methodologies that have obtained the most investment and focus.

6.5.1 Sensors

The utilization of IoT sensor nodes in the extraction of oil and gas helps in the restoration of continuous supply management regulation and the accelerated response to changing circumstances. Sensor-based techniques can be employed to supervise pipe pressure, the drilling procedure, device circumstances, and leak identification. Accelerated precision of concerns usually stems in billions of dollars in investment in this sector.

6.5.2 Smart Algorithms

Information and trials, registered and perceived through numerous detectors, are cross-referenced and analyzed by intelligent algorithms. They generate unique insights that assist the management in making critical decisions, such as when to switch and halt a drilling process to evade problems.

6.5.3 Prognostic and Preemptive Maintenance

Upgrading technology could quickly determine when the requirements of such compact equity transformation necessitate repetitive or incident servicing. By definition, on-demand preservation is much more structured than regular tests and ensures safety regulations.

6.5.4   Robots and Drones

Drones and robotic systems seem to be pivotal components of machinery as in oil and gas production process, according to the IoT. They permit more effective location investigation processes, regularly occurring information gathering, three-dimensional geomatics of dumping sites, as well as the ability to withstand the harsh environment encountered at the drill sites.

6.5.5   Smart Accessories

In this sector, wearable technologies have been increasing efficiency and sometimes even saving people's lives. Controller fits, fitness bands, Google glass, and headgear allow for ongoing surveillance of laborers' unsafe processes, reliable connectivity with the platform, and augmenting of workforce abilities by offering useful guide, alerts, or precautions. A smart helmet equipped with various devices is shown in Figure 7.

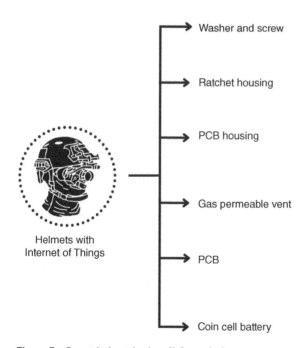

Helmets with
Internet of Things

→ Washer and screw

→ Ratchet housing

→ PCB housing

→ Gas permeable vent

→ PCB

→ Coin cell battery

Figure 7   Smart helmet in the oil & gas industry.

6.5.6 Wearable Watches

Just few oil service industries use wearable watches when their field engineers commute to distantly located offshore fields. The wearable watch can interact with the command center for precise positioning and security purposes. For example, it notifies the Command Center in real-time basis as to how so many workers are visible in a critical zone when in an oil spill or rig fire. Both voice and text notifications can indeed be recounted from the control room on a one-to-one or one-to-many basis, allowing workers to emigrate to a secure place.

Workers in mineral deposits are barred from having to carry mobile phones as they may lead to accidents. Laborers who use these wearable watches can be instantaneously notified to the potential of rock implosion. Individuals also can prevent a device in the event of a possible failure by hitting a button just on the watch and sending the data to the Control Room.

6.5.7 Wearable Glasses

They are electro-optical head-mounted spectacles that are handy. They show information on the goggles in a hands-free layout, similar to a mobile phone. Such spectacles offer workers with factual facts. Apart from visual representations, they also provide data on the gathered items.

They as well produce three-dimensional images that may be recounted instantaneously to specialists for improved cooperation. Workers working in remote regions can also obtain useful inputs from coworkers long distances away in real-time. Real-time guidance can occur in the form of portion replacement and maintenance, security protocols to be followed, mentoring, timecard and enrollment authorizations, and so forth.

6.5.8 Drones

They are incredibly helpful when visually inspecting transmission lines for spills, and inspecting systems are situated subsea for flare energy consumption. Pipelines can span thousands of meters across ruddy steep regions. They could indeed perform in unreachable, aggressive, or remote areas and capture superior pictures. For preventive analytics, the video feed can be measured in real-time. Such unmanned drones have been shown to be extremely cost-efficient, and it will replace remotely controlled helicopters when performing the same work, possibly saving oil and gas firm funds. Individuals can help in trying to locate the GPS location of a piping system with fissures that enable hot oil products to leak.

Leaks that would take days or weeks to be identified are now discovered in seconds by infrared heat detecting cameras, resulting in investments on conventional checkpoints and maintenance services. Even though warm body systems transmit an IoT of infrared light, the smart gadget upon that drone follows the principle of identification of infrared searching. Such drones significantly increase work efficiency and facilitate pipeline producers in enhancing recruitment and selection processes, who would otherwise need to utilize personnel for pipeline prowling. The drone implementation will indeed lead to increased pipeline decency and effectiveness.

They are utilized in mineral extraction to transmit visuals from conveyors for the proof of identity of foreign substances in mineral metals. Individuals also can cause significant downturns of mineral metals that may go undiscovered, leading to high energy usage and faster wear out from the belt conveyor. Drones can also use infrared (IR) light emission to monitor the position of fire incidents due to coal burning in storage facilities, which would be a natural occurrence.

6.5.9 Monitoring Critical Systems 24/7

Remote surveillance systems provide an affordable remedy for constantly monitoring infrastructure. Users acquire an extra layer of security for their boreholes and equipment, despite those who lack a phone service, mobile signal, or network connection.

When any of the hardware screws up, individuals would be informed immediately by mobile, emails, or messages. The following environmental datasets could be strictly monitored:

- Temperature
- Pressure
- Rate of Flow
- Voltage
- Level of Tank
- Amperage
- Torque on a motor

6.5.10 PLC Emergency Alert Notification Systems

Environment-sensitive processes could be operated in a wide range of situations to track oil and gas pipes. Sensaphone gadgets can be utilized as simple callout equipment in PLC-equipped terminals. The same instrument is interconnected into the PLC's alerting connections. A PLC shall notify the relevant individuals whenever it recognizes an alert system.

Every detector interaction is identified uniquely, enabling professionals to show up on-site realizing exactly what is a miss. Although the PLC is a serial port, the

Ranger Plus can immediately extract information from it to confirm that the remote location is legitimate for updates, alarming, confirmation, and reporting and recording data.

6.5.11 Independent Verification

A few oil and gas firms utilize Sensaphone processes to confirm or refuse environmental circumstances at remote locations. Wet well limits, temperature, pressure, internal devices, and power failures are all monitored using sensors. To determine the current status of their station, they must visit the Sensaphone site or contact the system directly. Sensaphone devices can easily show adherence and analyze trends because they record data. Users should meet an environment-sensitive equipment or inform a process to track the actual situation of their terminal. Even though Sensaphone devices record, individuals could also easily demonstrate adhesion as well as identify patterns.

Despite the fact that satellite technology continues to stay cost-prohibitive, the globalization of 4G LTE coverage allows oil and gas businesses to set up a broad pipeline flow surveillance system using devoted meters which deliver power information via industrial procedures. In many cases, serial monitor via RS-485 and the Modbus manufacturing procedure are used.

The information as from the flow meter must be gathered and transmitted to command centers or SCADA systems for center accumulation and perception. The TRB145 Serial Access Point from tectonic networks is ideal for such implementations, with an RS-485 functionality, Modbus RTU Master implementation, and 4G LTE Cat1; this can read the flow measurement file and transmit it via MQTT to virtual HTTP/HTTPS data centers or emerging IOT systems. Eventually, there is a broad power distribution range and low consumption of energy.

6.5.12 Oil and Gas Survey and Manufacturing Process

The two major parts of oil and gas processes are as follows:

- Upstream (exploration and manufacture)
- Downstream (production)

Downstream refers to the industry that deals with the processing and extraction of crude oil and natural gas product lines, in addition to their distribution and promotion. Businesses in the sector might well be completely integrated or may focus entirely on such a particular section, including such checking and manufacturing, which is usually referred as an E&P organization, or merely on tugging and brand management.

6.5.12.1 Immersion Monitoring

In the early stage of the exploration for oil mainstay developments, environmental graphs are evaluated on workstation research to identify significant drainage sinks. Satellite pictures are employed for recognizing enticing site configurations such as flaws or else extensional. A geographical evaluation research is employed to gather additional data, which are then utilized to select one of the three choices.

The three primary evaluation types are listed as follows:

- Magnetic
- Hydrometric
- Seismic

Magnetic techniques are based on calculating the differences on the force of the attractive field, which can replicate the magnetic landscape of the numerous existing mainstays, whereas the Hydrometric Method is based on the measurement of small differences in the gravitational pitch on the ground's exterior. Measurements are taken with an aircraft or a survey ship, respectively, on land and in the sea.

One of the most widely accepted assessment methods is a seismographic review, which would be the commonly and the very first field action conducted. The Seismological Procedure, which is based on the various introspective characteristics of seismic waves to different geological formations beneath underground or ocean circulation substrates, is employed to identify geological formations. The oil and gas industry sends a sound energy pulse into the ground, which also needs to travel as a wave. At every point in which various geological hierarchies occur, the fragment of the signal has been transferred to the lower part within the Earth, whereas the remaining be mirrored to the exterior. Then, it could be identified by a system of the submerged geological systems, which are sensitive receivers.

The procured electronic signals are transferred via special cables. In a mobile research lab, they have been magnified and screened before even being digitalized and saved on magnetic media for prospective interpretation.

6.5.12.2 Exploration Drilling

Only one method for confirming this same existence of oil and gas as well as the circumference inside and compression of a stream after trying to identify a good potential geological feature is to drill an explorative compartment. Only those gas wells to discover petroleum have been categorized as "exploration" wells, furthermore recognized as wildcats by drilling rigs. The same environmental management accounting firms have an effect on the position of a drill site. Overall, environmental requirements can be compensated with logistical constraints and the requirement for systematically drilling. After digging and initial testing, the rig is generally dismantled and relocated to the next spot.

6.6 Advantages of IIOT for the Oil and Gas Industry

IoT offers numerous business recompenses for the oil and gas firms. Consider the following reimbursements with IoT in oil and gas producers.

6.6.1 Monitoring Pipelines

Automation in the oil and gas sector allows effective surveillance system, which results in reduced upstream inspection reports and greater oil refining throughput. IIoT systems can be set up to send notifications to handsets and perhaps other mobile platforms. These assets are justified throughout the upstream pipeline in which a pump speed malfunction might cost $1 million a day during shutdowns. Correspondingly, downstream suppliers could use IoT to record ship fleets and delivery vehicles throughout real-time basis, trying to pinpoint every device's exact position.

6.6.2 Risk Mitigation

The deployment of IoT in oil and gas processes lowers the risk by enabling complicated and exploratory drilling to be watched and controlled more effectively. Not only does this diminishes inaccuracy and the threat of machinery and asset destruction but also decreases the chance of human fatalities and injuries.

6.6.3 Environmental Impact

A further essential utilization of IoT will be to facilitate thorough protection of the environment because all energy firms become more involved with social and ecological issues. The application of Industry 4.0 in the oil and gas sector tends to minimize the carbon emissions of drilling rigs. Yet another illustration in using the concept of IoT is to track wastages and harmful spillages. Therefore, business decisions can be taken more rapidly and allow access to less expensive sensor technology with low power consumption, which is most reliable for large-scale manufacture in the oil and gas businesses.

- Employment of IoT in the industry improves energy efficiency as well. It distributes the data related to energy on the device level to be analyzed more efficiently.
- The IoT could also cast-off to progress product excellence. Construction time of high-quality products can be reduced by managing machine downtime.
- The decision-making process is accelerated, resulting in more accurate results.

6.6.4 Managing Emergency Conditions

Most oil drilling and extraction industries are situated in far-flung regions from any adequate human presence. As a result, it is incredibly hard for aid and assistance to attain the oil and gas production location during such urgent situations. The oil and gas industry will be unable to monitor and control and address conditions that may lead to catastrophic situations, except if they endorse integrated surveillance systems. It can lead to unexpected shutoffs, preceded by unexpected maintenance costs, as well as charges power companies a significant amount of cash due to non-productive time (NPT).

The IoT enables energy firms to easily track numerous on-field mechanisms, allowing them to manage circumstances (similar to oil spills) that really face consequences in these kind of emergency shutdowns. This allows them to anticipate unnoticeable glitches and organize regular inspections to minimize NPT.

6.6.5 Establishing Workers Healthy and Safety

The oil and gas sector does not offer an optimum workplace culture, with metric tons of relocating working systems as well as an enormous amount of hazardous and flammable liquids. Moreover, they are often situated in distant and unreachable places, making things much more hazardous and challenging.

Even the most highly skilled employees could face challenges with maintenance and examination responsibilities in such dangerous conditions. An IoT-connected architecture allows a business to clearly track the performance of machineries and other on-site and off-site processes, minimizing human maintenance and inspection tasks.

Furthermore, IoT-based mobile sensors have the potential to monitor laborers in hazardous environments. As a result, in the scenario of an emergency, the actual number of submerged laborers, as well as their position, could be recognized, enabling relocation and rehabilitation processes.

6.6.6 Supply Chain Management

With fuel costs constantly rising, oil and gas sectors are searching for places in which they can reduce expenses while keeping annual spending.

The IoT enables oil sectors to maintain their arranging, organizing, as well as sourcing methods. All through the logistics system, the IIoT would then facilitate individuals in collecting and interpreting the information gathered from various terminals to establish important ideas for different stages. For instance, an oil refinery could indeed regulate and supervise its sourcing methods based on the volume of oil preserved inside the storage tank.

6.7 Conclusion

The IIoT enforces to enable companies to become more efficient, productive, and safe globally and across all oil and gas industries. Regardless of the implementation field, the IoT can ultimately assist operators in optimizing processes, resulting in profits. Given that decision-makers are eager to embrace innovation and IoT implementation, the sooner organizations adopt IoT, the sooner they can optimize systems and stay ahead of the competition. Enterprises also can facilitate a responsiveness culture and empower a smooth virtual transformation besides instructing and informing workforce upon the benefits of IIoT.

The IoT instigating momentous variations in the oil and gas manufacturing transforms it into a well-structured sector with automated working capabilities. It has numerous advantages that can be used to improve business performance by making work more efficient. The ability of IoT to improve business performance by focusing on the most important business factors makes it the perfect fit for this industry. Biz4Intellia proposes complete IoT solutions for real-time tracking in the oil and gas industry. It also offers customized on-demand services for IoT solutions in a variety of industries.

Bibliography

1 Jeong, Y.-S. (2022). Secure IIoT information reinforcement model based on IIoT information platform using blockchain. *Sensors* (12): 4645. https://doi.org/10.3390/s22124645.

2 Lobyzov, V. and Shevtsov, V. (2021). The concept of a hardware-software system for protecting IIoT devices. *NBI Technologies* (2): 16–21. https://doi.org/10.15688/nbit.jvolsu.2021.2.3.

3 Gopalakrishnan, S. and Senthil Kumaran, M. (2022). IIoT framework based ML model to improve automobile industry product. *Intelligent Automation & Soft Computing* (3): 1435–1449. https://doi.org/10.32604/iasc.2022.020660.

4 Boyes, H., Hallaq, B., Cunningham, J., and Watson, T. (2018). The industrial internet of things (IIoT): an analysis framework. *Computers in Industry* 1–12. https://doi.org/10.1016/j.compind.2018.04.015.

5 Priyashan, W.D.M. and Thilakarathne, N.N. (2020). IIoT framework for SME level injection molding industry in the context of industry 4.0. *SSRN Electronic Journal* https://doi.org/10.2139/ssrn.3750588.

6 Hussain, Z., Akhunzada, A., and Iqbal, J. (2021). Secure IIoT-enabled industry 4.0. *Sustainability* (22): 12384. https://doi.org/10.3390/su132212384.

7 Fernández-Caramés, T.M. and Fraga-Lamas, P. (2020). Use case based blended teaching of IIoT cybersecurity in the industry 4.0 era. *Applied Sciences* (16): 5607. https://doi.org/10.3390/app10165607.

8 Shakulikova, G.T. (2021). New challenges in oil education. *Kazakhstan Journal for Oil & Gas Industry* (1): 99–107. https://doi.org/10.54859/kjogi88924.

9 Sotoodeh, K. (2021). Introduction to the subsea sector of the oil and gas industry. In: *Subsea Valves and Actuators for the Oil and Gas Industry*, 1–36. Elsevier https://doi.org/10.1016/C2020-0-04035-1.

10 Baxter, M. (2005). Global challenges facing the oil and gas industry. *Journal of Petroleum Technology* (04): 14–15. https://doi.org/10.2118/0405-0014-jpt.

11 Jaidka, H., Sharma, N., and Singh, R. (2020). Evolution of IoT to IIoT: applications & challenges. *SSRN Electronic Journal* https://doi.org/10.2139/ssrn.3603739.

12 Mayer, B. and Hartner, R. (2021). Multi-layer IIoT architecture for autonomous vehicles: a proof of concept. *SSRN Electronic Journal* https://doi.org/10.2139/ssrn.3868537.

13 Paytayeva, K. (2013). Environmentaland economocaspecte of environmental management in oiland gas industry. *The Russian Academic Journal* (4): https://doi.org/10.15535/14.

14 Papavinasam, S. (2014). Oil and gas industry network. In: *Corrosion Control in the Oil and Gas Industry*, 41–131. Elsevier https://doi.org/10.1016/C2011-0-04629-X.

15 Sonawane, K. and Bojewar, S. (2019). IIOT for monitoring oil well production and ensure reliability. *SSRN Electronic Journal* https://doi.org/10.2139/ssrn.3370201.

16 Dias, J.P., Restivo, A., and Ferreira, H.S. (2022). Designing and constructing internet-of things systems: an overview of the ecosystem. *Internet of Things* 100529. https://doi.org/10.1016/j.iot.2022.100529.

17 Singh, K.J. and Kapoor, D.S. (2017). Create your own internet of things: a survey of IoT platforms. *IEEE Consumer Electronics Magazine* (2): 57–68. https://doi.org/10.1109/mce.2016.2640718.

18 Okay, F.Y. and Ozdemir, S. (2018). Routing in fog-enabled IoT platforms: a survey and an SDN-based solution. *IEEE Internet of Things Journal* 6: 4871–4889. https://doi.org/10.1109/jiot.2018.2882781.

19 Chettri, L. and Bera, R. (2020). A comprehensive survey on Internet of Things (IoT) toward 5G wireless systems. *IEEE Internet of Things Journal* (1): 16–32. https://doi.org/10.1109/jiot.2019.2948888.

20 Aly, M., Khomh, F., and Gueheneuc, Y.-G. (2019). Is fragmentation a threat to the success of the Internet of Things? *IEEE Internet of Things Journal* (1): 472–487. https://doi.org/10.1109/jiot.2018.2863180.

21 Al-Hawawreh, M. and Sitnikova, E. (2021). Developing a security testbed for industrial Internet of Things. *IEEE Internet of Things Journal* (7): 5558–5573. https://doi.org/10.1109/jiot.2020.3032093.

22 Al-Rubaye, S. and Kadhum, E. (2019). Industrial Internet of Things driven by SDN platform for smart grid resiliency. *IEEE Internet of Things Journal* (1): 267–277. https://doi.org/10.1109/jiot.2017.2734903.

7

The Power of Edge AI Technologies for Real-Time Use Cases in the Oil and Gas Domain

7.1 Introduction

Simply put, edge AI is a strategically sound combination of edge computing and artificial intelligence (AI). Without an iota of doubt, it is a fact that the enigmatic AI domain is being empowered through a dazzling array of path-breaking machine and deep learning (ML/DL) algorithms. On the other hand, there are billions of network-embedded devices (alternatively termed as the Internet of Things (IoT) edge devices). Increasingly and interestingly, trained, tested, and refined AI models are being deployed on resource-intensive IoT edge devices (some term these devices as fog devices or edge servers). With the faster maturity and stability of AI algorithms, frameworks, and libraries, AI-specific processors, etc., creating, evaluating, and optimizing AI models to run on IoT edge devices have gained momentum.

This direct deployment facilitates the capture, storage, processing, and analytics of the IoT device data locally. That is, IoT data get captured, stocked, and processed in that IoT device itself. With generation of more IoT device data, there is a transition toward the leverage of multiple heterogeneous IoT devices for proximate and complex IoT data processing and analytics. In other words, several IoT devices in the vicinity get integrated to quickly perform edge data analytics by sharing the computed and storage resources of IoT edge devices. This clustering of multiple devices is typically termed as device clusters/clouds for enabling real-time and complicated edge data analytics. There is a new era of forming *ad hoc*, dynamic, purpose-specific, and small-scale cloud environments by networking of different and distributed IoT edge devices. These edge cloud environments have

The Power of Artificial Intelligence for the Next-Generation Oil and Gas Industry: Envisaging AI-Inspired Intelligent Energy Systems and Environments, First Edition. Pethuru Raj Chelliah, Venkatraman Jayasankar, Mats Agerstam, B. Sundaravadivazhagan, and Robin Cyriac.

laid down a stimulating and scintillating platform for accomplishing edge computing and analytics. Especially, streaming analytics platforms are being deployed in edge devices to perform real-time streaming analytics.

However, with the advanced and automated analytics capability being provided by AI algorithms, there is a twist. Increasingly, AI frameworks and libraries are being deployed in IoT edge devices to simplify and streamline next-generation edge analytics. The ML and DL algorithms deployed on the edge devices process the data generated by the devices themselves to emit out predictive insights in time. Devices can make independent and insightful decisions in a matter of milliseconds. There is no need for the edge devices to connect with faraway cloud platforms, applications, and databases all the time. Occasionally, edge devices can connect and synchronize with cloud-hosted services/applications and datastores to be on the same page. Such a change in thinking through the power of edge AI leads to the visualization and realization of a plethora of powerful business, technical, and user cases. With a host of edge AI products, accelerators, and solutions, a variety of sophisticated edge AI applications have come up to remarkably enable institutions, individuals, and innovators in their everyday assignments. The traditional and widely talked smart cities, buildings, retail stores, healthcare, defence, manufacturing floors, logistics, etc. are being significantly improved through the power of edge AI. In this chapter, we will deliberate about the practical details of edge AI and how it can be artistically leveraged for the betterment of industries.

7.2 Demystifying the Paradigm of Artificial Intelligence (AI)

The mesmerizing AI capability is penetrating every industry vertical these days. AI is all set to become a popular, pervasive, and persuasive choice for every worthwhile system to be adaptive and adroit in its functioning and service delivery. AI is termed as a collection of powerful technologies and tools, which come handy in smoothly replicating human brain capabilities in our everyday gadgets and gizmos, including smartphones, robots, drones, cameras, vehicles, medical instruments, and manufacturing floor machineries. That is, the distinct and distant goal of AI is to bring in human brain capabilities into everyday devices in our lively and lovely environments, business workloads, and IT services running in enterprise and cloud IT infrastructures. It is argued that such a dramatic transformation is being made possible through the smart leverage of machine and deep learning (ML/DL) algorithms.

There is an overwhelming acceptance that data get transitioned into information, which, in turn, get converted into knowledge through a host of AI technologies and tools. Thus, any enterprising business must collect all kinds of data meticulously and subject them to a variety of deeper investigations to extract actionable insights in time. In short, we are heading toward the era of data-driven insights and insight-driven decisions/deeds with the immense contribution of AI algorithms and models. AI is all set to become the new normal for the technology world. AI will be treated as the central and core capability for all kinds of business augmentation, acceleration, and automation requirements. With the faster adoption and adaptation of AI advancements, hitherto unknown use cases will be disclosed for the betterment of business and IT worlds. The role and responsibility of AI for setting up and sustaining the dreamt intelligent world are growing steadily.

7.3 Describing the Phenomenon of Edge Computing

Market researchers and analysts came with a solid forecast that there will be billions of IoT edge devices in the days ahead. There is resource-constrained and intensive IoT devices. Typically, resource-constrained devices are termed as edge devices. Resource-intensive devices are called as edge servers/fog devices. Edge devices are the new computing elements for facilitating real-time and real-world services. Edge devices are being used for data storage, processing, and analytics. Edge computing involves data collection, analysis, and processing, which are happening at edge devices. This means that the computing power and data storage are located where the actual data collection happens. Edge devices can be a smartphone, robot, drone, toaster, consumer electronics, medical instruments, defence equipment, manufacturing machineries, self-driving vehicles, kitchen utilities, telecom tower, etc.

The edge computing levels: Generally, AI training for model creation happens in the cloud due to the necessity of huge computational and storage resources. However, data collection and inferencing take place at the edge of the network. Thus, trained, tested, and refined AI models are made to run on edge devices to enable real-time edge data analytics. That is, inferencing takes place at the edge. On fresh data, the inferencing module running on edge devices takes competent decisions and embarks on the decided actions quickly. AI-specific processors are increasingly embedded in edge devices to quicken the inferencing activity.

Low edge computing level: With low edge computing, applications only send a small amount of useful data to the cloud. This significantly reduces data transmission time, saves precious bandwidth, achieves energy savings, etc. An arm-based platform without AI accelerators can be used on IIoT devices to collect and analyze local data to make quick inferences. Relatively speaking, compute complexity at the edge is less.

Medium edge computing level: This involves a good amount of computing. IP cameras stream videos for instantaneous analytics. This is a computer vision (CV) application. Face detection is a popular use case. Access control systems in our office locations are using the face detection module. Similarly, CCTV cameras are being installed in important junctions such as airports. For meeting high-performance computing, the Intel Core i7 Series CPUs offer an efficient CV solution with the OpenVINO toolkit and software-based AI accelerators.

High edge computing level: This involves a heavy load of data to be processed at the edge. AI models must do complex pattern recognition. The prime examples include behavior analysis for achieving automated video surveillance. Such a capability helps proactively detect security incidents or visualize threatening events. This is getting accomplished with the help of AI accelerators including a high-end GPU, neural processing unit (NPU), vision processing unit (VPU), tensor processing unit (TPU), and FPGA. A macro-level edge computing architecture is pictorially represented in the below diagram.

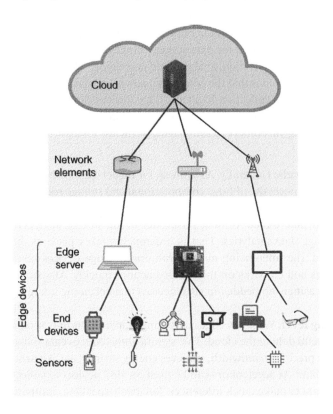

Thus, depending on use cases, the quantity of local computation needed varies drastically. Similarly, the amount of data getting generated, gathered, and crunched also significantly differs. Accordingly, edge devices are empowered with adequate resources such as memory and storage capacities and processing capability. Deeper edge analytics through AI models typically need bigger computing power.

7.4 Delineating Edge Computing Advantages

Edge computing enables bringing data processing tasks from the cloud to edge devices. As proximate processing is being facilitated through the edge phenomenon, a significantly lower amount of edge data traverses to nearby or faraway cloud platforms to be stocked and subjected to a variety of deeper and decisive investigations. Such a transition opens several noteworthy possibilities as illustrated below.

7.4.1 The Formation of Edge Device Clouds

Both small and large-scale computing activities are comfortably happening in enterprise and cloud servers. The cloud journey thus far is simply amazing. We are hearing, reading, and even experiencing hybrid and multi-cloud facilities. Multiple cloud service providers in conjunction communication service providers are in the scene these days to deliver business workloads and IT services to cloud consumers. To avail a variety of cloud services and applications being offered by multiple cloud providers, enterprises and governments across the world are jumping into the multi-cloud bandwagon. Further on, business workloads are being built as a collection of microservices, which are independently developed and deployed in geographically distributed cloud environments. Thus, multiple cloud environments must be stitched together to fulfil complex business processes. In other words, not only application components spread across different cloud environments, but also applications deployed across clouds must be linked up to perform specific assignments successfully. Today, we have public, private, and hybrid clouds. With multiple cloud environments getting integrated on need basis to run specific workloads, we now eventually enter the multi-cloud phenomenon.

With edge devices joining in the mainstream computing, some specific computing activities are being accomplished through edge devices and their clusters. That is, IoT edge devices emerge as the new computational environment to enable real-time service development and delivery. Edge device clouds are being formed dynamically and integrated into mainstream cloud environments. That is, the much-trumpeted multi-cloud capability now gets strong boost with the incorporation of edge clouds. Such a transition leads to the visualization and realization of a plethora of real-world applications (personal as well as professional). Edge clouds enable edge computing and analytics. There are lightweight versions of Kubernetes to facilitate a simplified setup and sustenance of device clusters/clouds.

7.4.2 Real-Time Computing

Transmitting edge data to faraway cloud environments takes time. That is, latency, which is usually about 100 ms, is unacceptable for certain problems. For example, new Porsches are equipped with hundreds of sensors that continuously produce data on various operational aspects of the car's components. Porsches are fitted with an NVIDIA GPU processor and Kinetica analytics software. This empowerment leads to the much-expected proximate data processing, which, in turn, helps take appropriate decisions in real-time. For example, if the car's speed is 200 km per hour, latency of even milliseconds is unacceptable. The decision to brake must be made before the car is plunging into a ditch.

7.4.3 Real-Time Analytics

With edge computing, it is possible to perform real-time analytics. Data get captured and crunched quickly to emit out actionable insights in time. The multifaceted machineries and robots in a manufacturing floor must be activated in time so that quality products are generated. Self-driving cars and robot-assisted surgeries are the other well-known activities that squarely rely upon real-time analytics capability. Real-time analytics is the most sought-after feature for real-time and intelligent enterprises.

7.4.4 Scalable Computing

There are market research and analysis organizations forecasting that there will be billions of IoT edge devices in the years to come. The research organization IDC has predicted that there will be 41.6 billion IoT devices generating 79.4 zettabytes of data in 2025. This can result in tremendous amount of edge data getting produced, collected, stocked, and analyzed. Streaming analytics toolkits are being deployed in edge devices to perform data analytics to produce timely insights. The device ecosystem is continuously growing. There are slim and sleek, trendy, and handy, and multifaceted devices in plenty. There is a faster maturity and stability of technologies and tools that come handy in forming edge device clouds to tackle specific tasks at hand. Due to the surge in device deployment in our home, relaxing, wandering, and working places, edge-native applications can get an enormous number of device infrastructures to perform improved processing and storage. AI models and frameworks can be deployed in a group of devices to local analytics of edge data to extract useful and usable information and insights. In short, additional devices can be leveraged quickly to match up with the amount of edge data. Due to the increasing abundance of edge devices in any mission-critical environments, edge devices contribute for scalable computing.

7.4.5 Secured Computing

Edge data typically stay within the edge environment, thereby data security and privacy goals are getting simply fulfilled. Less amount of data gets transmitted to the cloud through the Internet, which is the world largest communication infrastructure. When edges are offline, cyberattacks are comparatively less. The edge network is closed, and hence stealing confidential, customer, and corporate data becomes a tough affair. Accessing and assessing edge devices and data remain a challenge for outsiders. When data integrity and confidentiality are guaranteed, the data-driven insights will also be accurate. Sensitive data in edge environments get processed immediately and hence disappears quickly from the scene to escape from prying eyes and evil hackers.

7.4.6 Automated Analytics and Action

As indicated above, real-time analytics of edge device data is gaining speed and sagacity. With streaming platforms being deployed in edge servers, real-time streaming analytics is getting the attention of many these days. For such analytics projects, analytics professionals must write the data-processing logic.

However, with well-trained and optimized AI models getting deployed in IoT edge devices, the analytics process gets accelerated and automated. That is, inference and decision-making processes are being hugely simplified and speeded up. AI models can automatically learn from data and make right predictions in time. The knowledge gained immensely helps software systems and actuators to plunge into remedial and rewarding actions with all the confidence and clarity.

7.4.7 Reduced Costs

Data processing gets distributed with the steady availability of heterogeneous edge devices. Edge data gets local storage and processing. Thereby, the quantity of data gets transmitted to cloud environments (private and public) over any shared and dedicated networks decreases. The expensive network bandwidth therefore gets saved, and the network congestion also gets avoided to a greater extent. With the ready availability of pools of edge devices, horizontal scalability required to tackle data and user spikes is made simple and swift. Easily and affordably additional devices can be deployed to accomplish large-scale data processing. Mission-critical decisions can be made momentarily, and counter-measures can be initiated fast according to changing situations. There is no need for transmitting all the data getting generated to faraway cloud storage appliances. Thus, edge storage and processing are being touted as the one for the future of the IT.

In summary, edge computing is laying down a progressive and productive environment for the flourishing of business-critical, people-centric, and edge-native applications and services. With edge devices joining in the mainstream computing, efficient data processing and analytics are destined to happen at the network edge. As reported above, newer business use cases are bound to emerge and evolve.

7.5 Demarcating the Move Toward Edge AI

As accentuated above, the combination of two strategically sound paradigms is going to be the momentous change for the entire world. Edge AI is being viewed as advanced and automated analytics that takes place locally and logically in IoT edge devices. Edge devices can get the vision capabilities and the intelligence to exhibit adaptivity and sagacity while executing and delivering deliveries to their users. That is, the themes of machine vision and intelligence gain the attention of many with the penetration of AI into everyday devices. Video analytics, once the prime thing through powerful cloud servers, is now being activated and accomplished in edge devices. There are algorithms and implementations for sensor data fusion; thereby complex analytics could be performed at the device layer. With the microelectronics and nanoelectronics fields growing fast with the dedicated and disciplined contributions of semiconductor players across the globe, the pace of the growth of edge computing is to accelerate further in the years to unfurl. Edge analytics or in-device data processing contributes for extracting context intelligence. Any service being developed and delivered dynamically with the enriched context details can be more appropriate and assistive to users.

The edge AI capability is being methodically embedded in IoT edge devices to empower them to be cognitive in their decisions and deeds. Some well-known devices getting strengthened through edge AI include smart speakers, smart phones, robots, self-driven cars, drones, and surveillance cameras. With intelligence ingrained, edge devices join in the mainstream computing. Business processes are being readied to include edge devices to bring in deeper business automation. Data and process pipelines are accordingly prepared and preserved to elegantly energize business actions. Modern data platforms emerge to take care of copious amounts of edge device data to make real-time computing

7.5.1   How Edge AI Helps to Generate Better Business

In short, when edge computing is seamlessly combined with artificial intelligence, we get an unbeatable and versatile combination. Edge AI speeds up knowledge discovery and dissemination; makes data capture, storage, and processing more secure, smooth, and swift; improves user experience with many hyper-personalization features; and lowers costs (capital and operational). In short, real-world and real-time applications get realized to smoothen and speed up the journey toward intelligent real-time enterprises.

A popular example for an edge AI device is a hand-held tool, which is extensively used in a factory. This tool is stuffed with a microprocessor that intrinsically utilizes edge AI software. There are several business, technical, and user cases. The battery life increases; there is no need to send edge data to the cloud; the tool locally collects, stocks, processes, and analyzes data to extract actionable intelligence in time; the tool can get synchronized with the cloud at a scheduled time; the cloud server can do comprehensive analytics on historical and current data, etc. A tool embedded with the AI feature could turn itself off in the event of an emergency. The manufacturer receives all sorts of decision-enabling information about the health, performance level, etc. to bring in a series of innovations in product development.

Cisco has introduced an analytics software for Meraki MV cameras that tracks the location of an inventory in warehouses, machinery on factory floors, and the use of protective gear in hospitals.

7.6   Why Edge AI Gains Momentum?

For creating viable and venerable machine learning (ML) models, we need to leverage one or more proven ML algorithms on a chosen dataset to train an ML model. Training the model means that it is programmatically empowered to precisely pinpoint useful patterns in the training dataset. Subsequently, the trained model is evaluated on a test dataset to verify and validate its performance and accuracy levels. Once the trained ML model reaches a state of stability in predicting correct results on test data, then the ML model gets deployed in cloud environments. There are AutoML platforms (open source as well as commercial grade) for assisting data scientists to create appropriate ML models in an automated manner. There is a galaxy of assisting tools for feature selection, optimization, and engineering. There are feature stores and data pre-processing tools in plenty toward lessening the workload of data scientists.

MLOps platforms are to deploy AI models in production environments in a simplified and speedy manner. With the arrival of fresh data, the ML model gets retrained and redeployed. Thus, the fast and frequent deployment of ML models that is needed to match up with fast-changing business sentiments gets realized through competent MLOps solutions. For example, Kubeflow is an MLOps implementation to deploy, run, and manage ML models in Kubernetes-managed cloud environments. The ML model can be expressed and exposed as a microservice. That is, the model works via an API. The ML model makes predictions. The model output can be an input to a software application or visualized through a 360° dashboard for business executives and users. Thus, derived insights are supplied to the concerned person or system to ponder about the next course of actions in time.

As mentioned above, cloud platforms and infrastructures are highly affordable through a host of automation features and the much-celebrated aspect of sharing. The large-scale computing needed for creating ML models is being succulently fulfilled through cloud storage facilities. Running ML models on cloud infrastructures has opened fresh possibilities and opportunities. However, the worrying factors of cloud IT are the network latency and the wastage of precious network bandwidth. So, the prevailing trend is to train and test ML models using cloud infrastructures, whereas the inference on fresh data gets accomplished through edge devices. That is, trained, tested, and refined models are getting deployed in edge devices to enable fast inferencing and decision-making.

7.6.1 The Growing Device Ecosystem

With the surge of Industry 4.0 machineries, autonomous vehicles, connected gadgets and gizmos, embedded systems, and IoT sensors and devices, real-time inferencing through the edge gains dominance and prominence these days. If an ML model lives in the cloud, then edge data must be transmitted to the cloud to make inferences. In this mode, gaining real-time insights is an arduous task. Resource-intensive edge devices have brought in a paradigm shift in realizing edge-enabled real-time AI capabilities.

7.6.2 Federated Learning

This is an interesting method for training an ML model using multiple edge devices. ML models are locally trained by leveraging several devices, and only the model updates are communicated to the central server, which, in turn, aggregates the updates and sends the updated ML model back to the edge devices. Thus, complex model training gets simplified through the simultaneous leveraging of many edge devices. There are cluster enablement and management software solutions to make use of multiple devices for efficient model training. This is known as federated learning, which comes handy when an edge device is incapable of doing data processing alone. Thus, edge AI becomes a mainstream subject of study and research with the emergence of powerful techniques such as federated learning.

7.6.3 Optimization Techniques to Run AI Models in Edge Devices

Artificial intelligence (AI) models are being produced and used to solve a variety of business, scientific, and technical problems. AI model engineering processes, platforms, and products are acquiring special significance across industry verticals. Due to deeper automation, the number of features being used for model generation is numerous, and hence the resulting AI models are bulky. Therefore, AI researchers have come out with several powerful optimization techniques and tools to compress AI models. This section explores a suite of pioneering methods for achieving AI model optimization. Pruning and quantization techniques are being used in combination to reduce the size of complex AI model architectures and make them optimized and performant so that they can be easily deployed in the IoT edge devices.

In the field of computer vision (CV) and natural language processing (NLP), neural networks are being used in innovative and impactful ways. However, abundant computational resources are required for implementing neural networks (NNs) (realized through machine and deep learning (ML/DL) algorithms). Further on, the energy consumption of these artificial intelligence (AI) models is also high, while the heat dissipated by them into our fragile environment has a hugely damaging effect. Additionally, deploying such feature-rich neural networks on the Internet of Things (IoT) edge devices is beset with several technical challenges and concerns. Thus, AI models ought to be optimized to reduce the usage of huge computational resources. Also, real-time analytics is the need of the hour for producing real-time services, applications, and enterprises. The optimization typically involves the identification and elimination of redundant and irrelevant features. Especially, there is a surge in edge devices, robots, drones, instruments, equipment, appliances, wares, machineries, and gadgets. There is a requirement for producing and deploying edge-native intelligent applications. Highly optimized AI models are therefore insisted for intrinsically empowering edge devices with all the data-driven insights.

The choice of AI algorithms used in edge devices is crucial as these devices lack substantial power sources. Selecting the right machine learning algorithm will help edge devices process enormous amounts of data and significantly decrease the complexity of the network. These choices will cut down on parameters such as inference latency, power consumption, and memory use. This will also improve performance parameters of the chosen model. By doing so, we will reduce the number of redundant calculations in neural networks during the training phase. There are powerful methodologies to create lightweight (compressed) models, and in this section, we will focus on two such methodologies to create lightweight and superior models. Quantization and pruning are the two common techniques using which ML algorithms can be made to work well.

7.6.4 Neural Network (NN) Pruning

Pruning is a proven way to reduce the size of NNs through compression. After an NN is pre-trained, it is then fine-tuned to determine the importance of connections. This is done through the ranking of the neurons from the neural network.

The basic principles of pruning include removing unimportant weighted information using second-derivative data. This results in better generalization results, improved speed of processing the results, and a reduced size as well. Pruning is usually done in an iterative fashion to avoid the pruning of important neurons. Neural networks are a black box, and hence the vital step is to determine which neurons are important and which are not. After this, the least important neuron gets removed. This is followed by the fine-tuning of the model. At this point, a decision must be made to continue the pruning process or to stop pruning.

Pruning can come in various forms. Which pruning method must be chosen depends on the kind of output that is required. In some cases, speed is more preferred, whereas in some other cases, the storage requirement must be reduced. One of the finest methods of pruning is pruning entire convolutional filters. Using an L1 norm of the weight of all the filters in the network, the filters get ranked. This is then followed by pruning the "n" lowest ranking filters globally. The model is then retrained, and this process is repeated.

There are methods for implementing structured pruning for a more lightweight approach of regulating the output of the method. This method utilizes a set of particle filters, which are the same in number as the number of convolutional filters in the network. Then, the network's accuracy is determined by a test on a validation set. At its basic level, this method involves pruning each filter and observing how the cost function changes when the layers are changed. This is a brute force method, and hence the ranking method is highly effective and intuitive. It utilizes both the activation and gradient variables as ranking methods, providing a clearer view of the model.

Quantization is a machine learning technique that involves converting data from floating point 32 bits to a less precise format, such as integer 8 bits, and then using integer 8 bits to do all the convolutional operations. In the last stage, the output with lower precision is changed back to the output with higher precision in floating point 32. This process allows models to be represented in a compact manner and allows vectorized operations to run quickly on a wide range of hardware platforms.

Regularization is a set of techniques which can help avoid overfitting in neural networks, thereby improving the accuracy of deep learning models when it is fed entirely new data from the problem domain. There are various regularization techniques, and some of the most popular ones are L1, L2, dropout, early stopping, and data augmentation. The characteristic of a good machine learning model is its ability to generalize well from the training data to any data from the problem domain. This allows it to make good predictions on the data that the model have never seen.

If the model is not generalized, a problem of overfitting emerges. In overfitting, the machine learning model works on the training data too well but fails when applied to the testing data. It even picks up the noise and fluctuations in the training data and learns it as a concept. This is where regularization steps in and makes slight changes to the learning algorithm so that the model generalizes better.

7.6.5   L2 and L1 Regularization

Regularization works on the premise that smaller weights lead to simpler models, which in turn helps in avoiding overfitting. So, to obtain a smaller weight matrix, these techniques add a "regularization term" along with the loss to obtain the cost function.

$$\text{Cost function} = \text{Loss} + \text{Regularization term}$$

The difference between L1 and L2 regularization techniques lies in this regularization term. In general, the addition of this regularization term causes the values of the weight matrices to reduce, leading to simpler models. In L2, we depict cost function as

$$\text{Cost function} = \text{Loss} + \frac{\lambda}{2m} * \sum \| w \|^2$$

Here, lambda is the regularization parameter, which is the sum of squares of all feature weights. The L2 technique forces the weight to reduce but never makes them 0. This technique performs best when all the input features influence the output, and all the weights are of almost equal size. In the L1 regularization technique,

$$\text{Cost function} = \text{Loss} + \frac{\lambda}{2m} * \sum \| w \|$$

Unlike in the case of L2 regularization, where weights are never reduced to 0, in L1, the absolute value of the weights are penalized. This technique is useful when the aim is to compress the model. In this technique, insignificant input features are assigned zero weight and useful features with non-zero.

There are a few more popular regularization methods. Thus, model optimization along with regularization is being termed as essential for producing highly optimized AI models that can be easily and quickly run on edge devices for fulfilling the varying goals of edge AI.

There are other model optimization approaches such as knowledge distillation and federated learning. There are enabling frameworks, libraries, and tools for performing model optimization. Without embracing the proven model optimization techniques, AI models are bound to consume a lot of computational and power resources unnecessarily. There are numerous research papers exclusively focusing on the importance of model optimization and various techniques to reducing the model size. The irrelevant and repetitive features and parameters are identified and removed to arrive at lightweight models. As edge devices are being empowered to be intelligent through the dynamic deployment of AI models, AI model compression acquires special significance. Having understood the importance, product and tool vendors are bringing forth several enabling techniques and tools to simplify and streamline model optimization. With this movement, the idea of edge AI is bound to flourish in the days to come.

7.7 The Enablers of Edge AI

As indicated above, the field of edge AI is receiving significant attention these days. There are several enabling technologies and tools for fulfilling the ideals of edge AI. In this section, we are to discuss the prominent and dominant technologies that contribute immaculately for the flourishment and nourishment of edge AI. Let us start with the 5G-advanced communication

7.8 5G-Advanced Communication

5G networks are exceptionally reliable and blessed with a lot of bandwidth. Thus, transmitting a sizable amount of data is being speeded up through the 5G paradigm. 5G networks can accommodate one million edge devices in a kilometer surrounding. That is, without any hitch and hurdle, thousands of IoT devices can be comfortably tackled through the 5G communication capability. Resultantly, the 5G phenomenon contributes immensely for the desired success of edge AI. There are industrial and private 5G network facilities. It is expected that by the year 2024, 5G-Advanced specifications will be readied.

It is expected that 5G-Advanced communication will bring out the richest capabilities of 5G for borderless information flow. It is to break down all kinds of sickening boundaries to provide extreme and deep connectivity. From immersive extended reality (XR) experiences to high-precision location, presence, and timing technologies, 5G-Advanced will profoundly transform the way we interact, collaborate, correlate, and do our everyday activities. AI-inspired automated and advanced data analytics is to strategically empower core, RAN, and network management functionalities. Expertly designing production plants and intralogistics to be flexible, futuristic, autonomous, and efficient requires the right competency of connectivity. The 5G communication capability, which is being extended to private places and industrial environments, is to open fresh business prospects.

7.8.1 Industrial 5G

The 5G communication capability is being expanded and embedded into industrial establishments to bring in much-needed reliable and real-time actions. 5G-Advanced is the next evolutionary and revolutionary step in the reverberating 5G journey. 5G-Advanced will bring in a spate of enhanced capabilities beyond connectivity. Newer use cases are being published to get the support for 5G-Advanced. The critical advancements in the artificial intelligence (AI) algorithms and accelerators are to insightfully improve network performance. There will be noteworthy improvements in the areas of spectral efficiency and energy savings. There are articles and blogs illustrating how the emerging 5G-Advanced is to change how we live, work, interact, decide, and act.

Factory automation requirements ranging from production to intralogistics and transportation will be easily accomplished. The green computing paradigm will thrive across. Other applications include the leverage of mobile robots in production environments, autonomous vehicles on the road, the Industrial IoT-inspired industry 4.0 and 5.0 applications, and augmented reality (AR) applications for service and maintenance technicians. All these applications yearn for the 5G-Advanced communication capability.

Thus, 5G and 5G-Advanced are all set to transform the world. Our private and public spaces along with manufacturing floors are to be dextrously empowered to be beneficial for common people and industry workers in their everyday assignments.

7.8.2 Edge Computing

This is becoming the most popular computing model due to a variety of causes and reasons. Edge computing is a distributed and differentiated computing paradigm.

This is gaining a surging popularity as distributed computing services and applications are becoming the new normal in the IT industry to comfortably tackle complicated business problems. Without an iota of doubt, ours is becoming a deeply connected world. Businesses are expanding their operations across continents, countries, counties, and cities. The computing aspect therefore must cope up with the changing business sentiments to be right and relevant. That is, computing becomes distributed and decentralized. As the traditional cloud computing is centralized, there is a greater scope for edge computing, which intrinsically facilitates the much-expected decentralized computing.

Thus, edge computing is a vital cog for the intended success of edge AI. There are emerging and evolving purpose-specific and agnostic devices. These devices are integrated with one another in the vicinity and linked up with the Internet. Thus, the voluminous production and deployment of connected edge devices and servers are being seen as a positive sign for edge AI. Edge devices, already described above, are being clubbed together to form device clusters/clouds. Large-scale and data-intensive processing is getting accomplished through device clouds. Data analytics leveraging the traditional analytics platforms and in the recent past, through highly optimized AI models, is being enabled through edge clusters.

7.8.3 Massive Amounts of Edge Devices Data

As indicated above, there will be trillions of IoT sensos and devices capable of producing zettabytes of data. For example, the latest aircrafts are fitted with 50,000+ sensors that collect several terabytes of data every day. Thus, with the ready availability of Big Data, the power of AI is bound to increase sharply. Thus, collecting, cleansing, and crunching at the edge through on-device processing gain the attention of many these days. All kinds of operational data get gathered and subjected to a variety of investigations through analytics and AI platforms deployed in edge device clusters to extract operational intelligence. AI algorithms can produce higher accuracy in their decisions/conclusions/inferences when there is a tremendous amount of data.

7.8.4 The Emergence of Accelerators and Specialized Engines

On one side, there are IoT sensors, actuators, equipment, machineries, instruments, robots, drones, appliances, etc. getting produced and installed across our personal, professional, and social environments. On the other side, high-performance processors such as GPU, TPU, and VPU accelerators are being manufactured in enormous quantities. They are increasingly affordable due to the voluminous production. With these two breakthroughs, the versatile concept of edge AI becomes penetrative and pervasive. Also, the lightweight versions of various AI frameworks, toolkits, libraries, and platforms are being directly deployed in edge devices and clusters to fast-track the mystic edge AI paradigm. Edge AI use cases are being overwhelmingly explored and experimented across industry verticals. The manufacturing sector is benefiting out of all the distinct advancements in the edge AI space. Smart manufacturing/factory automation and Industry 4.0/5.0 applications such as pre-emptive and proactive care of high-value assets and machineries through predictive and prescriptive insights, monitoring of people and properties, etc. are termed as the prioritized contributions of edge AI.

Thus, the spread of resource-intensive IoT edge devices for local data storage and proximate processing, the emergence of powerful AI-specific processors to fulfil real-time computing and analytics, the faster maturity and stability of edge device cluster/cloud formation technologies, etc. collectively contribute for the intended success of the edge AI paradigm. Thus, with the widespread availability of 5G communication, local computing through edge devices, AI processing at the edge level, and an abundant supply of AI-specific processors have speeded up the domain of edge AI.

7.8.5 Cyber Physical Systems (CPS)

Digitization, as accentuated above, is fast catching up. All kinds of physical, mechanical, and electrical entities get digitized methodically through a host of digitization and edge technologies (sensing, perception, vision, communication, computation, alerting, identification, and actuating modules). Data capturing, storing, and processing and knowledge discovery and dissemination and decision-making components are also being attached on to produce multifaceted yet miniaturized digitized assets.

Digitized elements can acquire and exhibit extraordinary capabilities. They are being designed and developed to find, bind, and leverage each other's unique capabilities in their vicinity. Further on, digitized systems can directly or indirectly get linked up with the Internet. All kinds of web content, services, applications, and datastores can be accessed and availed by digitized entities to be distinct in their operations, offerings, and outputs. Physical systems at the ground level are being integrated with cyber-applications through connectivity infrastructures. The idea of cyber-physical systems (CPS) originated with digitization. Physical systems are artistically and adequately empowered through cyber-capabilities to be intelligent and autonomous in their service deliveries. Thus, besides electronics gadgets, physical systems are enabled to join in the mainstream computing. Such a technologically inspired transformation can result in producing, deploying, and delivering context-aware and cognitive services in real-time.

7.8.6 Digital Twins (DT)

This is another futuristic field of study and research for the knowledge-driven society. Building a digital/logical/software version of a physical system to automate some of the crucial activities has accelerated in the recent past. Precisely speaking, digital twins are the virtual replicas being meticulously built for all kinds of physical assets, devices, and processes. All kinds of mission-critical and multifaceted physical systems will benefit through their digital twins. Researchers, product engineers, and data scientists can run product simulations by leveraging its digital twin.

Digital twins are primarily used to detect and prevent problems in products, predict their performance, and optimize processes through real-time analytics to enhance the productivity. There are research publications, case studies, knowledge guides, white papers and e-books clearly articulating how digital twins can bring forth a suite of business transformations. Physical devices can be integrated with their respective digital twins, which are typically running at cloud environments. Digital twins need a few additional supporting systems such as data analytics platforms, AI model engineering platforms, data lakes, and digital twin development toolkits. Data integration tools also are seen as an important cog in realizing highly competent digital twins. Blockchain technology is being explored to safeguard digital twins' data.

Through digital twins, building, verifying, monitoring, measuring, managing, and enhancing overly complex physical systems become simpler, faster, and risk-free. AI-enabled digital twins are being viewed as the future of digital twins. AI-based data processing comes handy in expertly solving some of the prediction, recognition, correlation, classification, etc. Digital twins are expected to be an important ingredient in elevating the power of edge computing in the years to come.

From the initial stage of a chipset design or layout of a circuit through the end-product validation, emulators are indispensable for building complex interfaces and environments. Digital twins are the advanced version of these emulators. A digital twin can be a simple signal generator or a complete environment emulator. AI simplifies the design, development, and management of feature-rich digital twins.

7.9 Why Edge AI is Being Pursued with Alacrity?

We have studied about edge AI in detail. Edge AI is being vigorously and rigorously pursued by many as it is to make strategically sound impacts on businesses and people. The vision and mission for edge AI are clearly demarcated. With the outstanding growth of artificial intelligence (AI) algorithms, frameworks, platforms, libraries, toolkits, and models, the phenomenon of edge AI is bound to grow and glow in the years to unfurl. There are model optimization and compression techniques to enable AI models to be made to run on resource-constrained edge devices.

There are distinct advantages being made from edge AI. The long-pending use cases such as setting up and sustaining smarter cities, homes, hospitals, manufacturing, hotels, etc. are being readied through the power of edge AI. People-centric, service-oriented, event-driven, situation-aware, process-optimized, knowledge-filled, mission-critical, and proximate-processing services and applications.

7.9.1 The Need for Customer Delight

Customers, clients, and consumers want real-time and real-world experience in all things they do. Service delivery, therefore, must be highly personalized and timely. A kind of smooth and seamless experience is being demanded by people. Any kind of delay and deviation is not acceptable for people anymore. Edge computing is the way forward to fulfil this critical demand. Real-time services and applications can be readily built and delivered in time. Such a unique ability can result in real-time and intelligent enterprises. The transition from business efficiency and empowerment to people empowerment will be streamlined through the smart leverage of edge AI capabilities. Service delivery will be adaptive and adroit; thereby, customer delight requirements can be easily and elegantly accomplished.

7.9.2 Unearthing Fresh Use Cases for Edge AI Across Industrial Verticals

AI can cope up with an increasing number of complex situations by finding optimal decisions. It can learn and adapt quickly. It can pinpoint useful patterns in datasets. It is precise and perfect in recognition, detection, and language processing. Having understood the strategic significance of edge AI across business and functional domains, thought leaders and industrial veterans are publishing a dazzling array of future-proof use cases. Readers can find prominent use cases in the subsequent sections.

With more real-world and real-time use cases, the scope for edge AI is bound to increase remarkably. Businesses and end-users will get the confidence and the boost for embracing the idea of edge AI.

7.10   Edge AI Frameworks and Accelerators

There are several deep learning (DL) frameworks facilitating AI engineers and researchers to build DL models more easily and quickly.

PyTorch (https://pytorch.org/) is an open-source machine learning framework that accelerates the path from research prototyping to production deployment. This is used for applications such as computer vision and natural language processing. It has a rich ecosystem of libraries, tools, and more to support development. It is being supported through a growing developer community, which continuously contributes, learns, and gets your questions answered.

TensorFlow (https://www.tensorflow.org/) is the core open-source library to help you develop and train ML models. This enables fast prototyping, research, and production with TensorFlow's user-friendly Keras-based APIs, which are used to define and train neural networks. TensorFlow is being well-supported through a growing ecosystem of tools, libraries, and community resources that lets researchers push the state-of-the-art in ML and developers easily build and deploy ML-powered applications.

Real-time data analytics methods and AI algorithms on edge devices have led to the realization of industrial applications such as predictive maintenance, digital twins for complicated and sophisticated physical assets at the ground, and anomaly detection. There are several other enabling ML and DL frameworks for democratizing AI model generation, deployment, and improvement. There are corresponding lightweight versions of these frameworks to empower the ideals and goals behind the paradigm of edge AI.

Hardware-based accelerator toolkits: AI accelerator toolkits emanate from hardware vendors, and these accelerate producing advanced AI applications.

Intel OpenVINO Toolkit (https://www.intel.com/content/www/us/en/developer/tools/openvino-toolkit/overview.html): The Open Visual Inference and Neural Network Optimization (OpenVINO) toolkit is to help developers build robust computer vision (CV) applications on Intel platforms. OpenVINO also enables faster inference for deep learning (DL) models. This toolkit converts and optimizes models trained using popular frameworks like TensorFlow, PyTorch, and Caffe. This also deploys across a mix of Intel hardware and environments (on-premise enterprise servers and on-device, off-premise clouds). This is vividly illustrated in the figure below.

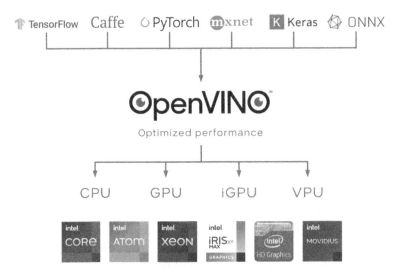

NVIDIA CUDA (https://developer.nvidia.com/cuda-toolkit): The NVIDIA CUDA toolkit provides a development environment for creating high-performance GPU-accelerated applications. With the CUDA Toolkit, you can develop, optimize, and deploy your applications on GPU-accelerated embedded systems, desktop workstations, enterprise data centers, cloud-based platforms, and HPC supercomputers. The toolkit includes GPU-accelerated libraries, debugging and optimization tools, a C/C++ compiler, and a runtime library to build and deploy your application on major architectures including ×86, Arm, and POWER.

Keeping mass transit on track: Trains are made to run on metal tracks, which need to remain upright and properly spaced. If there is any small deviation or unevenness, there is a possibility for trains to derail. Herein, track fasteners play a very vital role in ensuring smooth and safe rides for travelers.

However, due to constant friction and vibration between fast-moving train wheels and the tracks, as well as damage from the natural environment, track fasteners degrade and break over time. Therefore, it is essential to detect and repair track fasteners. Usually, after train service ends on one of the railway lines, human maintenance engineers perform manual visual inspection of the tracks and check for loose fasteners. Since visual inspection of railway tracks is error-prone, edge AI solutions come to the rescue by accelerating track fastener inspection with computer vision.

Moxa's V2406C Series rail computer is preferred due to its compact size with an Intel Core i7 processor that provides ample computing power for running the trained AI inferencing model. The V2406C supports the Intel OpenVINO toolkit and features two mPCIe slots for Intel Movidius VPU modules to accelerate image recognition computations and edge AI inferencing. This can improve efficiency and reduce maintenance expenses.

Autonomous Haulage Systems (AHS): These systems also rely on computer vision and navigation technology to enable autonomous haul trucks to pinpoint obstacles and move in the right direction and in proper position to collect excavated rocks from excavators and dump the debris in designated locations.

By installing a high-performance edge computer such as the Moxa MC-1220 series to connect PTZ cameras and sensors on each autonomous haul truck in the fleet, mining companies can obtain real-time video data from the excavation site as well as the exact position of each truck. The MC-1220 provides high-performance Intel Core i7 processors for video analysis and self-driving systems, as well as Wi-Fi and cellular connectivity to transmit pre-processed field data to the control center.

1) **Smart manufacturing**: Edge AI plays a very vital role in the aspect of quality control in manufacturing and transporting quality products across the planet Earth. Video analytics is a prime example of Industrial Edge AI, and such an earned ability can diligently monitor various product quality attributes reliably and precisely. Video analytics can detect and articulate even small deviations and defects that are almost impossible to be noticed by the human eye. Predictive maintenance of mission-critical assets in factories is being enabled through edge AI. Such a capability ensures the much-insisted business continuity. All kinds of sensors and devices data get meticulously collected and crunched to detect any abnormal behaviour/outlier/anomaly. This extracted knowledge comes handy in timely prediction of any impending failure of electronic gadgets and gizmos. That is, such advanced analytics indicates whether there is any need for rest or repair of any component or system. This increases the product life.

   AI is to reduce costs, improve safety, and increase productivity. Automated steering of transport robots to avoid obstacles and visual quality control in production and expert support for complex maintenance processes are the renowned use cases of AI for the manufacturing domain.

2) **Smart city, transportation, and traffic:** Deep learning (DL) allows edge AI devices to be aware of prevailing traffic situation and to detect abnormal behaviors through smart cameras. ML models are being used for license plate recognition on the road and for finding parking places.

   Integrating edge AI into the fleet of vehicles makes it easy to remotely monitor the vehicles via portable devices like smartphones and digital assistants. Autonomous vehicles are becoming popular these days. As known widely, edge AI plays an indispensable role there. As described elsewhere, thousands of sensors are being used in every passenger aircraft to eloquently monitor, measure, and manage the operational part of aircrafts. Real-time analytics of aircraft engines' data through edge AI goes a long way in smartly and securely flying us across the world. The promising edge AI technology is being used to calculate the number of passengers and to locate fast vehicles with extreme accuracy. Thus, the scope of edge AI is bound to grow and glow consistently in the days to unfurl.

3) **Energy:** Due to the purposeful participation of a copious number of sensors, a smart grid produces substantial data. As widely articulated, there are a myriad of advantages of smart grids. A smart grid guarantees adequate supply of electricity as per changing demands, power energy consumption monitoring and forecasting, renewable energy utilization, and decentralized energy production. All these get comfortably fulfilled only if local communication is possible between devices. Thus, local communication and collaboration empower smart grids artistically in their service deliveries.

4) **Retail:** This is a flourishing domain. We have supermarkets, hypermarkets, malls, retail stores, etc. in our residential areas. CCTV cameras are plentifully used in these environments for a variety of reasons. Point of sale (PoS) devices are in large-scale use here. Robots are being used to smoothen the purchase activity for shoppers. Drones are also explored for product delivery at doorsteps. Thus, there are multiple connected devices being leveraged for precise retailing and delivery. Herein, real-time analytics of edge device data is insisted for personalized service offerings. Facilitating a smooth and smart shopping is essential for retaining the loyalty of current customers and for attaining new customers. Inventory and replenishment management activities are smoothly handled up through the edge AI power.

5) **Campus positioning:** The precise identification of objects' location is being seen as a critical aspect to achieve solid advancements in productivity, safety, efficiency, and cost savings for factories, warehouses, etc. Seamless and accurate indoor–outdoor positioning is a prerequisite for holistic autonomous intralogistics. Further on, low-latency edge positioning is necessary for autonomous drones for warehouse inventories or video from 5G-enabled drones that monitor potential security issues.

6) **Immersive experience:** As articulated above, digital technologies such as AI, Augmented and Virtual Reality (AR/VR), blockchain, and 3D printing in association with the accumulation of IoT devices and sensors in manufacturing floors have gained the wherewithal to totally transform the product lifecycle management and drastically to change consumer behavior.

   AR can be experienced via wearable smart glasses, smartphones, etc. These edge devices internally use AI models to assist users in performing complex tasks and getting real-time insights to take timely and correct decisions. Any AR-capable edge AI device allows visualization of computed results with low latency. A holographic AR cockpit with real-time data of all production machines reduces complexity for workers and enables on-site decision-making and planning. In a nutshell, AR devices stuffed with the edge AI facility allow organizations to speed up productivity, enable hands-free operation, and leverage real-time data analytics. The trick is that the data transmission between IoT edge devices and AR systems in any environment is quickly done through the advanced communication technologies such as 5G.

7) **Automated guided vehicles:** Such advanced vehicles have vast potential to lower overhead and cost, particularly in warehouses and Industry 4.0-enabled factories. However, these are simply powerful and path-breaking when enabled with edge AI capability and 5G communication. There are distinct benefits getting accrued out of the combination of the industrial IoT, AI, edge analytics, AR devices, etc. The well-known advantages are seamless use of autonomous vehicles for indoor material transport, improved performance and reliability, and reduced operational risks along with lowered costs.

8) **Gaming fun with edge computing:** ecentralized data processing ensures that gamers can use augmented reality (AR) games everywhere without any interruption. With edge computing, data processing is brought closer to where it is needed.

9) **The emergence of rtificial intelligence of things (AIoT) applications:** Increasingly mission-critical industrial assets such as machineries, instruments, equipment, wares, utensils, and appliances are digitized and connected with the Internet. These assets are being integrated with web- and cloud-based applications, services, and databases. These are typically termed as the Industry Internet of Things (IIoT).

   IIoT devices generate substantial data and send it to faraway cloud storage and processing. This wastes a lot of network bandwidth. To reduce latency and storage costs, and increase network availability, businesses are taking AI algorithms (machine and deep learning) to the edge for real-time knowledge extraction, decision-making, and actions. The empowerment of industry IoT devices with AI capabilities results in a flurry of sophisticated AIoT applications.

   The advent of AIoT helps in exploring new avenues for fresh revenues for enterprises. By obtaining performance, operational, security, health condition, and environmental data from field equipment and machinery in time, companies have actionable insights at their disposal to make informed and intelligent decisions. For example, if human vision is used to manually inspect tiny defects on golf balls on a manufacturing assembly line for eight hours each day, five days a week, it is going to be a tough and time-consuming affair. People are liable for fatigue and error-prone. Thus, AI-inspired automation is being widely recommended. In a nutshell, multiple IoT devices collect and aggregate data, whereas AI crunches the data to emit out insights. Thus, this combination is being termed as deadly for institutions, individuals, and innovators.

10) **Edge AI use cases in healthcare**: Healthcare applications such as remote surgery and diagnostics are being advanced through edge AI platforms. Monitoring of patient vital signs is being performed through feature-rich edge devices.

    The need for on-device data processing ceaselessly gains momentum as there are frequent needs for insight-driven decisions to be made in time. Intensive care is an area that could immensely benefit out of edge AI. Real-time data collection, processing, decision-making, and action are often insisted to deliver sophisticated care services. If there is any threshold break-in in any of the prominent physiological parameters, then appropriate and automated actions must be initiated in nullify any catastrophes in time.

With astounding and all-round advancements in producing high-performing and AI-specific processing units such as GPUs, TPUs, and VPUs, the much-celebrated edge computing and analytics are bound to see a huge uptake and uplift. On the other hand, there are pioneering machine and deep learning algorithms to artistically support the edge AI journey. This venerable and versatile combination of flourishing edge data and the AI-based data processing at the edge has paved the way for the surge in adopting and adapting edge AI in the healthcare domain. Besides simple, a bevy of complex parameters are being elegantly monitored, measured, and managed by AI-activated edge devices. Experts point out that AI devices natively gain the power to pinpoint any neurological activity and cardiac rhythms. Precisely speaking, AI-inspired diagnostics and even prognostics are seeing hope to save precise lives.

11) **Ambient intelligence (AmI)**: This is an incredibly old concept belonging to the era of ubiquitous, pervasive, and sentient computing models. That is, AmI is all about producing and experiencing technology-inspired intelligence everywhere all the time in an environment such as a home, hotel, and hospital, etc. Edge AI is a serious contender for implementing the ideals of AmI. Edge devices will gain the ability to sense and respond (S&R) intelligently. Ambient assisted living (AAL) is one renowned use case.

12) **Mining, oil, and gas use cases:** These areas are mostly remote, rough, and tough. Workers on these sites are prone to a variety of dangers and life-threatening risks. Edge sensors and actuators are, therefore, popular in bringing in automation of several activities in these places. These meticulously collect much data and share it to local edge servers to enable accurate prediction and to prescribe of what to do for what symptoms and signals. Anomaly detection and predictive maintenance are the well-known use cases getting fulfilled through edge AI.

Thus, without an iota of doubt, the versatile edge paradigm is a new catalyst in visualizing and realizing next-generation digital life use cases and experiences by seamlessly and spontaneously incorporating the magical power of artificial intelligence (AI) at the edge. Further on, a growing array of business-critical and people-centric devices such as augmented and virtual reality (AR/VR) devices, humanoid robots, autonomous drones and industrial internet of things (IIoT) sensors, and drones have facilitated worldwide enterprises and people to embrace and experiment the edge AI phenomenon with all the clarity, cognition, and confidence to offer scores of premium and path-breaking offerings and services to their constituents, clients, and consumers. In maintenance use cases, AI is integrated with AR glasses to visually aid maintenance personnel through work steps. At the same time, AI checks whether all work steps have been carried out properly.

The size of images is rapidly increasing. Sending local images to faraway cloud platforms to be processed is to waste precious network bandwidth resources. Therefore, image analytics for gathering real-time insights is to be done at the edge. That is, cameras are being stuffed with AI models and frameworks to perform proximate data processing. This is a change in thinking. AI on the edge enables local analytics of the visual scene in various flavors, such as understanding the scene for context information, simultaneous multi-object detection and recognition for obstacle avoidance, and people identification. There are other allied use cases as illustrated below.

13) **Surveillance and monitoring**: Lately, we have a plenty of AI-enabled surveillance cameras in critical junctions and locations such as airports, warehouses, and retail stores. These empowered camera sensors could capture and process captured images instantaneously to track and trace people, objects, etc. If there is any incursion of terrorists into any specified area, cameras can understand and inform nearby police personnel directly through one or other channels. Or a warning message can be broadcast through any public address system. This acquired knowledge can help security officers swing into appropriate counter-measures to stop any nefarious and nonsense activity proactively and pre-emptively. Thus, gaining real-time insights is indispensable to ensure security and safety for high-value properties and people.

As widely understood, AI-activated cameras come handy in recognizing incoming vehicles, traffic signs, snarls and signals, and road-crossing pedestrians. Thus, edge AI is the prime contributor for the immense success of autonomous vehicles.

14) **Video analytics**: This is an interesting use case getting elegantly accomplished through the power of edge AI. Video surveillance for security, facial recognition for access control, and flow analysis for flawless and fabulous productivity are being implemented through local video analytics.

In retail, video analytics can be used to track customer footfall and analyze the buying patterns. Inventory and replenishment management, real-time customer servicing, product placement, etc. can be optimally accomplished through edge video analytics. Thus, there is a growing list of use cases for AI-based real-time video analytics. Smart city applications will be drastically advanced through such video analytics.

15) **Audio analytics:** Edge AI contributes for audio analytics as well. Audio event detection is a solid use case. Detecting sounds such as the cry of a baby, glass breaking, or a gunshot can trigger a series of connected and corrective actions, including sending notifications to the concerned in time. AI at the edge is the way forward for effective and elegant recognition of an audio event among numerous overlapping sound sources. Recognizing a car or truck approaching or screeching brakes of any vehicle in front of us can be lifesaving.

Human–machine interactions (HMIs) and machine-to-machine (M2M) interactions are being simplified and streamlined through the enhanced NLP capability of machines. Thus, manufacturing process steps can be automated to produce high-quality goods. Text to Speech (TTS) and Speech to Text (STT) are the two well-known examples of complex tasks. These are fully accomplished through advanced deep learning algorithms. in which deep learning algorithms are used to bring these functionalities on the edge. Examples are hands-free text read and write functions in automotive, where the driver can keep attention on his main task (drive the car) while interacting with the infotainment system. Finally, DL-based Speech Recognition is used in conversational user interfaces (CUI) where the distinct abilities of NLP are drastically augmented by allowing, for example, a chatbot to interact (dialogue) with a human grade conversation.

16) **Sensor analytics**: Smartwatches and fitness bands, as well as smart buildings, homes, and factories extensively exploit inertial and environmental sensors. A DL-enabled on-device processing allows quicker analysis of local situations and faster response.

17) **Predictive maintenance in factories**: This is a well-known use case for edge analytics. Sensors attached to a machine can measure vibration, temperature, and noise levels, and local AI-based data processing can infer the state of the equipment, potential anomalies or damages, and early indications of failure. If there is a need for a repair or rest for a particular component of an industrial asset, edge AI is capable of articulating that well ahead of time. In addition, the local DL inferencing model could also communicate with cloud-based services and digital twins to get data for specific analyses and corrective actions.

    Smart factory applications, such as smart machines, aim to improve safety and productivity. For example, operators can remotely operate, monitor, manage, and maintain heavy machines, which are in rough and tough places. Precision monitoring and control of manufacturing machineries are also ensured through edge AI.

18) **Body monitoring**: There are excellent write-ups about building and deploying body area networks (BANs) comprising several infinitesimal sensors and actuators such as insulin pumps to closely monitor various body parameters and how to insightfully counter if there is any emergency. These wearable devices collect a lot of data about our body conditions such as sugar and pulse levels in real time, the intensity of our physical activities, where are we put up, and what is the time. These details can be correlated with health, stress levels, diet, etc. and if needed, the BAN system can immediately alert wearers or their spouses or doctors to a potential health issue before it becomes critical.

Thus, sensors and their network data can be captured and analyzed instantaneously to ponder about the best course of actions with speed and sagacity. Artificial neural networks (ANNs) can be further exploited for multimodal context analysis by receiving and fusing data from a variety of data sources. There are pre-trained and advanced neural network (NN) models to recognize video, image, and audio data. Also, the models can blend all these data to better understand what is happening around the user. Such an understanding furthers to automate the necessary actions.

19) **Smart AI vision:** Computer vision applications, which are overwhelming at the cloud AI environments, are now replicated in edge AI environments. AI-inspired video analytics is becoming the new normal across industry verticals. Visual processing units (VPUs) are enabling high-performance computer vision services. With edge AI, it becomes possible to power scalable, mission-critical, and private AI applications.

20) **Smart energy:** Today, there is a steady movement toward renewable and green energy sources such as solar, wind, and green carbon. Wind farms are formed and connected. This is to facilitate data collection and to do knowledge discovery through cloud-based AI platforms. As wind farms use several data-generating sensors and actuators such as video surveillance cameras, security sensors, access sensors for employees, and sensors on wind turbines, edge AI capabilities are recommended to arrive at real-time insights with less expense.

21) **Smart entertainment:** The promising applications include virtual reality (VR), augmented reality (AR), and mixed reality (MR) for streaming video content to VR glasses. The size of such glasses can be sharply reduced by off-loading computation from the glasses to edge devices and servers. For example, Microsoft recently introduced HoloLens, for a holographic computer built onto a headset for an AR experience. HoloLens assists in Manufacturing (https://www.microsoft.com/en-us/hololens). It reduces downtime, transforms your workforce, and builds more agile factories. With HoloLens 2, employees can quickly learn complex tasks and collaborate in the moment from anywhere.

22) **Smart security:** Traditional surveillance cameras record and stock videos for hours. But, they could not process the captured videos to take intelligent decisions in time. With AI-enabled cameras, appropriate decisions can be taken in real time to strengthen the security and safety of people at crowded places.

Edge AI gives the wherewithal to generate automated responses/alerts to audio–visual stimuli in robots. It can also be used for real-time recognition of spaces and scenes. Multiple industries consciously explore and experiment differentiated edge AI use cases. For example, during the global COVID-19 crisis, AI-powered IoT products and solutions such as advanced X-ray machines, scanning and monitoring devices, and medication electronics and instruments were deployed in malls, hospitals, and other crowded places to do automated and real-time testing and treatment for people.

There is a wide range of edge AI applications including facial recognition and real-time traffic updates on semi-autonomous vehicles, etc. The emerging domains of connected intelligence and machine vision will get spruced up. Video games, robots, speakers and toasters, machines with human and machine interfaces, drones, wearable health monitoring devices, and security cameras will be astutely and aesthetically amplified.

7.11 Conclusion

Edge computing helps expertly empowering our everyday devices to join in the mainstream computing. That is, edge devices can be communicative, sensitive, perceptive, responsive, computation and vision-enabled, active, etc. Edge devices can collect data from their environments and process the gathered data instantaneously to emit out actionable insights. Thus, the renowned and reverberating aspect of data-driven insights and insight-driven decisions, deeds, and deals becomes a real-time affair with all the advancements and accomplishments in the edge space. By tackling personal, social, and professional problems through edge devices together, network latency decreases considerably, network bandwidth gets saved, data security gets beefed up, etc. With the deployment of refined AI models on edge devices, AI-enabled data processing results in the transition of data into information and into knowledge quickly. Thus, knowledge discovery by edge devices individually and collectively facilitates the visualization and realization of next-generation and people-centric AI applications.

Thus, with the emergence of a few resilient, robust, and versatile AI-centric technologies and tools, real-time and intelligent applications for setting up and sustaining smarter environments, establishments, and enterprises will arrive and thrive in the days to unfurl.

Bibliography

1 Gilmore, M. Edge computing in oil and gas: driving efficiency in digital transformation. https://stlpartners.com/articles/edge-computing/edge-computing-oil-gas-efficiency.

2 Edge computing for oil and gas companies. https://www.nagarro.com/en/blog/edge-computing-oil-gas-companies.

3 Oil and gas: digital transformation using edge computing https://www.pwc.in/assets/pdfs/emerging-tech/oil-and-gas-digital-transformation-using-edge-computing.pdf.

4 Edge analytics - making oil drilling more efficient. https://www.ltimindtree.com/wp-content/uploads/2022/02/Edge-Analytics-Making-oil-field-drilling-more-efficient-PoV.pdf.

5 Shifting operations to the edge in the oil and gas industry by Eli Daccach. https://www.rtinsights.com/shifting-operations-to-the-edge-in-the-oil-and-gas-industry/

6 Edge AI in manufacturing industry benefits and use cases. https://www.xenonstack.com/blog/edge-ai-in-manufacturing-industry.

7 High-value applications of computer vision in oil and gas (2023). https://viso.ai/applications/computer-vision-in-oil-and-gas/

8 The growing role of generative AI in the oil and gas industry. https://blog.se.com/industry/energies-and-chemicals/2023/07/21/the-growing-role-generative-ai-oil-and-gas-industry/

8

AI-Enabled Robots for Automating Oil and Gas Operations

Robots are increasingly becoming an intimate and insightful part for the oil and gas industry to survive and thrive in the midst of many business and technical challenges. With the artistic assimilation of potential and promising digital transformation technologies, the robotic industry is all set to grow manifold and glow in the days to unfurl. This chapter aims to analyze and articulate the various aspects of current and emerging robotic solutions and how the dawn of artificial intelligence (AI) technologies and tools is to produce game-changing robots for the struggling oil and gas sector. There is no dispute as far as the power of the AI paradigm. With the cool convergence of AI and robotic technologies, the world can easily expect sophisticated, versatile, and resilient robots that can work wonders for the critical oil and gas industry.

8.1 Briefing the Impending Digital Era

We have been fiddling with many information and communication technologies (ICT) for long. These technologies are primarily dealing with information. They do information collection, transmission, cleansing, aggregation, storage, processing, analytics, and visualization tasks. For example, there are analytics methods for transforming information and knowledge. A growing array of business requirements and operations are being accelerated and automated through a host of IT systems and services. Further on, there is a galaxy of business workloads such as enterprise resource planning (ERP), supply chain management (SCM), customer relationship management (CRM), knowledge management (KM), and

The Power of Artificial Intelligence for the Next-Generation Oil and Gas Industry: Envisaging AI-Inspired Intelligent Energy Systems and Environments, First Edition. Pethuru Raj Chelliah, Venkatraman Jayasankar, Mats Agerstam, B. Sundaravadivazhagan, and Robin Cyriac.
© 2024 The Institute of Electrical and Electronics Engineers, Inc.
Published 2024 by John Wiley & Sons, Inc.

manufacturing execution system (MES) toward business empowerment. Thus, without an iota of doubt, IT-enabled business transformation is happening fast.

For ensuring the people empowerment, information and communication technologies contribute in multiple ways. For example, we could do e-commerce and e-business transactions to buy groceries, fruits, electronics goods, eatables, books, etc. Websites throw a lot of useful information about establishments, enterprises, and entities. Through bank websites, we could do money transfer. Web applications and services are providing information, transaction, brokerage, management, concierge, and empowerment services to their consumers. Now, the world is heading toward the era of Web 3.0. We can expect hyperautomation in the days to unfurl. Thus, technologies and tools are flourishing toward bringing in the much-needed convenience, choice, care, and comfort for people.

The fast maturity and stability of digital technologies (the combination of digitization and digitalization technologies) are promising much more for business behemoths and commoners. That is, all kinds of physical, mechanical, and electrical systems are digitally enabled and transformed into digital artifacts through digitization and edge technologies. These digital entities in any personal, social, and professional environments, on purposefully interacting with one another in the vicinity, generate a massive amount of multi-structured data. By smartly leveraging digitalization technologies, all kinds of digital data get subjected to a variety of investigations to extract actionable insights in time. Thus, digitization and edge technologies transform ordinary items in our everyday environments into extraordinary objects. Dumb things turn into dynamic objects.

The Internet of Things (IoT) is the key digitization technology for transitioning any tangible object into digital material. Then, the internet communication infrastructure comes handy in networking digital objects. The connectivity is not only with local objects but also with cloud-based software services and applications. Such connected digital entities empower experts to envisage a dazzling array of fresh use cases. The buzzwords are digital innovation, disruption, transformation, and intelligence. We are looking forward to digitally transformed cities, offices, manufacturing, homes, hotels, hospitals, warehouses, retail stores, campuses, eating joints, entertainment plazas, airports, railway stations, etc.

8.2 Depicting the Digital Power

We discussed the digitization, edge, and connectivity technologies and their unique contributions especially in modernizing all kinds of things into digital elements. The era of networked embedded systems capable of bringing forth fresh possibilities and opportunities is now with the arrival of digitalization technologies and transitioning of digital data and intelligence simplified and speeded up.

That is, the target of knowledge discovery and dissemination gets streamlined, thereby the long-standing goal of setting up and sustaining intelligent products, solutions, and services is being fulfilled. With the ready availability of digitalization technologies such as artificial intelligence algorithms, data analytics methods and platforms, message brokers, buses and queues, databases, data warehouses and lakes, and storage appliances and networks, making sense out of digital data becomes the new normal. With proper utilization, digital technologies (the unique and strategically sound combination of digitization and digitalization technologies) are gaining the confidence of enterprises to embark on digital transformation initiatives and implementations.

The key distinction of the digital power is to facilitate physical services for multiple industry verticals such as manufacturing, supply chain, and healthcare. Several manual tasks are automated through digital intelligence. For example, ambient assisted living (AAL) is possible with IoT edge devices and sensors. Diseased, debilitated, and bedridden people can order their coffee maker to make a cup of coffee through their smartphone and ask their humanoid robot to go to kitchen room to fetch the coffee near the bed side. Fall and fire detection are getting automated. Every gesture, event, mood, snoozing, orientation, etc. get meticulously captured and understood to deliver context-sensitive, people-centric, real-time, and real-world services. Digital entities, sensors, actuators, devices, appliances, equipment, instruments, wares, utensils, robots, drones, etc. work together to understand the varying needs of people clearly and concisely in time and fulfil them with all the clarity and alacrity. Briefly, context-aware techniques in association with a host of digitized entities come handy in visualizing and realizing context-aware services and applications. In short, we will be succulently empowered through a bevy of situation-aware and sagacious physical services in our everyday life through the application of digital technologies.

As per the leading market watchers and analysts, it is estimated that there will be millions of microservices, billions of connected devices, and trillions of digitized elements. These connect, correspond, collaborate, correlate, and corroborate to great zettabytes of poly-structured data. By utilizing digital technologies with care, the blooming idea of data-driven insights and insight-driven decisions is to see the light sooner than later.

8.2.1 The Emergence of Advanced Drones

Unmanned aerial vehicles (UAVs), widely called as drones, are becoming an important contributor for inspection toward fulfilling a variety of use cases. Drones are attached with multifaceted cameras and state-of-the-art sensors to collect visual data about mission-critical assets across industry verticals. For the oil and gas industry, the prominent assets include oil rigs, platforms, tanks, and robots. Similarly, there are several important machineries in oil refineries and plants. In between, there are pipelines, which also need to be meticulously monitored, measured, and managed to bring in the much-needed optimization, prediction, and automation in the oil and gas sector. Such an aerial inspection eliminates the huge cost of installing and maintaining elevated structures. Such an unmanned flying inspection leads to the enhanced safety of field workers. Lately, by equipping drones with powerful sensors such as radiometric thermal sensors, optical gas imaging, and cameras, additional capabilities are being derived through drones.

Newer applications and services are being supplied by advanced drones, whose application scope expands to monitoring construction areas. Also, environment monitoring is also picking up with drones. E-commerce and business companies use drones extensively for product delivery. People and product security and safety are being ensured through the growing and glowing power of drones. Self-managing drones across industry verticals are understanding something clearly at last with the precise embedding of artificial intelligence (AI) models.

8.2.2 The Grandiose Arrival of the State-of-the-Art Robots

Robots are exceedingly popular and pervasive these days. They get used in many important junctions and locations. Robotics is being proclaimed as the game-changing technology for industry automation. Several manual activities are precisely and perfectly accomplished through intelligent robots. Robots work with industrial systems to significantly boost the productivity. The industrial operations are being speeded up through the careful utilization of pioneering robots. There are humanoid robots being placed across personal, social, and professional environments. Robots are self-sufficient enough to contribute as individual contributors. Robots are autonomous, so they do routine inspection and support any emergency situations with aplomb. Robots are being continuously empowered through self-analyzing and self-managing technologies. With a greater understanding of robots' broader and deeper benefits, businesses are ecstatic about robots and their immense contributions toward factory automation.

Robots gain advanced capabilities as they gradually get embedded with feature-rich sensors, motors, and navigation systems. Especially, the oil and gas industry is benefiting immensely through such multifaceted robots, which automate and accelerate oil and gas operations. High-quality robots are agile, adaptive, and autonomous. They assist field workers in completing complex tasks. Robots ensure the highest safety and security for assets and employees. Robots are the most rightful and relevant ones for rough, remote, and risky environments. Not only oil fields but also oil plants and refineries benefit out of path-breaking robots. With the faster maturity and stability of digital technologies such as digital twins, cyber physical systems (CPS), the Industrial Internet of Things (IIoT), 5G, artificial intelligence algorithms and pre-trained models, cybersecurity, and software-defined cloud environments, the scope for robots brightens across industrial environments.

8.2.2.1 Real-Time and Real-World Robots

With the concept of edge computing gaining the much-insisted prominence and dominance in the recent past, the upcoming days and years are going to be bright for robots in achieving Industry 4.0 and 5.0 applications. With the technology assimilation, robots in our environments are set to become self-contained to guarantee autonomous behavior. Robots collect their environment data and crunch the collected and cleaned data to bring forth useful insights in time. The on-robot data processing is being facilitated through edge computing. Increasingly streaming analytics platforms are installed in edge devices to emit out real-time insights. Further on, with the arrival of lightweight AI frameworks, models, and libraries, edge devices can do proximate data processing to squeeze out knowledge.

The aspect of edge AI is flourishing with the voluminous production of AI-specific processing units, which could be accommodated in edge devices such as robots and drones. AI models, as written in the previous chapter, are being subjected to a variety of optimization and compression phenomena to bring forth optimal AI models, which are suitable for resource-constrained edge devices. Thus, with the convergence of multiple enabling technologies, real-time and real-world robots are being produced and deployed. These robots are made cognitive and competitive in their operations, outputs, and offerings. Real-time and intelligent physical services will be gradually acquired and articulated through such breakthrough robots. The long-pending notion of real-time enterprises is to be accomplished through the delectable advancements in the field of robotics and all its implementation technological spaces.

8.3 Robotics: The Use Cases

Here are the top perceivable benefits from the inculcation of robotics into the oil and gas industry.

- **Safer rigs**: We have read about dreadful incidents that claimed hundreds of precious lives and several oil rigs got destroyed. Thus, people and property safety and security are vital for the oil domain to march ahead. Such catastrophic calamities ought to be proactively and pre-emptively pinpointed and stopped. The focus, therefore, turns toward producing safe rigs. Such advanced and automated rigs come handy in minimizing any unwanted occurrences.

Safe oil rigs can act autonomously and artistically across all situations and spots. Intelligent automation is the need of the hour. Capturing and crunching context details to extract viable intelligence on surroundings and situation are being seen as the way forward. Going forward, insightful industrial rigs are being recommended as the best bet. Rig manufacturers produce safer and smarter oil rigs to be deployed across safety-critical environments to reduce any agonizing moments. The noteworthy advancements in the robotics space are being attached into rigs; thereby, rigs function in a safe and smart manner.

- **More with less**: Offshore oil rigs are expensive due to various causes such as extreme weather. Further on, ensuring the highest safety for field workers and safeguarding them to escape unscathed if there are any emergency aggregates and aggravates the operational costs. The suggestion is to reduce human resources in such rough and tough environments. By deploying trendsetting AI-enabled robots there, it is possible to cut down operational costs along with the fulfilment of the much-needed safety and sagacity. Furthermore, such sophisticated robots use high-end AI models to understand the location and varying requirements unambiguously before making insight-driven decisions in time. All kinds of error-prone activities can be surmounted through the leverage of breakthrough robots. The much-needed perfection and precision get accomplished by intelligent robots. The productivity increases, the much-discussed concentration lapse gets eliminated, etc. In short, accomplishing more with less resources is the mantra getting replicated in the evergreen oil and gas industry.
- **Improved operational efficiency**: With the mesmerizing improvisations in the AI and edge computing spaces, oil rigs, robots deployed in oil fields, and other critical assets such as drilling machineries are going through a tectonic shift. By incorporating AI and local computing capabilities, machines are ingrained and inspired with sagacity. They instantaneously find and interact with one another purposefully to increase the operational efficiency. Machines learn fast and do better in their subsequent tasks. Robotics-infused rigs are smart and swift at spotting any inconsistency and potential shortcomings. Thereby, rigs can contemplate and complete course corrections quickly if necessary. Thus, robot-assisted oil rigs guarantee greater operational efficiency.
- **Deeper inspection**: We know that there are different robots to inspect different things in terrestrial and aquatic environments. Robots are enabled to be connected even when they are inside the water. Humans would find it too difficult to go the areas, where these robots comfortably function. Thus, with robots, inspection can be done anywhere anytime with all the precision and without harming the marine ecosystem.
- **Improved data generation**: With more assets and robots being involved and invoked for oil and gas exploration, production, transmission, refining, and retailing, there is a massive amount of multi-structured data getting generated.

With Big Data in custody, there is a bright and big scope for data-driven insights. With insights at hand, all kinds of participants are bound to get the much needed sophistications. Thus, we can safely expect there will be multiple intelligent artifacts getting stuffed in oil plants, pipelines, and refineries. All kinds of bottlenecks and barriers get resolved easily and elegantly with such knowledge-filled robots and equipment.

8.3.1 Upstream Oil and Gas

This primarily represents the exploration and production phase. Here, the oil companies earnestly search for identifying oil and gas fields. Then, they start the drilling activity to extract the oil or gas. Oil companies employ ROVs that are stuffed with a high-resolution video camera, lights, and artificial arms. The empowered ROVs can measure the alignment and penetration of the suction anchor of the drillship and then feed what is captured into the control system for arriving at accurate results. Precisely speaking, by leveraging advanced robots, there is a sharp reduction in human involvement. Industrial robots are good at accurately figuring out the areas where they can find oil resources.

8.3.2 Midstream Oil and Gas

The midstream component deals with the transporting, through pipelines and storing. Due to the repeated occurrences of oil leakage and pilferage on the way to refinery, there is an increased usage of path-breaking robots lately. With the steady deployment of purpose-specific and agnostic robots, the midstream inspection and maintenance are being tackled in an efficient fashion. Robots are being stuffed with additional capabilities to expand their scope and contributions in the days to come. Aging yet mission-critical infrastructures are being minutely monitored by robots for the sake of their continuous functioning.

8.3.3 Downstream Oil and Gas

This involves refining and distribution of crude oil and natural gas, resulting in several by-products. It is still difficult for humans to work in offshore oil refineries. Herein, robots come handy in expertly handling materials and in automated refueling. Robots remove errors and help optimize complicated processes. Thus, with lean and light processes, things become smooth. Human-caused errors are being eliminated through the leverage of insight-driven industrial robots. Robots are not wearisome and hence are expected to be precise in their assignments and engagements. Autonomous robots will be pervasively used to monitor, measure, and manage various facets and aspects of the oil and industry sector.

Due to the widely circulated society and sustainability goals, the oil industry must traverse through a significant transformation to be right and relevant to its constituents, clients, and consumers.

With the growing power of digital technologies, the oil and gas (O&G) sector is methodically prepared and poised for real transformation. The prime focus areas include a deeper and decisive understanding of reservoir resource and production potential, on how to boost operational efficiencies, and improve health and safety factors. As the industry is undoubtedly data-intensive, digitalization technologies (data analytics through traditional analytical platforms and the recent AI algorithms), there is a bright future for the oil industry. That is, all that are required to explore and establish oil fields, pipelines, plants, and refineries are digitally enabled and transformed for the betterment of the society.

Robotic technologies are in the fast track mode. With the emergence of evolutionary and revolutionary technologies, sophisticated robots are being built and released for activating and accomplishing ground-breaking activities. Robots are fabulous and famous in performing and providing a variety of automation, acceleration, and augmentation capabilities. Robots are AI-enabled, resilient, and versatile. Robots relate to the nearby and remote cloud-based business applications and IT services. 5G connectivity is available for enabling device-to-device (D2D) integration and device-to-cloud (D2C) integration. The much-touted edge computing is booming with proximate data processing. The newly conceptualized edge AI is being concretized with the arrival of AI-centric processing units, lightweight AI frameworks, and models. Thus, cognitive, and competent robots are being visualized and realized toward business and people empowerment.

The advancements in the robotics space bring in the potential to transform the oil and gas industry. Industry 4.0 and 5.0 applications mandate for the seamless and spontaneous convergence of the information and operational technologies. As indicated elsewhere, all kinds of industrial assets and artifacts are being synched with cloud-based applications and datastores. The emerging field of metaverse incorporating blockchain; decentralized applications; smart contracts; and augmented, virtual, and mixed realities (AR/VR/MR) technologies and tools is being seen as a positive indication for the oil and gas industry to survive and thrive.

8.4 Real-Life Examples of Robotic Solutions in the Oil and Gas Industry

Well-established oil companies across the globe are showing a keenness in beneficially leveraging the specific and praiseworthy powers of robots toward the much-trumpeted efficiency and elegance. Robotic solutions gain the wherewithal for disrupting the oil and gas industry, especially offshore operations, with a myriad of subsea and deep-sea applications. Toward this, modern-day robots are being stuffed with maintenance tools, multifaceted sensors, 360° cameras, etc. Further on, they are provided with on-robot data processing, heavy lifting, and edge AI capabilities to cope up with emerging needs. Specifically talking, underwater robots need such competencies to reduce labor. Machine power is bound to increase with the number of distinct advancements in the machine vision and intelligence spaces. In this section, we are to discuss a few popular robotic solutions.

For example, Saudi Aramco (https://www.aramco.com/en/creating-value/ technology-development/in-house-developed-technologies) is using many in-house developed technologies to gain efficient and safer inspection capabilities. In addition, their technologies guarantee appropriate response for any emergency. They do security-enablement, environmental and structural monitoring, aerial mapping, etc. Especially, robots and drones play a vital role in shaping up the oil and gas industry. Fire detection and fighting in time ensure enhanced safety. There is no discontinuity in providing assigned services. Robots are specially focused on three areas: air, ground, and subsea.

- **Shallow water inspection and monitoring robot**: This robot does pipeline inspection in an accelerated, safe, and efficient manner. This robot can be remotely operated to perform underwater inspection tasks in shallow water environments. This robot can be deployed from shore or by using a low-cost inflatable vessel. Thereby, the inspection time and the cost are reduced. The inspection can be done fast.
- **Saudi Aramco inspection robot**: This is a magnetic robotic crawler capable of doing visual inspection, ultrasonic thickness gauging, and gas sensing on hard-to-crack steel surfaces. This robot remarkably reduces potential hazards, inspection times, and costs. If we follow the traditional method for the inspection activity, then the cost would have been on the high side.
- **Robotic dust mitigation**: Dust accumulation on solar panels can have a detrimental impact on the output of solar arrays. However, manually cleaning solar panels consumes a lot of time and effort. To surmount this persisting problem, this robot cleans the panels in an efficient fashion.

Air robots: These hover and maneuver in air using propelling techniques such as wings or spinning propellers. There are mini-air robots for short-range applications such as visual inspection, surveying and mapping, and construction monitoring and management. Similarly, there are air robots for long-range operations such as pipelines and offshore operations for surveys, emergency response, and marine oil-spill monitoring. This robot is stuffed with special sensors to detect and measure methane emission.

Ground robots: These are typically mobile or fixed and operate primarily in a manual manner. There are ground robots working semi-autonomously to support field operations, which include maintenance and visual and thermal inspection. Further on, they read gauges, sniff and understand gas leaks, and hear and understand surrounding and mechanical vibrations.

Saudi Aramco has deployed an explosion-proof multipurpose firefighting robot, which is equipped with multiple visual and thermal cameras, heat tracking sensors, gas detectors, and 12 bar water connections to support safe firefighting operations. This robot has the innate capability to send live-video streaming to the command-and-control center.

Saudi Aramco is also building ground and legged robots to facilitate remote and unmanned field operations. This robot is empowered through top-mounted robotic arms with multiple axis freedom. This can control valves and local panels.

Deep-sea robots: The Earth's oil and gas resources primarily lie beneath its oceans. Therefore, the deep-water wells and pipeline systems rely on unmanned underwater vehicles (robots) to perform inspections, installations, monitoring, and maintenance. The challenge here is these robots ought to be transported to the offshore site using a staffed vessel topside. Deep-sea robots have an underwater docking station and are controlled remotely from the rig. These elegantly move through the water and get into areas, which are typically inaccessible to humans. With a connectivity module attached, these robots can be controlled from anywhere. The number of accidents has decreased sharply with the smart leverage of new-generation robots.

Underwater robots spend most of their time underwater to perform repair and maintenance tasks of machineries. This kind of robot has an underwater docking station and is controlled from the rig, which is above the water. Some underwater robots are enabled with an Ethernet connectivity. These can be controlled from anywhere through a buoy on the ocean's surface. These continuously perform inspections by reducing the damage for the fragile environment (in an environment-friendly manner).

By methodically using robots to do frequent inspections, the number of accidents can be reduced to a bare minimum. Humans need not get into those dangerous spots to do inspections; thereby, any human loss can be avoided completely. Thus, robots deployed on the ground and underwater do several vital tasks in an accelerated and automated way. As more powerful robots are invented and involved, industry experts and engineers can focus more on digital innovations, disruptions, and transformations. It is proven beyond any bit of doubt that robots significantly boost productivity, solidly improve cost-efficiency, and intrinsically guarantee safety.

Spot from Boston dynamics (https://www.bostondynamics.com/products/spot): This robot has been deployed once in the Norwegian Sea to collect sensory readings, scans, and images and seamlessly share them to the user's network. It was also deployed in the deep-water Gulf of Mexico to scan for methane. This robot could identify potential leaks in deep water and allow the team to read gauges.

Robots decreases downtime. Downtime of an oil rig can cause irreparable loss. Thus, by deploying robots in place of human workers, drilling activities and the functions of oil rigs can go ahead without any slowdown and breakdown.

Taurob inspector robot (https://taurob.com/taurob-inspector/) gathers data with unprecedented versatility and objectivity. It takes ultra-high-definition (UHD) photos and videos, records sounds, detects gas leaks through thermography, and oversees its environment with a 3D LIDAR scanner. The Taurob Tracker is a good example of the specialized robots for tracking oil and gas facilities. This is designed to perform tasks that include the reading of dials, valve positions, and level gauges; monitoring concentrations of harmful gases; and moving up and down stairs.

ANYmal (https://www.anybotics.com/): This robot can inspect offshore sites and is equipped with visual and thermal cameras, microphones, and gas detection sensors. This generates a 3D map of its surroundings to carry out inspections and operations in an efficient manner.

Eelume (https://eelume.com/): This is designed to "live" permanently underwater and perform subsea inspection, repair, and maintenance tasks.

Oceaneering's liberty E-ROV (https://www.oceaneering.com/rov-services/next-generation-subsea-vehicles/liberty-e-rov/) is a self-contained and battery-powered vehicle that allows you to perform routine tasks with fewer deployments. It can carry out inspection, maintenance, and repair (IMR); commissioning; and underwater intervention activities.

Reach robotics (https://reachrobotics.com/products/subsea-light/): The Reach Light is a tough and reliable underwater light, designed for ROV manufacturers and operators who are tired of leaking lights. The distinction of this robotic solution is its configurable control, temperature monitoring, and high luminosity. Everything is accommodated in a tiny package. This solution can be an entire ROV, or it can be added to give additional spotlight illumination.

uWare robotics (https://uware.io/) is a start-up company specializing in data-driven engineering solutions for coastal ecosystems. They use their own mobile autonomous underwater vehicle (AUV) and modular sensor systems for data gathering and provide a software platform for the processing and analysis of that data.

Rigarm (https://www.rigarm.com/robotics) is a drilling systems automation (DSA) robot consisting of drill floor robots, robotic roughnecks, multi-size elevators, and robotic pipe handlers to improve the performance of the drilling operations in an efficient manner. This robot connects drill pipes in challenging places like oil-bearing rock and the ocean. These make life easier and safer for people working on the rig. Rigarm offers a patented autonomous rig as a part of a DSA solution. The autonomous rig comprises smart robotic pipe handlers, which work together to transport pipe stands. Other prominent solutions include downhole sensor systems, vision-enabled autonomous robots, LIDAR, and radar.

Aquanaut (https://nauticusrobotics.com/aquanaut/) is an all-electric subsea robot that transforms from a long-range AUV to an untethered ROV with two robust arms. In the AUV mode, it can cover over 50 nautical miles in one mission, thanks to an onboard lithium-ion battery and thrusters for propulsion. Along its journey, Aquanaut's onboard high-precision geophysical instruments allow it to survey the seabed and collect data.

Once it reaches its destination, then it gets transformed to play the role of a ROV. The top half of the hull raises and the head swivels into place to expose stereo cameras and powerful 3D sensors. It can do important manipulation tasks such as waterjet cleaning, inspection of cathodic protection, and flooded member detection (FMD). This robot can assess the condition of oil and gas assets on the seafloor.

Communication is facilitated through an onboard acoustic modem. A small, unmanned surface vessel relays signals between the robot and communication satellites. This allows the robot to be controlled from anywhere.

Aerial robotics company Apellix (https://www.apellix.com/): The Apellix power wash drone is a safer and faster way to clean elevated structures. It can rapidly clean water towers, storage tanks, etc. This is for non-destructive testing (NDT), which is an inspection and analysis technique to evaluate the properties of pipelines without destroying the serviceability of the original system. Testing includes ultrasonic testing (UT) and dry film thickness (DFT) measurements.

For ultrasonic testing, the drone uses a probe to send an ultrasonic pulse through the wall of a steel structure to measure its thickness. These readings gauge corrosive wear and remaining service life of the asset. The DFT testing measures the thickness of coatings on ferrous and non-ferrous metals.

Guardian S (https://www.sarcos.com/products/guardian-s/) is a remote visual inspection and surveillance robot reliably capable of traversing challenging terrains and facilitate two-way, real-time video, voice, and data communication, all from a safe distance. This snake-like ROV is designed to inspect areas where humans cannot or should not go.

This robot is equipped with 360°-view cameras, an LED light package, two-way audio, and an array of other sensors including GPS. This is used for pre-commissioning inspections of oil and gas assets to efficiently and easily look for everything from debris to structural integrity issues. By incorporating other powerful sensors, it is possible to accomplish more like sensing gas, radiation detection, etc.

The platform Oseberg H (https://www.offshore-technology.com/analysis/ inside-the-first-fully-automated-offshore-platform/) is the fully automated oil and gas platform, which is entirely unmanned and requiring only one or two maintenance visits a year. It is a remotely operated platform. Digitization and digitalization technologies and tools are accommodated in this platform to be future-ready.

Inspection robots are indispensable for the oil and gas industry. Robots do a variety of tasks without any fatigue. The CIRRIS XI and CIRRIS XR robots by ULC Robotics (https://ulctechnologies.com/technologies/cirris-xi-inspection-robot/) are two fine examples of inspection robots for gas pipelines. These can move throughout a gas pipelining system and extensively use measurement sensors to find wall thickness and stress. These robots radically reduce the amount of excavation required to inspect a pipeline system. The data collected by them can be subjected to deeper investigations to ensure the structural integrity of the pipelines. This empowerment alleviates any hidden risks.

Further on, with the introduction of "Sludge cleaning robots," the dangerous tank cleaning is fully automated. These specialized and miniaturized robots are capable of handling power tools while navigating the small entrances of storage tanks. Gas distribution utilities face the excessive costs and disruption associated with replacing the pipe as well as the challenges of maintaining these assets. CISBOT enables gas networks to safely extend the life of these assets with no disruption of gas service to customers and reduced impact to the public and the environment (https://ulctechnologies.com/services/cisbot-robotic-cast-iron-joint-sealing/).

Submersible robots are growing in sophistication. In addition to the Eelume, Equinor employs its empowered remotely operated vehicle (E-ROV) to inspect its underwater infrastructure. This unit is battery-powered and operated using an Ethernet connection. This is well-supported through a dedicated subsea network. There are a few charging stations; thereby, the need for a mother ship is eliminated. These are known as autonomous underwater vehicles (AUVs). The Aquanaut (https://nauticusrobotics.com/) AUV can cover up to 200 km in a single mission and designed to transform into an ROV if the situation demands it.

Robots in services, security, and surveillance: Robotic solutions are becoming ubiquitous these days in many industries. For example, remote and precision surgeries are being accomplished through breakthrough robots. Retail stores, warehouses, national borders, and other sensitive environments are seeing a lot of pioneering robots cognitively catering to varying demands. Industrial robots are becoming popular for their productivity, power, precision, and perfection.

Collaborative robots (alternatively referred to as cobots) assist human beings in their daily assignments and engagements. Smart sensors are being attached with cobots to automate the data collection process. All the collected data get cleansed and crunched in time to squeeze out venerable and viable insights. Operational and log analytics, anomaly detection, correlation analytics, root cause analysis, and predictive and performance analytics are being done on data gathered to derive actionable intelligence in time.

Robots are used for assembling products, handling dangerous materials, spray-painting, cutting, polishing, etc. Product inspection is another prominent contribution of robots. Robots are also utilized for cleaning sewers, detecting bombs, etc. We nowadays have robotic lawn mowers. Robotic solutions are being explored and experimented across several verticals. Robots incredibly abound in retail stores for delighting shoppers, immensely assist in agriculture production, and aid in handling and transporting logistics. As per news reports, humanoid robots are being experimented in fast food restaurants in advanced countries. Thus, the scope for robotic solutions brightens as robots are becoming intelligent in their actions and reactions.

Microbots are miniaturized robots gaining surging popularity as emergency responders. Because of their shrunken size, they can penetrate places where humans or larger robots could not reach and work. These micro robots play a vital role while exploring small environments. Microbots can work in areas, which are being categorized as dangerous for humans. Where larger robots do find it difficult, miniaturized robots do well. Robots are being powered with alternate energy sources like solar, wind, and wave energy to work in locations, wherein the microgrid is not made available. Inspired by living creatures, modern robots are being designed and leveraged for a dazzling array of real-world and real-time applications.

In mission-critical environments, autonomous robots are being deployed to collect and react to planned and unplanned incidences in those environments. These robots carry multiple visual and thermal cameras, gas sensors, two-way communication systems, etc. to be vigilant and venerable in their assignments.

8.5 The Advantages of Robotic Solutions

We discussed many categories of robots for cognitively tackling different tasks associated with oil and gas exploration, drilling, production, storage, pipelining, refining, and retailing. With the diminished human resources, robotics usage has picked up sharply these days. A plethora of digitally transformed robots have arrived on the scene to automate and accelerate several specific activities in the oil and gas industry. With the mesmerizing advancements in the digital space, the world is experiencing hugely trendy robots across industry verticals.

The solidity of artificial intelligence (AI) for deeper data analytics and 5G connectivity for deeper penetration, the continuous explosion of networked embedded systems, the IT metamorphosis through the cloud paradigm, the arrival of edge AI, etc. have profusely contributed for the upliftment of the oil and gas industry, which gradually and glowingly utilizes state-of-the-art robots to automate repetitive, risky, yet rewarding tasks. Robots eliminate perpetual inefficiencies and reduce operational costs. Robots help explore fresh avenues to reap additional revenues. With trendsetting robots being deployed in large numbers, the return on investment (RoI) increases, whereas the total cost of ownership (TCO) for the oil and gas sector reduces remarkably.

The distinct automation being brought in through breakthrough robots is really understood and applauded across. Besides guaranteeing operational efficiency toward enhanced performance and productivity, robotic solutions eliminate the need for deploying professionals and skilled workers in remote and rough environments. That is, time-consuming and tedious tasks are being taken over by humanoid robots. Aerial drones and unmanned underwater vehicles can easily explore locations, which are quite unreachable for humans. Robots can handle life-threatening jobs such as underwater welding and repetitive drilling tasks. Thus, technologically sound robots fitted with versatile cameras and handlers are being viewed as the bright future for the oil and gas domain. Robotic technology brings the following unique advantages.

Robots eliminate safety hazards: Oil and gas companies produce toxic substances in hazardous locations such as subsea wellheads and undersea pipelines. Thus, there is a lingering security risk. As articulated and accentuated before, remotely controllable robots can accomplish drilling tasks and eliminate the risks of humans doing these life-threatening tasks. For example, robots can inspect offshore sites using sensors and create 3D maps of the environment so that humans can safely navigate these environments.

Such types of remotely operated robots can move around on four legs and use microphones and visual and thermal cameras to obtain the environment information with all precision and perfection. They also use gas detectors and thermal cameras to accurately assess the viability of such locations. Robots eliminate the risks associated with humans inspecting the sites. Robots minimize deaths, injuries, and roughnecking for humans. Material losses can also be curtailed. Such roughnecking robots can install drill pipes in tough environments including oil-bearing rocks and even in the ocean.

Robotic technology has gained the strength and the sagacity to automate entire oil rigs. There is no physical intervention in operating these rigs. There are specific robots that can carry out maintenance tasks and assess repairs in underwater. Such robots are empowered to have docking stations. These can be controlled remotely from the oil rig. They are designed and developed to have snake-like bodies to move through the water and to work in areas, which are impossible for humans to reach.

Some of these robots are embedded with an Ethernet module and a battery system to be connected and controlled remotely. Therefore, robots are self-contained to be autonomous in their activities. Robotic solutions can do monitoring and inspection without damaging their operating environments. Robots work with small carbon footprint. Thus, advanced robots are seen as a boon for the oil and gas industry.

Robots can be the "first responders" to safety incident: Robots are critical in explosive atmospheric conditions. It is well known that any large oil field is stuffed and surrounded by a multitude of heavy drilling equipment. Further on, there would be many pipes for rapidly siphoning hydrocarbons back to the coast. All these clearly indicate that there are bigger and brighter chances for a plenty of unpleasant happenings for field workers. Oil companies consciously use powerful robots to minimize risks for engineers and workers. By deploying robots to do these things, the safety of professionals is guaranteed. Cleaning robots are tasked with cleaning exchangers on platforms.

Robots reduce costs: Robots can be securely and swiftly deployed in rough and remote locations, which are too risky and sometimes catastrophic for humans. Crawlers can investigate deep within oil and gas pipelines. Underwater robots can capture and send images about the pipe conditions. Robots are good in inspection of assets and activities. Constant inspection is to ensure the correct functioning of machineries and platforms. In short, robots can work in places wherein people find it difficult to navigate.

There are robots inspecting offshore sites with a variety of sensors to create 3D maps of the surroundings. These robots use four legs to roam around and use their cameras and microphones to see and hear. They use thermal cameras and gas detectors to check whether the explored sites under exploration are of some use or not.

As robots automate several aspects and activities, the operational costs are bound to decrease sharply. With digitally transformed robots adaptively operating in oil fields, setting up and sustaining additional safety facilities and features are not needed. The safety costs, therefore, decrease. By removing humans from these specific environments, high-end robotic solutions eventually save a lot of capital and operational costs. While humans are error-prone, robots are stringent and straightforward in their assigned tasks. Robots work continuously without any breaks and barriers.

Robots increase productivity: As we know, technologically advanced robots perform far better than humans. Robots are precise and perfect in carrying out inspections of assets and their environments. Human resources with appropriate education, expertise, and experience can operate robots from a safe distance.

Robots execute things clinically. Through the predictive maintenance capability, robots can be continuously watched for their weariness and any flaws or failures. Thus, without any slowdown and breakdown, robots contribute well for the cause.

With a series of praiseworthy advancements in the AI space, any forthcoming robotic solutions will be deeply driven by competent cognitive technologies and tools. Robots will be adaptive and accommodative in their service deliveries. Their survivability and sustainability powers during challenging situations are something to boast about.

The renowned automation capability being delivered by innovative robotic solutions comes handy for energy companies in enhancing efficiency. Robots keep production continuity by insightfully scheduling repairs, pre-emptively detecting leaks, constantly inspecting machinery and infrastructure, appropriately managing pipelines, and carrying out repairs. In short, without an iota of doubt, robots enhance insight-driven industrial productivity.

Robots emit out real-time data: As indicated above, the oil and gas (O&G) industry leverages multi-faceted robots stuffed with visual and thermal cameras, lasers, and sensors to see and capture visual and operational data from assets and places, wherein humans find it difficult to enter. Robots, then, share the captured data with the server machines in the control centers in real-time. We have a variety of streaming analytics platforms to make sense out of the data in time. Further on, the captured data can be utilized by domain and technical experts to create 3D maps of pipelines and other mission-critical assets. Such knowledge empowers operators to proactively pinpoint any sickening problem emerging and to nip any deviation and deficiency in the budding stage itself with all the clarity and alacrity. The much-coveted AI technology contributes immensely here.

In summary, robotics technology is trend-setting and beneficial for several business domains. Robots are becoming miniaturized, feature-rich, and cognitive. Robots become sophisticated with the direct involvement of AI-enabled digital twin technology. The high bandwidth and exceptionally reliable connectivity solutions do bring in a lot of improvisations to robots. With digitally transformed robots, several kinds of physical activities are being accomplished in a reliable manner. With the growing presence of software-defined cloud environments across the world, robotics capabilities are being provided as a service. Thus, the new paradigm of robotics as a service (RaaS) is emerging and evolving fast. With the field of edge AI picking up fast, robots can capture, process, analyze, and decide what to do next in real-time. On-robot data processing is being facilitated to take timely decisions to plunge into quick actions with speed and sagacity.

8.6 The Dawn of the Internet of Robotic Things

We are well-aware that the power of the Internet of Things (IoT) brings in the convenience, comfort, choice, and care for us in our everyday life. The IoT paradigm has resulted in a dazzling array of multifaceted connected devices, services, and things. In other words, the IoT phenomenon has led to the proliferation of networked embedded systems. With our everyday equipment, instruments, machineries, wares, utensils, consumer electronics, handhelds, wearables, and implantable getting digitized and linked up with one another in the vicinity and with the fast-growing internet, the long-standing goal of producing and deploying self-, surrounding- and situation-aware systems is to understand something clearly at last sooner than later. The IoT paradigm is penetrating every worthwhile domain. We are already comfortable with the Industrial Internet of Things (IIoT), the Internet of Medical Things (IoMT), etc. Now with the flourishing of connected robots, the new field of the Internet of Robotic Things (IoRT) is emerging and evolving fast. With the surge in the number of software-defined cloud centers (SDCCs), the aspect of providing products and solutions as cloud-based services is gaining momentum. That is, the concept of everything as a service (XaaS) has become the new normal.

As reported earlier, robots are becoming digitized, connected, and leveraged for a variety of automated services. Especially for the oil and gas industry, the role and responsibility of robots are getting greater. With the faster maturity and stability of digital technologies, the scope, speed, sagacity, and strength of robots are becoming better. Therefore, the distinct contributions of modern robots are awesome. Thus, the combination of connected robots and scores of fabulous and futuristic robotic services through software-defined clouds is all set to instigate the inspired minds to visualize and realize next-generation services for the oil and gas domain.

The solidity of IoRT is to generate a massive amount of multi-structured data. With the smart application of digital technologies, all kinds of robotic data get captured and subjected to a variety of deeper and decisive investigations to extract actionable insights in time. Such knowledge discovery and dissemination make robots intelligent in their operations, offerings, and outputs. In a nutshell, the overwhelming idea of data-driven insights and insight-driven decisions and deeds is fast happening in the robotic space too. The new phenomenon of the Internet of Robotic Things is being seen as a game-changing facet and factor for the oil and gas sector.

As articulated above, there has been a multitude of noteworthy advancements in the fields of robotics and the Internet of Things (IoT) in the past few years. Both are complementary to guarantee heightened efficiency, cost reduction, and deep automation. We have written extensively about the breakthrough implications of the notion of the IoT paradigm in this book.

Precisely speaking, everything is becoming smart, every device is smarter, and every human being is all set to be the smartest with the artistic leverage of digitization and digitalization technologies.

Connected entities join in the mainstream computing to deliver hitherto unknown and unheard capabilities. The IoT concept expands the influence of the internet beyond compute servers, desktops, laptops, and smartphones to all types of physical, mechanical, electrical, and electronics systems. With abundant manufacturing of highly miniaturized processors, communication, and sensing modules, ordinary articles become digital entities. Thus, all mission-critical environments are being stuffed with a lot of digital artifacts. Experts proclaim that our everyday environments such as homes and offices will be digitally transformed to provide not only information and commercial services but also to accomplish context-aware physical services. Devices talk to one another to share their data and capabilities. Devices also interact with cloud-based software applications, services, and datastores directly or indirectly. There are human–machine interfaces (HMIs) for enabling humans to interact with machines and devices in a natural way. With the emergence of lightweight Kubernetes platforms, forming device clusters and clouds is being streamlined and speeded up. Devices forming clusters are to perform specific, simple, and dynamic tasks in real time. Such a real-time computation comes handy in envisaging real-time services and applications, which are being pronounced as the most essential competency for setting up and sustaining real-time intelligent enterprises.

Robotics data get accumulated in cloud databases or data warehouses or data lake houses to be analyzed by cloud-based analytics platforms to extract actionable insights, which can be fed back into ground-level robots to exhibit adaptive behavior. Nowadays, AI-based data analytics platforms are being made available in cloud environments. Thus, personalized, predictive, and prescriptive insights are being derived out of data through the AI platforms. Thus, the target of ambient intelligence (AmI) is being fulfilled through the evolutionary and revolutionary technologies and tools.

This new domain of the Internet of Robotic Things (IoRT) will open fresh possibilities and opportunities. A variety of intelligent automation and orchestration can be achieved through the noteworthy accomplishments happening in the IoRT space. The distinctive goals of robotic vision and intelligence can be realized. Robots will be intrinsically empowered to be cognitive and competitive in their processes. This paradigm of IoRT will facilitate the realization of autonomous robots. Remote monitoring, measurement, management, and maintenance of robotic solutions can become the new normal. Robots will acquire the vision and intelligence competencies so that a plethora of sophistications can be availed by oil companies. Robots will form local, transient, and purpose-specific clusters to tackle complicated tasks quickly and locally. Through the respective digital twins, the operational status of each robot can be obtained and understood.

The architecture of the Internet of Robotic Things (IoRT) can be divided into five layers.

8.6.1 The Hardware Layer

As enunciated above, latest robots are embedded with several action-packed modules. Highly miniaturized yet mesmerizing sensors and actuators are being attached in robots to capture different data and to implement appropriate actions in time. Cameras are being embedded in robots to acquire vision and perception capabilities. Through the much-maligned edge AI feature, real-time knowledge discovery, decision-making, and action are being facilitated. Besides, robotic solutions being integrated with cloud-based software packages and libraries, additional capabilities are being realized through modern robots. Thus, the internal attachment and the external integration empower modern-day robots to be innately innovative, disruptive, and transformative.

8.6.2 The Network Layer

Now with the installation of 5G communication infrastructures across the world, the realization of connected robots is happening fast. There are both short- and long-range communication protocols and technologies to empower robots working at ground and underground places appropriately. The Bluetooth low energy (BLE) is one among the growing array of communication protocols.

8.6.3 The Internet Layer

As we all know, the Internet is the world's largest communication infrastructure. The Internet is affordable, open, public, and flexible. HTTP is the prominent Internet protocol. There are other related protocols such as transmission control protocol (TCP) and message queuing telemetry transport (MQTT), etc.

8.6.4 Infrastructure Layer

Now clouds represent highly organized and optimized IT environments for hosting, delivering, and managing business workloads and IT services. Cloud centers are being formed through the consolidation of compute, storage, security, and network solutions in a centralized location. Nowadays, different and distributed cloud environments are being linked up toward federated clouds. A troublesome, time-consuming, and tough cloud IT operations are being automated through a host of pioneering software solutions. IT products and tools are plentiful for automating and orchestrating IT management activities. Also, cloud resources are being shared through the concept of compartmentalization (virtualization and containerization). The recent buzzword is serverless computing, which is deepening the automation characteristics of cloud computing. Integrated data analytics platforms, datastores, digital twins, knowledge visualization dashboards, platforms for AI model engineering, evaluation, optimization, and deployment, etc. are being installed and offered as services through cloud infrastructures. Thus, for the brewing idea of IoRT, the cloud concept is being viewed as the most competent IT infrastructure. Robot-enablement and management platforms are being modernized and migrated to cloud environments (public, private, hybrid, and edge).

8.6.5 The Application Layer

Pioneering robotic applications and services are being developed and run on cloud-based IoRT-centric platforms in order to epically and expertly manage various contributions of robotic solutions. Robot management and security platforms are being deployed in cloud environments.

The emerging field of IoRT is being seen as a change in thinking for the robotics field. With the IoRT technologies growing rapidly, there will be state-of-the-art robots to be built and used for fulfilling the different needs of the oil and gas industry. The cloud-enabled robots will be a game changer. With the combination of cloud AI and edge AI techniques, forthcoming robots will learn the changing situation and demands quickly and accordingly accomplish the assignments with all the clarity and alacrity. Robots in the neighborhood connect, collaborate, and correlate insightfully to accomplish bigger and better tasks with all the precision and perfection. AI will be the key contributing technology for futuristic robots to be right and relevant to their owners and consumers.

8.7 Conclusion

Robots are now extensively used in multiple industries to improve working conditions for field workers while increasing productivity. As discussed in this chapter, the oil and gas sector (O&G) heavily uses state-of-the-art robots. The industry demands high-end robots and on the other hand, with the emergence of pioneering robots, many mundane activities are being delegated to robots to achieve more with less. Robots guarantee the much-needed perfection and precision. Robotics open fresh possibilities and opportunities for companies. Robots are penetrative, pervasive, and persuasive. Robots advance automation, remove inefficiencies, enhance safety for workers, and cut costs. Robots are deployed and operated in places, wherein humans face danger. Robots can be used in land parcels and as underwater vehicles. They are also used for drilling and underwater welding. The prevailing trend for skilled people is to remotely control robots rather than risking their lives in extreme conditions.

With several distinct advancements in the field of artificial intelligence (AI), the next-generation robotic solutions are going to be trendsetting. A host of service-oriented, event-driven, knowledge-filled, cloud-native, process-aware, business-critical, context-sensitive, people-centric, and environment-friendly robots will be produced and deployed in rough and tough oil field environments to ensure the highest safety for people and properties. Robot management platform solutions will emerge and automate several robot lifecycle management aspects.

Bibliography

1 Robots will be the oil and gas industry's growth engine, says GlobalData. https://jpt.spe.org/robots-will-be-the-oil-and-gas-industrys-growth-engine-says-globaldata.

2 The essential guide: how to transform oil & gas industry with robotics. https://www.birlasoft.com/articles/oil-and-gas-industry-robotics.

3 Bradstock, F.The growing importance of robots in the oil and gas industry https://oilprice.com/Energy/Energy-General/The-Growing-Importance-Of-Robots-In-The-Oil-And-Gas-Industry.html (accessed 11 May 2023).

4 Deep dive: robotics in oil & gas, improve safety and productivity. https://www.automate.org/industry-insights/deep-dive-robotics-in-oil-and-gas-improve-safety-and-productivity.

5 Use of robotics in the oil and gas industry. https://copas.org/use-of-robotics-in-the-oil-and-gas-industry/

9

AI-Empowered Drones for Versatile Oil and Gas Use Cases

9.1 Introduction

With the greater understanding, the usage of drones is increasing steadily. Newer capabilities are being incorporated into modern drones to visualize and realize advanced applications. Especially AI-powered drones are gaining the attention of many these days. Drones are deployed in disaster-stricken environments, border areas, and war zones. Drones are becoming more reliable in their service deliveries. They can continuously work for longer hours and cover more areas quickly. With the widespread availability of 5G communication networks, resource-intensive drones are being stuffed with on-drone data processing capability. Thus, gaining real-time intelligence through localized data processing is seen as a change in thinking for drones to do improved things for industry verticals.

Drones are being categorized as a type of robots. Drones can fly over a place to capture the operational status of various assets and people in that place. The application domains for powerful drones are constantly expanding. Besides delivering parcels and products for e-commerce service providers, security agencies are exploring the possibility of leveraging drones for law enforcement. The vit search and rescue (SAR) is another popular use case of drones in the case of any disaster. Farming is another niche domain for drones.

Drones and robots will reduce the operation cost of assets associated with oil and gas exploration and production, transmission, refining, and retailing. Efforts and time needed to complete most of the activities will also reduce. Repetitive tasks are being elegantly completed by intelligent robots and

The Power of Artificial Intelligence for the Next-Generation Oil and Gas Industry: Envisaging AI-Inspired Intelligent Energy Systems and Environments, First Edition. Pethuru Raj Chelliah, Venkatraman Jayasankar, Mats Agerstam, B. Sundaravadivazhagan, and Robin Cyriac.
© 2024 The Institute of Electrical and Electronics Engineers, Inc.
Published 2024 by John Wiley & Sons, Inc.

drones. The future drones will be smaller in size, smarter, and faster in processing the data produced by the assets. The knowledge discovered gets disseminated to appropriate assets and drones to exhibit and expect an adaptive behavior in real time. With the surging popularity of AI-based data processing, oil and gas assets in association with robots and drones can be intelligent in their actions and reactions. Let us start with the three prime components in the oil and gas industry.

9.2 The Upstream Process

Experts say that this initial and important process primarily involves the identification of potential sites and plunging into drilling and exploration. Once wells are found, then crude oil and gas are methodically produced from those wells. Drones contribute to automating and accelerating this long, hazardous, manual, and arduous task. Drones aid in extracting and providing real-time data to support all the steps associated with this crucial phase.

Drones are being attached with thermal sensors and cameras that assist in pinpointing sites and identifying the best routes of taking and transferring appropriate infrastructure modules. Drones can continuously monitor the progress of any construction work by flying over them. If there is any need for any improvement to be done on the construction activity or if there is any deviation in the work being carried out compared to the originally accepted plan, then drones can clearly articulate that. Drones also assist in predicting any repair and maintenance requirement. Drones monitor oil rigs and other nearby assets carefully in accomplishing fault and failure detection.

Drones do collect a lot of operational, log, health condition, performance, and workload data (audio, image, and video) and transmit it to any competent cloud platform to subject it to a variety of deeper investigations. Thereby, all kinds of predictive and prescriptive insights can be obtained ahead of time to enable actuation systems and human experts to ponder and plunge into appropriate countermeasures with all the clarity and alacrity. Thus, the transition of data into actionable insights in time helps the companies increase the productivity and the brand value. Apart from that, drones check and survey oil wells, fields, and platforms to eliminate any kind of accidents. Thus, any loss of people and property gets subsided significantly.

On summarizing, this is an exploration to the production phase. The exploration phase mandates surveying of large areas to find the abundant presence of oil and gas. This is a time-consuming and tough assignment. These places are sometimes inaccessible and life-threatening for field workers. Drones extend a helping hand in surmounting the persisting and prickling issues.

9.3   The Midstream Process

In the previous phase, we discussed finding and getting natural resources to support the humankind. In the midstream phase, the collected crude oil and gas must be transported to refineries for refinement. The transportation happens via pipelines between oilwells and refineries. Huge tankers are also used for oil and gas transportation. The pipelines may exist over the ground or underwater. There are storage mechanisms to stock oil and gas. For such a process beset with numerous challenges and concerns, the arrival of high-end drones is being seen as a positive sign. Drones

- cut down the efforts, time, and cost needed to finish the activities in this midstream process and improve productivity.
- increase protection for professionals through automated inspection and supervision of pipelines and avoid downtime through early detection of corrosion and leakage.
- cover large distances in a short span of time, gather a lot of data in real time, and get it analyzed immediately to extract actionable insights.
- smoothen the transportation by pinpointing any roadblocks, vehicle breakdowns, weather condition, etc.

Drones also steadily contribute for the oil and gas sector by detecting and articulating any kind of pilferage and damage in refineries proactively. Any hazardous gas and chemical emission can be identified and addressed well before any untoward incident happens. In warehouses, drones also perform stock checks excellently by reading the RFID tags; thereby, the aspects of inventory and replenishment management are optimally done.

In summary, this is the second important phase. Pipelines are being laid down for hundreds and even thousands of kilometers to transmit precious oil and gas in a safe and secure manner. Even for remote and rough locations, pipeline infrastructure is being installed and managed remotely through competent technologies. These pipelines ought to be minutely monitored and maintained through automated software solutions. Any leakage or pilferage of these expensive resources must be proactively and pre-emptively pinpointed, and necessary counter-measures must be initiated and implemented through robotic and drone solutions. Any problem with these pipelines may degrade or even destroy the fragile environment, and any damage to the vegetation can be problematic for human living. Thus, leak detection and timely response are vital. Further on, pooling of liquid on the ground and dirt or debris blowing up from the ground can create a detrimental effect on pipelines. There can be other external sources of irritations for pipelines. Thus, this is a critical component for the oil and gas industry.

9.4 The Downstream Process

The collected crude oil and natural gas are meticulously refined to create multiple end-user products such as petrol, jet fuel, and gas. This process deals with taking the end-products to their retailers to be bought and used by consumers. The much-discussed delivery of petroleum products involves a large human force and is time-consuming. The transportation also must deal with several risks. Thereby, advanced drones are being leveraged to reduce risks and transportation time. The prominent and dominant contributions of modern drones in this phase are listed below for the benefit of our esteemed readers.

- Drones can move into humanly inaccessible spaces such as towers and tanks to inspect their inner walls. Drones can readily relay all the captured information to any centralized control servers to be processed to ponder about the best-in-class actions quickly. Thus, drones ensure real-time data capture, cleansing, and crunching to emit out venerable insights.
- Drones also extend their helping hand in expertly managing and maintaining required stocks so that the delivery process becomes timely and delightful.
- Drones are embedded with advanced thermal cameras, and such an empowerment enables drones to pre-emptively detect any kind of damages to pipelines and storage facilities. This keeps customers and consumers happy. Any wastage of precious fuels is nipped in the budding stage itself.

Refinery and petrochemical plants are hazardous. Workers and operators in these environments are liable for risks and sometimes catastrophic disasters. The downstream processes are equally complex, and this phase needs cognitive and pioneering digital technologies to guarantee the much-required transformation. In this chapter, we discuss how modern and technologically sound drones (alternatively termed as unmanned aerial vehicles (UAVs)) are to contribute magically for the automation and optimization of multiple activities spread across the three components.

9.5 Navigation Technologies for Drones

Drones for fulfilling the intended success ought to leverage highly accurate navigation systems for accurately and quickly surveying landscapes, monitoring search-and-rescue (SAR) operations, and building 3D maps. Drones can fly in both satellite and non-satellite modes. Drones, therefore, utilize the global navigation satellite system (GNSS) to track their position and navigate in self-driven mode. The GNSS utilizes a constellation of orbiting satellites, and each transmitting satellite transmits coded signals at fixed time intervals. This signal information is converted by the receiver into position, time, and velocity. This is used to calculate the correct position of the satellite and its distance.

On starting its arduous journey, a drone records the home coordinates once it receives enough signal from the GNSS. Some drones seek location through GPS satellites. In the beginning, drones operated independently. But, in the recent past, many drones have purposefully been synchronized to perform complicated tasks. Thus, drones need to communicate with one another to operate as a single unit to finish business processes. There are both wireless and wired communication technologies and topologies for networking drones and other communication-enablement systems. However, the communication mechanism to be selected for enabling drones to form an ad hoc, purpose-specific, and dynamic communication network solely depends on the underlying application. For an example, in outdoor communication, a simple line of sight and point-to-point (P2P) communication link between the flying drone and the device at the ground is sufficient for establishing a continuous signal transmission. Drone communication must maintain connections between drones and a ground station. In the case of surveillance, drones must communicate through satellite communication links. Satellite communication is primarily used for national and in-land security. But, for personal applications, the cellular communication is being recommended.

However, for indoor communication, Bluetooth and other P2P communication protocols are being preferred due to the need for meter-level accuracy. Thus, drone communication is a critical area of study and research. As we all know, drone communication contributes immensely for military and civilian applications. The communication infrastructure for drones must guarantee a high throughput, a large coverage area, and reliable communication. Drones can be used to stretch the coverage area by acting as a relay. With appropriate communication capacity and capability in place, drones are bound to shine and succeed in their missions.

9.5.1 5G Enables Seamless Communication for Drones

5G is being proclaimed as a game-changing phenomenon for autonomous drones. There is a possibility for the point-to-point communication to fail. That is, the much-worried signal fade at any time may occur when drones are flying. However, on 5G communication networks, drones experience the ultra-high reliability and the low-latency connectivity. That is, without any disconnect, drones can continuously receive and act on commands from the ground control system. Due to low latency, 5G helps drones in navigation scenarios, where drones fly in GPS-denied environments. With 5G, drones can work on environments, which are beyond the visual line of sight. In this use case, instead of the GPS, drones rely on visual inertial odometry (VIO) to navigate in areas, wherein the pilot's view is obscured.

With 5G communication, the drone's camera sends the pilot an accurate and aerial view of where the drone is flying.

Major players like e-commerce giants invest their treasures heavily in having a fleet of autonomous drones. Telecom players and business houses are setting up private 5G communication networks that are compatible and constructive to drones. These are to simplify and speed up the proliferation of drones across industry verticals. Drones are being equipped with 5G capability to achieve 3 Gbps upload and 7.5 Gbps download speeds. You will need a 5G modem compatible with your wireless carrier and then an appropriately configured SIM card to make the data connection.

With 5G network infrastructure and equipment in place, mobile operators are keenly strategizing and executing path-breaking and premium capabilities to users. With 5G-enabled drones getting deployed plentifully, there will be pioneering use cases. Network operators will gain additional material benefits with 5G drones.

9.5.2 5G Drones for the Oil and Gas (O&G) Industry

Oil and gas are the key fuels for most of the industries and nations across the world. Considering the declining fuel sources, the world is scouting for renewable energy sources to prop up the industry and economy. There are complicated processes involved in producing and distributing oil and gas. The processes involve state-of-the-art infrastructures and innovative technologies and tools. The capital and operational costs are high, whereas the safety and security of people and properties ought to be ensured at any cost. There are challenges. However, with 5G drones being involved, there arises several fresh use cases. In this section, we are to discuss how 5G communication networks are beneficial for the oil and gas sector.

As articulated above, for envisaging next-generation use cases, we need a suite of digital and communication technologies. These technologies help in digitizing ordinary items. And, they enable digital artifacts to interact purposefully. When digital entities collaborate, correlate, and corroborate, there is a massive amount of multi-structured data getting generated. Data, then, get gleaned and subjected to a variety of deeper investigations. This analytics process emits out intelligence to be fed into assigned actuation systems. There are drones, robots, and a bevy of equipment in need of competitive intelligence to exhibit self-, surrounding-, and situation-aware behavior. 5G through its native power of high-capacity, low-latency, and high-reliability features are to empower drones to accomplish better and bigger things for the oil and gas industry. 5G is also able to guarantee signal transmission over a larger area. 5G can penetrate indoor locations also.

For realizing Industry 4.0 and 5.0 applications, the well-known connectivity solutions such as Wi-Fi 6 or Ethernet are found to be inadequate for real-time, mission-critical, and reliable applications such as virtual spaces, autonomous vehicles, and remote surgery. Therefore, the world is tending toward 5G communication. Private 5G networks are gaining the speed and sagacity to empower new-generation industry use cases. In the recent past, the promising and potential phenomenon of edge computing has been gathering momentum. Edge devices in our everyday environments are being stuffed with AI-centric processing units and AI models to enable AI-based on-device data processing toward gaining edge intelligence. 5G is the key driver for such a paradigm shift. Such transformation being facilitated through a host of digital innovation and intelligence methods foretells a lot for service providers. The fast-emerging and evolving concept of edge AI is to bring forth a suite of breakthrough advantages toward business and people empowerment.

Seismic exploration is a prime task of oil and gas companies for identifying oil and gas reserves. However, the primary objective is to increase seismic exploration while reducing well drilling. Now, we all know that there is a vigorous and rigorous deployment of 5G networks and equipment in developed and developing countries. The oil and gas industry executives are keen on doing more with less by smartly leveraging the widely reported power of 5G communication, which guarantees a lot for optimizing industrial processes. Seismic data collection, crunching through a dazzling array of technologies such as big and streaming data analytics platforms and AI models, and extracting useful knowledge in conjunction with the contributions of 5G come handy in reducing the exploration workload. The faster proliferation of the IoT sensors and devices, data collection and transmission through an IoT gateway to a nearby or faraway cloud platforms, and data analytics at cloud or edge through AI models are conveying a substantial change for the oil and gas industry. Thus, digitization and digitalization technologies play a very vital role in shaping up the drone world.

Thus, the cellular connectivity has all the potential to transform industrial environments rapidly and rewardingly (both onshore and offshore facilities that are being empowered by fiber, satellite, and microwave backhaul). In addition, 5G enables both cloud and edge computing models. Edge computing strengthens edge devices (wearables, handhelds, hearables, mobiles, portables, and implantable, nomadic, and fixed gadgets and gizmos) to join in the mainstream computing. That is, localized computing along with in-device data processing is being viewed as a paradigm shift in the computing world. Edge devices are being embedded with computing, communication, sensing, vision, perception, decision-making, and actuation capabilities.

Edge devices keep mission-critical data on premises and minimize data transfer over the already clogged internet. Edge computing opens new possibilities and opportunities for established public cloud service providers. That is, edge computing empowers hyperscale clouds to benefit financially by entering new industrial spaces.

9.6 Drones Specialities and Successes

Drones are typically remotely controlled aircrafts with three-axis gimbals, fixed camera mount for high-definition video recording or still photos, and a host of assisted piloting controls. Experts point out that modern drones are blessed with transformative designs, 360° gimbals, higher-value instrumentation, autopilot, and intelligent piloting mode. Other high-end facilities and features include automated safety mode, platform and payload adaptability, full airspace awareness with auto-action like take-off, landing, mission execution, and intelligent piloting. Thus, the functionalities of modern drones are gradually becoming sophisticated. With the conscious addition of digital technologies, the scope and smartness of drones are bound to expand bountifully in the days to unfurl. In this section, we are to discuss the renowned contributions of smart and safe drones.

9.6.1 Drones Assist Human Workers

Drones have brought in real transformation in inspecting and maintenance of critical assets such as pipelines, oil fields and platforms, storage tanks, and refineries. Drones are being embedded with a variety of high-quality sensors and cameras to understand the prevailing condition of the assets and to respond to any incident accordingly. A lot of useful and usable visual data gets captured and streamed to data analytics platforms and databases hosted in remote or nearby cloud environments to make sense out of it in time. Thereby drones sagely gain the necessary capabilities to immeasurably assist human workers involved in exploration, production, transmission, refinement, and retailing.

With drones, the number of human efforts reduces sharply. Similarly, the capital and operation costs for erecting and taking down scaffolding are reduced considerably. Such an aerial manning of vital assets enhances the safety of the field workers, who need not enter those confined spaces. Lately, multifaceted sensors and actuators are being attached in drones. As experts pointed out, high-resolution sensors being incorporated into drones include optical gas imaging, radiometric thermal sensors, magnetometers, gyroscopes, accelerometers and GPS modules, and LiDAR. These externally or internally attached sensors can bring forth additional capabilities to drones to do more with less. Newer use cases can be elegantly visualized and easily accomplished through such flexible integration.

9.6.2 Drones Enhance the Safety and Security

As articulated in the previous chapter, the safety of mission-critical assets being deployed and used in oil wells and platforms is paramount for the oil and gas companies to achieve the intended success. Besides, the human workers must be saved and secured from any kind of incidents and accidents. Robots and drones play a very vital role here in replacing field workers. So timely detection and addressing of any problems is a crucial factor for ensuring the much-expected safety. Drones and robots have gained the distinct capability for replacing problematic and ancient systems and laborious humans from the scene. Drones are being empowered with path-breaking digital technologies such as highly miniaturized and AI-centric processing units, data analytics platforms using machine and deep learning (ML/DL) algorithms, cloud-based data stores, and digital twins. Thus, artificial intelligence (AI)-enabled drones are innately capable of providing more adroit services.

9.6.3 Drones to Analyze Surroundings and Avoid Obstacles

It is inadvisable to embed onboard systems into drones. This approach increases the weight of drones. Also, with additional modules, the energy requirement is to increase. Primarily drones leverage battery-powered power. All these clearly bat for capturing and streaming the videos back to cloud-based object-recognition platform, which analyzes the terrain details carefully and quickly. The knowledge discovered through the video analytics process gets disseminated to drones to take right decisions in time. Thus, drones can understand their working environments well. Because of the widely expressed concerns of drones getting crashed into people, critical assets, vehicles on the ground, or any other flying objects such as crewed aircrafts, drones must have the innate ability to understand their terrains well. Drones are being typically empowered with GPS modules and other location-enablement technologies to do their assignments and engagements with all the clarity and alacrity.

9.6.4 Positioning Capability to Achieve Failsafe Function

The aspect of highly accurate drone navigation contributes immensely to creating 3D maps, analyzing terrains, surveying landscapes, and performing search and rescue (SAR) operations with all the precision and perfection. The modern-day drones are therefore embedded with two global navigation satellite systems (GNSS). The satellite systems collectively help in identifying the current position, orientation, and location of drones. These details come handy in cognitively operating drones remotely. Besides the satellite systems, there are multifaceted sensors in arriving at the location details precisely. Accurate position information goes a long way in producing powerful use cases. The primary benefits of drones for oil and gas companies are many, including the ability to conduct inspections safely and quickly.

9.6.5 Increased Safety

Safety is indispensable for the oil and gas industry. Drones comfortably remove the growing risks associated with humans getting into rough and risky places. Drones can cleanly complete inspections of multiple locations and objects such as storage tanks, transmission pipelines, refineries, chimneys, and other dangerous areas.

9.6.6 Less Downtime

Manual inspections are typically error-prone, and hence there may be occasions for stopping productions. This directly affects the company's brand value. Drones come handy here in guaranteeing less downtime.

9.6.7 Cost Savings

Involving human professionals for oil and gas exploration, production, transmission, and refinement operations costs a lot for any energy company. However, drone-based operations cost less, whereas the precision increases significantly. Drones do inspections of multiple assets quickly. Thus, the mantra of achieving more with less is getting fulfilled through self-managing drones.

In summary, the future for drones is very bright. Smart drones arrive at the scene with powerful motors, a bevy of highly miniaturized yet smart sensors and actuators for effective monitoring, control and cruising, lightweight and long-lasting battery, multifaceted processing units, and self-managing software. Other noteworthy things include the incorporation of autopilot system and built-in compliance technology in new-generation drones. Drones are also vision-enabled to detect objects and avoid collision during inspection. Precisely speaking, AI is the main driver for next-generation drones. The futuristic drones can deliver data in real time once the destination or task is configured. These will be self-contained so that the aspects of self-monitoring of fuel levels and self-correcting of any damage to their propellers, cameras, and sensors can be fulfilled. To contribute handsomely for real-world use cases, drones must be equipped with a host of sophisticated technologies and tools. Such an empowerment makes drones to fly for hours and to come back to its source.

Drones monitoring and management platform software are getting hosted in cloud environments. And, smartphone applications are emerging and evolving to operate drones with correct configuration and control. Drones must be autonomic in their actions and reactions to be commercially viable. Drones are being used across industry verticals for fulfilling critical missions.

9.7 The Emergence of State-of-the-Art Drones

There are versatile algorithms and tools for collision avoidance. Such a competency enables drone operators and businesses to visualize and vigorously deliver sophisticated drone services. There are powerful sensors such as vision, ultrasonic, infrared (IR), Lidar, and time-of-flight. Drones can be fused with one or more of these sensors to sense and avoid obstacles. There are other requirements for drones such as the maximum height and the weather condition of the day and time. Also, drones are prohibited in areas near airports. In the years ahead, the drone technology is to go through several noteworthy innovations, disruptions, and transformations. Resultantly, the world will see and experience intelligent drones, which are miniature in size, quieter in operation, and quicker in response. With the field of edge AI all set to flourish, the next-generation drones are going to be hugely smart and safe. The machine vision, perception, and intelligence technologies contribute immensely for the design, development, and deployment of ground-breaking drones across industry verticals. Hydrogen power will energize futuristic drones.

9.7.1 Drones with Real-Time Flight Parameters

All the present-day drones are being activated and actualized through a controller at the ground. Nowadays, there are smartphone applications to keep track of drones and their trajectories. All drones have a ground station controller (GSC) or a smartphone app, allowing users to keep track of the drones' current flight telemetry. Drones should not accidentally fly into no-fly zones. They should not cross the path of aircrafts. Drones must be continuously monitored and managed to avoid any unnecessary incident. Firmware updates must happen consistently to keep up the condition of multi-purpose and commercial-grade drones. Considering the changing world, drones must be designed and developed by leveraging the purpose-driven digital technologies.

9.7.2 Connected Drones

With the surging popularity of the Internet of Things (IoT) and edge technologies, every concrete and common thing in our everyday environments is getting methodically digitized and hooked into the internet. That is, digitized entities when integrated with cloud-based software solutions and services can acquire special capabilities. Going forward, drones are also networked with others in the vicinity and over the internet communication infrastructure to communicate, collaborate, and correlate to finish off real-world business use cases. With such connectivity, drones are expected to contribute immeasurably in rough, remote, and risky areas. Thus, newer applications are being worked out for connected and cognitive drones.

The unprecedented growth of digital technologies makes present-day drones to be penetrative into any worthwhile domain to be pervasive and persuasive too. Experts and evangelists point out that a drone can elegantly march into dangerous places and guide a team of searchers through a dark forest. With the inherent connectivity capability, drones are getting ready to perform sophisticated processes across agriculture, national and in-land security missions, industry operations, product and package delivery, entertainment, etc. E-commerce service providers increasingly leverage drones these days to deliver packets and parcels. The agriculture industry benefits in the form of carrying seeds and fertilizers into fields for cultivation and nourishing in an optimal manner. The oil and gas industry benefits out of all the unique competencies of connected drones.

As articulated and accentuated above, powerful yet disappearing sensors and actuators, tags, pads, codes, chips, controllers, software agents, LEDs, beacons, etc. are being neatly embedded in drones. Such a transition enables modern drones to do powerful and peculiar tasks. Radio and GPS modules are included for location identification and data transfer. In addition, video cameras are attached for recording positioning of objects and people. Criminals on the run can be tracked with all the diligence. Drone technology is ideally fit for search and rescue (SAR) missions. Drones can easily fly to remote places faster than vehicles. Drones can fly over areas with lights and sensors including cameras to locate living beings using infrared (IR) technology. Search and rescue drones can deliver flotation devices and life-saving supplies, such as foo id, water, rope, life jackets, inflatables, or life preservers.

9.7.3 Cognitive Drones

With the arrival of cognition-enablement technologies, the days of cognitive drones are nearer. There are innovative technologies and state-of-the-art infrastructures emerging and evolving to transform data into knowledge, which, then, can be fed into drones to exhibit intelligent behavior in their working environments. All kinds of data getting collected by drones are being cleansed and crunched through cloud AI and edge AI platforms to extract actionable intelligence in time.

Without an iota of doubt, the AI field is going through a series of paradigm shifts with the solidity of several breakthrough machine and deep learning algorithms. The process of knowledge discovery out of data heaps is getting speeded up through AI algorithms, frameworks, processing units, and other accelerators. Thus, empowering drones with all the intelligence in real time is gathering momentum. Both cloud AI and edge AI processes contribute immeasurably toward intelligent drones. For an example, edge AI can help drones recognize and navigate obstacles to avoid collisions. This unique power put drones in rescue missions in difficult terrains such as mines, tunnels, and caves.

9.7.4 Integration with Artificial Intelligence (AI)

Any worthwhile system must be seamlessly and spontaneously empowered with AI capabilities to deliver next-generation functionalities. Drones are also not an exception to this hard rule. AI models are being deployed in cloud environments to propel drones remotely. In the recent past, even lightweight AI models are made, evaluated, optimized, and deployed in drones directly. Such on-drone data capture, storage, and processing promise building and releasing real-time drone services and applications. Thus, all kinds of innovations and improvisations occurring in the AI space are hugely impactful for the drone industry also. When paired with AI, drones can revolutionize equipment inspections in a risk-free and rewarding manner. The inception and incorporation of contemporary techniques have immensely contributed toward the upliftment of the oil and gas industry.

AI is being proclaimed as the most influential and inspiring technology for the future of the society. Especially, drones are being emboldened through the direct leverage of the unique AI capabilities. AI can speedily crunch any amount of data to emit out useful insights in time. Drones can be embedded with ultrasonic sensors and vision modules such as cameras and thermal imagers. These empowered drones are primarily used for performing non-destructive inspections of oil and gas assets to pinpoint flaws or defects in time. Drones can fly remarkably close to oil and gas transmission pipelines and other assets to capture the details of the interior and exterior parts of tanks and marine vessels. By merging, mining, and massaging all the gathered data, it is possible to clearly understand what is happening in and around. The knowledge discovered gets disseminated to actuating systems to embark on corrective works cognitively.

In summary, drones are hugely beneficial in upstream assets, midstream pipeline infrastructure, and downstream facilities.

Drones equipped with thermal imaging or optical gas imaging can safely contribute to locating problem areas. High-resolution images and videos of oilwells, machineries and robots deployed, and transmission pipelines supply all that are needed for experts to ponder about the best-in-class combating solution approaches and complexity mitigation techniques. In short, drones are gaining immense popularity in accurately doing infrastructure inspection in hard-to-reach locations.

Drones sufficiently enhance operational efficiency throughout the value chain. The recent technological advancements in accurate sensing, imaging, perception, pattern recognition, knowledge discovery, and action result in state-of-the-art drones.

Drones are being used across industries to assist human workers ably and artistically in their occupations, improve productivity drastically in complicated and risky activities, and lead to a bevy of newer creative interactions.

Drones have become an adept assistant for industrial workers. Lately, drones have been stuffed with AI to automate several manual tasks. Drones are empowered with AI-based data analytics capability to crunch data collected by their sensors during their flights. Multifaceted sensors are being attached with drones to collect visual data while flying over certain locations such as oil wells, fields, platforms, pipelines, and refineries. Drones duly powered by AI can do computer vision (CV) and natural language processing (NLP) tasks. As we all know, the primary role of AI is to empower hardware and software systems to gain the human brain-like capabilities such as learning, understanding, thinking, articulating, and proposing new hypothesis. Thus, drones are vision-enabled to comprehend the prevailing situation and accordingly act upon. That is, AI-enabled drones gain the vision and intelligence to avoid collisions. They can detect and track specific objects. There are certain purposes for drones to fulfil.

Surveillance: Increasingly drones are outfitted with high-resolution surveillance devices to gather high-definition videos and photographs even during nighttime. Drones can collect vehicle license plate information and data about specified targets such as people and assets. Drone surveillance facilitates monitoring and data collection without getting visibly noticed. Thus, any kind of covert operation can be identified and conveyed in time.

Weather forecast: This is another important contribution of drones. Drones, as we all know, help in the event of a natural disaster by assisting emergency workers with all the right details. For forecasting, drones can collect air samples, and this is being seen as a significant improvement over the conventional ways of gathering data. Drones increase the reliability of climate forecasting models; thereby, with 5G drones, the aspect of weather forecast is going to be more accurate.

Healthcare and food: On the delivery front, drones are gaining a lot of attention these days. E-commerce service providers leverage drones for product and parcel delivery to their customers. During the unforgettable COVID period, medical supplies such as COVID toolkits, tablets, medicines, and vaccines were delivered by drones. Food items are also being delivered through drones. The United States postal service (USPS) has announced that it is exploring using drones to advance mail delivery operations across.

The number of AI-enabled applications is growing rapidly. Similarly, the embedding of AI capabilities into drones is to accelerate in the days to unfurl. With more memory, processing, and storage power, complex AI models are run in drones. With computer vision (CV) and natural language processing (NLP) use cases being implemented through deep learning (DL) algorithms, the days of vision-enabled drones are near. That is, drones by gaining the vision capability can exhibit intelligence in their operations, outputs, and offerings.

First aid drones: As per a technical article, a drone has beaten an ambulance in race to deliver first aid to patients (https://spectrum.ieee.org/drone-vs-ambulance-drone-wins). Drone technology is being seen as a way forward to get lifesaving medical supplies to patients as quickly as possible. For a person experiencing sudden cardiac arrest, there is a need to rush automated external defibrillator (AED) to restart the heart. First-aid drones can do that swiftly.

Drones in agricultural applications: Crop spraying for fertilization and pest management is being accomplished through connected drones. The IoT drones can sow seeds along with nutrients to sustain them. Drones are popular here as they do it with precision. Further on, drones can produce 3D maps, which, in turn, provide actionable insights to farmers to plan their seed planting, fertilization, and irrigation with all the astuteness and alacrity. There are several noteworthy contributions of drones, which are increasingly prevalent in military operations. Drones offer high-resolution aerial photography and videos for sporting events and entertainment.

Drones stuffed with sensing and perception capabilities are to be used across engineering and scientific applications. Specialists across industry domains are passionately experimenting and exploiting the observability factors of drones to optimize hugely complicated industrial processes. Such a drone-enabled transition eventually results in operational efficiency and higher productivity. Drones are being applied for an assortment of activities, ranging from indoor and outdoor inspections to complex applications such as verifying the physical condition of pipelines.

With the penetration and pervasiveness of 5G-enabled drones, newer use cases emerge and evolve fast. The traditional use cases are getting strengthened with the ready availability of 5G communication networks. The use cases such as aerial inspection, entertainment, landscape mapping, product delivery and intelligence, surveillance, and reconnaissance (ISR) are getting bolstered through 5G. 5G-enabled drones could reliably deliver medical equipment like

COVID test kits to infected person safely. Thereby, the spread of infectious diseases can be stopped. The hugely popular search and rescue (SAR) operations can be minutely reported to the concerned by drones with all the visual input in real time with low latency.

Drones with 5G connectivity contribute immeasurably to collecting data easily, quickly, and affordably. It is undoubtedly clear that data are emerging as the strategic asset for any growing organization to plan and execute different things correctly. With cloud-based and edge AI platforms, unearthing viable and venerable insights out of data heaps becomes simpler and speedier.

Higher yields: For the rising population, higher food production is being insisted. Thus, digitally transformed farming is being recommended. With autonomous drones being availed, the farming community is to get tactical as well as strategical benefits to sharply enhance the yield. Drones gather high-resolution images of agriculture, sericulture, and horticulture fields. The insights getting generated out of these captured images come handy for farmers to focus on certain areas so that the overall production is bound to increase remarkably.

Drones have propelled the oil and gas industry to greater heights. Drones can do aerial photography and thermography. Thereby, drones guarantee improved image quality with precisely measured and angled shots. The noteworthy advancements in sensing, vision, perception, proximate processing using AI models toward real-time insights, etc. have enabled modern drones to be deployed for a growing number of use cases. Drones monitor, measure, and predict the operational, health condition, performance/throughput, and other critical aspects of oil and gas industry infrastructure assets precisely.

Data analytics through cloud and edge AI platforms comes handy in producing actionable insights for oil and gas industry consultants to visualize and realize newer powerful and pertinent applications. Drones are found to be hugely beneficial for remote surveillance. Drones can audit all kinds of principal assets, including oil wells, rigs, platforms, tankers, trucks and pipelines, machineries, specialized equipment, and robots in oil fields. Drones can provide a 360° view for constantly overseeing field operations. Drones also contribute to providing the progress of facilities, which are under construction or upgrade. If anyone is intruding into the oil facilities or if there is any encroachment, drones inform the concerned immediately. In short, for a mission-critical vertical such as oil and gas, drones automate and optimize several oil and gas industry processes. Thereby, the much-expected precision and productivity improvement are being relentlessly accomplished through advanced and autonomous drones.

9.8 Drones in the Oil and Gas Industry

With a greater understanding, the use of drones in the oil and gas industry is picking up. The industry people are steadily gaining everything about drones and how they are critically contributing for automating and optimizing oil and gas processes. Drones can reduce inspection time and cost. Most of the operational complexities are being addressed through the smart leverage of powerful drones. The drone technology can scale for more advanced use cases as the enabling technologies are maturing fast. The prominent use cases include the following.

- **Asset inspections:** Drones are attached with powerful cameras, thermal imaging, and modeling software. Hence, drones can create precise 3D models, which optimize the inspection of pipelines, storage tanks, marine vessels, and other important assets and aspects. There are drones capable of flying for several hours at the speed of 40 miles per hour. Drones can swiftly and securely carry out inspection activities. Otherwise, inspection takes a lot of precious time, talent, and treasure. Also, without drones, the inspection process would have put many employees at personal risk. A report says that drones have the intrinsic power to guarantee 33% heightened inspection efficiency and the inspection costs are to be reduced by 50%.
- **Aerial surveillance:** Drones are famous for remote monitoring of oil and gas fields, platforms, marine vessels, and refineries. Drones also give a 360° view for a variety of assets, equipment, tankers, trucks, platforms, sites under construction, etc. The visual data generated by drones come handy for experts and engineers to gain a deeper and broader understanding of the happenings in and around the vital junctions.
- **Inspection and proactive maintenance:** Gradually, modern drones are designed and manufactured with ultrasonic sensors and vision-enablement equipment. Such empowered drones do close yet non-invasive and non-disruptive inspections of oil and gas assets to look for any defects, deficiencies, or dangerous situations. With more breakthrough technologies, both exterior and interior portions of assets, pipelines, tanker ships, etc. can be assessed and articulated. With AI-based real-time data analytics platforms, any kind of slowdowns and breakdowns can be pre-emptively pinpointed and addressed.
- **Detecting leaks:** Methane leak detection and quantification are critical to reducing methane gas emissions from fossil fuel production. There are sensors that use spectral imaging technology to measure methane emissions. Such sensors are easily attachable and detachable and can be mounted at the bottom of drones. Such empowered drones are found to be relevant for detecting methane.
- **Emergency response:** There may be human-caused as well as natural disasters across industrial environments. For the oil and gas sector also, this

possibility persists. In such critical situations, drones are found to be highly suitable for aerially monitoring any kind of disasters and for arriving at quick and informed decisions and finally for providing the much-needed emergency response. Thus, drones are utilized to minimize property and people losses.

Drones provide an aerial view of all that is happening in any accident place. Image, audio, and video analytics on the data captured and streamed by drones can assist in mapping oil spills or fire incidents. This enables in earmarking correct and immediate response and sending resources to the right people and places.

- **Handling and moving materials:** Delivering critical equipment and machineries through drones is becoming penetrative and pervasive. Drones can deliver important assets between land-based facilities and marine vessels. Thus, oil and gas companies focus on leveraging the power of drones in transmitting vital materials between various locations spreading across land and sea. Drones are increasingly used to transport small payloads to production platforms located nearby.
- **Enhancement of workplace and workforce safety**: Safety is important for industrial environments located in rough, touch, and remote places. As seen above, drones play a handsome role in securing and saving human lives and critical assets. Drones can monitor most of the sites and spots including oil wells, platforms, and refineries spectacularly.
- **Methane management**: There is a sustained campaign for mitigating climate change and global warming across the globe. Therefore, oil and gas companies are making efforts to reduce methane gas emission. Organizations are investing in sensing technologies. These sensors can be mounted on drones for gathering real-time data.

Drones are being positioned and proposed as a precise, easily maneuverable, and cost-effective method for asset checking and management. Drones enhance the inspection quality and improve the safety field workers. Also, drones through automation and optimization lower the capital and operational costs for the oil and gas sector. For an example, drones, without an involvement of human workers, can conduct detailed inspections of essential oil refinery equipment, chimneys, smokestacks, jetties, and storage tanks. Thus, safe and secure inspection of many mission-critical assets is being fulfilled through drones.

In summary, drones do the inspection of various equipment and assets deployed in and around oil wells and platforms regularly; thereby, inspection specialists are taken off the equipment. Thus, the safety of human workers is fully ensured. Drones are unmanned and hence fly close to infrastructural components and assets; thereby, any associated risks are majorly avoided. Drones continuously collect a massive amount of equipment data. These data gives a lot of insights on the operational aspects of involved equipment and machineries. Drones automate most of the manual processes and facilitate the process optimization, which is crucial for any industry segment to survive and thrive in the excruciatingly knowledge-driven world. With less time, talent, and treasure, oil and gas companies can achieve more. Drones do everything efficiently; thereby, the productivity of the oil wells increases significantly.

9.9 AI-Enabled Drone Services

There is no doubt on the overwhelmingly accepted truth that AI is the domain of grandiose choice for building next-generation, mission-critical, and people-centric intelligent systems for most of the industry verticals. AI-enabled systems can sense, perceive, understand, decide, articulate, and actuate several things for humans. For example, within our home environments, AI-enabled sensors, actuators, instruments, equipment, kitchen utensils, wares, gadgets, and gizmos work individually as well as collectively to correctly understand the unique needs of homeowners and occupants and work on them to deliver the identified services to the people in an unobtrusive manner in time. That is, AI systems are destined to accompany and assist professionals and commoners in their everyday assignments and engagements. AI systems can think like humans, take decisions, plunge into appropriate actions, and improve peoples' decision-making ability by continuously learning from multiple sources including the environment. Today, AI models and services are being embedded into all kinds of digitized and connected devices (network-embedded systems) to unlock newer opportunities and possibilities for service providers as well as consumers.

Thus, AI is to help in learning something useful and usable from data heaps. There is a dazzling array of learning algorithms and approaches to facilitate the much-required automated learning from a growing amount of multistructured data. There are cognitive science techniques in plenty. Similarly, the paradigm of cognitive computing is emerging and evolving to enable computing and communication devices to perform tasks that we can perform (understanding, proposing new hypotheses and reasoning, articulating the findings naturally and unambiguously, etc.) in our everyday life.

AI-enabled systems gain the ability to correctly analyze data. Such analysis helps the system understand what is required to accomplish certain things. Such understanding helps the system achieve the target and finish the assigned tasks in time. Drones leveraging the power of AI can do several hitherto unheard things for businesses and people. Drones can traverse in the right path toward their destination by surmounting any obstacle on their route. They cognitively navigate through crowded places and areas with poor visibility. They can monitor their surroundings and understand their situations perfectly and precisely. Thus, self-contained and AI-activated drones tend to be adaptive in their decision-making and actions.

Drones with appropriate enablement through AI algorithms (machine and deep learning) automate several manual and risky tasks such as inspection, surveying, monitoring, delivery, and capturing environment and operational data. AI capabilities such as computer vision and natural language processing (NLP) teach drones to do as per the changing needs and situations.

Cameras mounted on drones help capture visually rich data in the form of still and dynamic images. Their image and video analytics software solutions are useful in the market. These are being adequately empowered through AI algorithms and models. That is, AI-powered image and video analytics software plays a key role in elevating drones to perform path-breaking tasks in real time. AI-enabled drones can identify and locate different objects in an image, count the number of objects (class instances) in an image, and classify pixels in an image into several finite segments to ensure simplified representation. These drones can identify any perceptible changes between two temporally spaced images and can perform image classification. Thus, understanding and learning from images and videos are being automated through the power of AI.

9.9.1 Drones with AI Power

Drones gain immense power through the noteworthy innovations and disruptions happening in the domain of computer vision. Drones are being vision-enabled to detect different objects accurately while flying.

As we all know, computer vision is one of the prominent applications of AI. There is a myriad of deep neural network (DNN) architectures. Images and videos are being expertly processed by deep neural networks (DNNs) to detect, recognize, and track various objects hidden in input image files. Such DNNs provide the power to drones to perform object detection and classification. All sorts of captured information are fused together to arrive at insights, which enables drones to avoid any collisions with other drones in the vicinity with aplomb. Thereby, drones can easily reach target environments and assets without any hurdle and hitch. There are deep learning (DL) algorithms in plenty to assist AI engineers and data scientists to train, retrain, and refine AI models with efficacy to achieve the above complex tasks. Further on, there are libraries and toolsets to do hyperparameter optimization (HPO) tuning to improve the prediction accuracy with less resource consumption.

9.9.2 Sensors for Data Collection

As illustrated above, multifaceted sensors play a very vital role in shaping up the drone technology. Sensors can sense, capture, and transmit all that is happening in and around the area of inspection and monitoring. Cameras are visual sensors and hence can transmit visual data. As written above, there are highly refined and optimized ML algorithms to educate drones to act according to the underlying situation. Also, sensor data get used for non-flight-related analyses. This is used to find potential mining locations or to evaluate the water quality of reservoirs.

AI models can process a huge amount of data quickly. Also, AI's prediction accuracy increases drastically if there is a big data to be analyzed. Also, there are several scenarios that need real-time insights. AI is the way to go for producing real-time services and applications. AI possesses the capability of guaranteeing real-time processing of data streams. Thus, embedded sensors collect and share data with the drone's AI platform to facilitate the generation of real-time insights. Thus, the overly dissected and deliberated concepts of data-driven insights and insight-driven actions are seeing the reality with the blending of sensors data and AI models.

9.10 AI Platforms for Drones

A litany of breakthroughs in the AI space promise to supercharge new-generation technology platforms. As widely known, AI adds capabilities like prediction and prescription to drones. These sophisticated features are being provided to drones through these platform solutions. The AI-centric platforms help drones be context-aware. With the context-awareness functionality, drones are succulently and sagaciously enabled to be autonomous in their contributions. Therefore, having understood the long-term benefits of AI platform-supported drones, researchers explore this phenomenon to enrich the drone technology. As described above, drones are already doing wonders in the military domain. The scope for the drone technology is gradually growing and glowing. It is penetrating agriculture, construction, critical infrastructures and assets, security and surveillance, etc. Using feature-rich sensors, drones can continuously collect and transmit data to AI platforms to crunch the captured data. The combination of software-defined cloud centers, sensor-attached drones for data collection, 5G connectivity, big data on data lake, machine and deep learning (ML/DL) models running on AI platforms, etc. has laid down a stimulating foundation for designing, developing, and deploying competent, cognitive, and conscious drones for the future of the oil and gas industry. The details about AI-enabled drone applications can be found at https://consortiq. com/uas-resources/drone-ai-technology-how-it-works-why-it-matters.

FlytBase AI is a cloud-based and AI-enabled platform for drones (https://flytbase. com/ai-drones/). Because drones are fit with so many multi-purpose sensors, drones can capture a lot of data from multiple sources such as their areas of inspection. Also, data come in disparate formats (images, videos, audios, etc.).

Any visual data can tell a lot about the operational domain to domain experts and data scientists. As the world is steadily realizing the fact that data are a strategic asset for any enterprising business to be on the right side, all sorts of data get meticulously gleaned and subjected to deeper and decisive analytics processes.

The FlytBase AI platform is specifically for drone-enablement by leveraging the elasticity feature of cloud infrastructures. Several AI models are getting trained, retrained, refined, and deposited in this platform for activating and automating many use cases. Especially real-time visual data captured by drones are analyzed through this platform-based AI models. Such an empowerment comes handy in making drones intelligent in their actions and reactions. The workflow diagram given below illustrates how this platform accomplishes various drone-centric use cases.

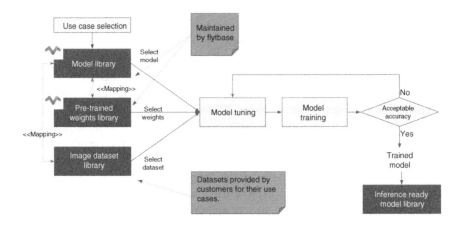

There are strategically sound benefits with real-time data processing capability, which is being realized through the seamless and spontaneous fusion of the AI power. With this feature, drones can function insightfully in locations, where people dread to go. AI assists drones to make adjustment and adaptivity to conditions in rough and tough locations.

In short, drones are internally and externally empowered to be AI functional. Several new use cases are being proposed and piloted. Construction companies increasingly employ AI-enabled drones to scan and monitor construction sites and activities continuously. Similarly, agricultural companies use drones for monitoring operations for getting higher yields. Drones are good at planting and nourishing crops by optimally scattering seeds and fertilizers.

Equipped with computer vision, AI-based drones can continually assess their areas under inspection and do the appropriate counter-measures in time. Thus,

AI-enabled drones with a huge volume of correct data can contribute more intuitively and inspiringly for the total society.

Folio3 (https://www.folio3.ai/) is vigorously conducting advanced research and producing many AI-centric solutions for tackling a set of business problems. Especially AI is being applied in empowering drones to be cognitive and conscious of their environments and activities. A growing array of tasks is being automated and accelerated through AI-enabled drones.

- **Livestock management**: Manually counting cattle is time-consuming and error-prone. But AI-enabled drones can do a better job here. Folio3 has demonstrated its unique capability in automated livestock management. Identifying sick animals and separating them from healthy ones to stop any disease proactively can be elegantly handled by drones.
- **Terrain mapping**: Folio3 has developed and released a competent AI solution for exemplary terrain mapping. AI-backed drones equipped with 3D cameras and LIDAR detectors can record the appropriate area, and the recorded data can be used for building useful 3D models, which come handy in precisely automating the terrain mapping need.
- **Precision agriculture**: AI-enabled drones can overcome several challenges of the agriculture industry. This company's solution assists farming companies in taking right decisions at the right time to enhance the farm yields.

In short, AI makes machines intelligent in their actions and interactions. By artistically embedding drones with AI power, it is possible to bring forth pioneering use cases for the total society. Different industrial sectors such as energy, construction, agriculture, and oil and gas are to leverage this unique power to be self-sufficient and sagacious in their service delivery. There are blogs, magazine articles, and research papers illustrating how the AI conundrum fuses with machineries to provide exemplary services to the humankind.

DRONE VOLT (https://www.dronevolt.com/en/) is developing drone (enabled through computer vision) solutions for the energy, construction, civil engineering, and security industries. The prominent applications are

- Object detection, counting, segmentation, and tracking
- Person or animal detection and tracking
- Thermal detection
- Verifying the adherence of face mask rule
- Detection of the use of protective equipment (glasses and helmets)
- Face detection and recognition
- Fire and smoke detection
- License plate reading
- Crack damage detection on surfaces

Digital Aerolus (https://digitalaerolus.com/) makes drones, which can safely and stably fly into areas where other drones cannot go. They can move into areas that are GPS-denied, such as inside power plants and storage tanks.

Qii.AI (https://qii.ai/) is to solve real-world problems. The company has entered the drone inspection space. They came across a big problem. They could collect massive volumes of data during inspections. But there is no solution to manage and share this information. Qii.AI is to help inspection teams find and fix defects faster and better.

The Dronehub docking station (https://dronehub.ai/) allows rapid and remote inspection of the refinery objects, without stopping the worker's job. Drone-based routine inspection improves efficiency and optimizes the time inspection. Drones could get closer to inspect parts of the infrastructure than traditional methods. This solution has brought in a series of benefits for the oil and gas industry.

The incorporation of efficient, edge, and explainable AI enables drone vendors to collect, cleanse, and crunch data from multiple sensors attached to the drone. The AI-enabled data processing leads to the realization of autonomous drones. Drones have become a grandiose part of the growing array of smart mobility offerings. As indicated above, AI drones depend upon the aspect of computer vision. Thus, asset detection, recognition, monitoring, tracking, and management are being hugely simplified.

9.11 Conclusion

Drones are unmanned, remotely controlled, and aerially flying devices that accomplish a variety of tasks, which are typically tiresome, troublesome, and time-consuming for human beings. With the solid improvement in the AI technology space, drones are being innately embedded with a suite of AI competencies to perform complicated and long-running business use processes in an accelerated and automated manner.

With the incorporation of AI facilities, it is possible to collect all kinds of data (visual and environment) emanating from drone-attached sensors and subject it for a deeper investigation quickly to extract actionable insights, which, in turn, empowers drone vendors to use data from sensors attached to the drone to collect and implement visual and environmental data. The knowledge discovered gets disseminated into drones in time to exhibit an adaptive behavior during their flights. Thereby, AI-enabled drones become autonomous with less intervention, instruction, and involvement of human experts. With all the noteworthy enhancements being brought into modern-day drones, there are a host of smart mobility offerings being made available to individuals and industries.

Bibliography

1 10 Major advantages of using drones in oil and gas industry. https://www.zenadrone.com/10-major-advantages-of-using-drones-in-oil-and-gas-industry-in-2023/

2 The importance of drones in the oil and gas industry. https://consortiq.com/uas-resources/the-importance-of-drones-in-the-oil-and-gas-industry.

3 Unleashing the full potential of drones in the oil and gas industry. https://viper-drones.com/industries/oil-gas-industry-drones/

4 5 Major benefits of drones in oil and gas. https://thedronelifenj.com/benefits-of-drones-in-oil-and-gas/

5 ideaForge drones can effortlessly capture and provide in-depth details of flare stacks, oil rigs, and pipelines to diagnose faults. https://ideaforgetech.com/mapping/application/oil-and-gas.

6 How are drones used in the oil and gas industry? https://copas.org/how-are-drones-used-in-the-oil-and-gas-industry/

10

The Importance of Artificial Intelligence for the Oil and Gas Industry

10.1 Introduction

What is AI?

As per Wikipedia, Artificial intelligence (AI) is intelligence – perceiving, synthesizing, and inferring information – demonstrated by machines, as opposed to intelligence displayed by animals and humans. Example tasks in which this is done include speech recognition, computer vision, translation between (natural) languages, as well as other mappings of inputs.

The theory and development of computer systems is in their ability to perform tasks that normally require human intelligence, such as visual perception, speech recognition, decision-making, and translation between languages.

AI applications include advanced web search engines (e.g. Google), recommendation systems (used by YouTube, Amazon, and Netflix), understanding human speech (such as Siri and Alexa), self-driving cars (e.g. Tesla), automated decision-making, and competing at the highest level in strategic game systems (such as chess and Go). As machines become increasingly capable, tasks considered to require "intelligence" are often removed from the definition of AI, a phenomenon known as the AI effect. For instance, optical character recognition is frequently excluded from things considered to be AI, having become a routine technology.

Artificial intelligence was founded as an academic discipline in 1956 and in the years since has experienced several waves of optimism, followed by disappointment and the loss of funding (known as an "AI winter"), followed by new approaches, success, and renewed funding. AI research has tried and discarded many different approaches since its founding, including simulating the brain, modeling human problem-solving, formal logic, large databases of knowledge,

The Power of Artificial Intelligence for the Next-Generation Oil and Gas Industry: Envisaging AI-Inspired Intelligent Energy Systems and Environments, First Edition. Pethuru Raj Chelliah, Venkatraman Jayasankar, Mats Agerstam, B. Sundaravadivazhagan, and Robin Cyriac.
© 2024 The Institute of Electrical and Electronics Engineers, Inc.
Published 2024 by John Wiley & Sons, Inc.

and imitating animal behavior. In the first decades of the twenty first century, highly mathematical–statistical machine learning has dominated the field, and this technique has proven highly successful, helping solve many challenging problems throughout industry and academia.

The various sub-fields of AI research are centred around goals and the use of tools. The traditional goals of AI research include reasoning, knowledge representation, planning, learning, natural language processing, perception, and the ability to move and manipulate objects. General intelligence (the ability to solve an arbitrary problem) is among the field's long-term goals. To solve these problems, AI researchers have adapted and integrated a wide range of problem-solving techniques – including search and mathematical optimization, formal logic, artificial neural networks, and methods based on statistics, probability, and economics. AI also draws upon computer science, psychology, linguistics, philosophy, and many other fields.

The field was founded on the assumption that human intelligence "can be so precisely described that a machine can be made to simulate it." This raised philosophical arguments about the mind and the ethical consequences of creating artificial beings endowed with human-like intelligence; these issues have previously been explored by myth, fiction, and philosophy since antiquity. Computer and philosophers have since suggested that AI may become an existential risk to humanity if its rational capacities are not steered toward beneficial goals.

In short, AI refers to the intelligence that is demonstrated by machines with the data input provided to them. The machines will learn from the data (called machine learning) and get trained in taking decisions in lieu of humans when similar incidents occur and when a decision must be taken.

In the O&G industry, AI is used in multiple areas like hazard detection and prevention, reducing equipment downtime, detecting anomalies by enabling automation in assets, predictive maintenance to improve asset integrity and increase productivity, prevention of workplace safety risks, and surveillance of safety deviations.

Below are some use cases focusing on AI in the O&G industry

10.2 Reducing Well/Equipment Downtime

As oil prices are dynamic, there is a need for oil companies to remain profitable. One of the key areas where oil companies can target for saving cost is reducing well/equipment downtime. For this purpose, oil companies rely on predicting unplanned shutdown/downtime/equipment failures and taking corrective action using data analytics, machine learning, and AI.

For instance, in this use case, we use historical customer data to build a demand forecast and based on the forecast modify the downtime of the wells. There could be an expected surge in demand, which means the company can reschedule its maintenance activities to a later date or drop in demand where the company can afford to close a well for maintenance

Downtime costs are highly expensive for oil and gas companies. Whenever an oil well or a refinery shuts down for unplanned reasons, production halts, workers sit idle, parts and equipment must be urgently purchased, and the flow of materials is interrupted. Christensen et al. (2013) cited a study conducted by the Hydrocarbons Publishing Company based on data provided by the United States Department of Energy that reveals that between 2009 and 2012, there were over 1700 refinery shutdowns in the United States, equivalent to 1.2 shutdowns per day. The same study indicates that 46% of these shutdowns were caused by mechanical breakdowns, 23% due to maintenance, 19% because of electrical power outages, and 12% for other reasons.

Preventive and predictive maintenance can help salvage the unexpected equipment downtime to a certain extent, but the issue is when there are unplanned downtimes impacting the production. Despite companies doing predictive maintenance, a large part of downtime happens due to unexpected events. So companies try to find ways to save cost by predicting equipment downtime and hence preventing the downtime as the analysis indicates the importance of maintaining the assets in an O&G company. The other way organizations maintain their equipment is to replace the equipment or its parts at regular intervals, irrespective of whether they are working or not. To replace parts or perform preventive maintenance, ML can be used. Sometimes, there may not be a need to change or replace a part, and hence based on historical data from this company or another company with similar operating conditions, the part replacement can be planned.

One of the key areas of focus to save costs by applying ML is in the supply chain area. The figure below shows the supply chain for O&G.

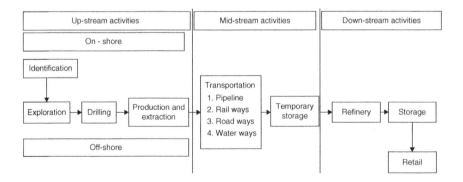

As indicated in the figure above, the O&G industry comprises upstream, downstream, and midstream activities. So, when we are planning a supply chain for the complex industry demands, the following points are to be taken into consideration:

1) Availability of the right component/part at the right time. Eg: Which component needs to be replaced? Where is the asset? Does the engineer present have the knowledge to replace the component?
2) Integrating vendors: Does the vendor have the component or part that needs to be installed? Will the part arrive on time for the engineer to fix it?
3) Is the removed component or part disposed safely?
4) How long will the engineer need to do the maintenance? And what are the tools that would be needed for the maintenance?

If you can see, supply chain (SC) is a data-intensive process for O&G companies. The key here is to get the right data at the right time without data quality issues, and integrating supply chain, tools/solutions to manage transport and logistics, resource tracking, and delivering the goods/components as per the commitment is key. Creating an ecosystem of product vendor consumers and the tracking system are a very important part of a supply chain management process. If there are slippages in the SC process, there could be unplanned downtime of assets, the cost of which would be huge. Total unplanned downtime is a big money burner in the O&G industry.

So how was AI/ML used for reducing well/equipment downtime? Below are the common types of maintenance in the O&G industry

a) Reactive/corrective maintenance
b) Periodic maintenance
c) Proactive/preventive maintenance
d) Predictive maintenance
e) Condition-based maintenance
f) Risk-based maintenance

One of the major O&G companies enhanced its ability to predict well collapses before they occur, reduced maintenance and efficiently operated the wells, and extended their remaining use of life.

How did they do it?

The asset installed sensors for getting the details of each unit and its components. The plant's engineers built a traffic light system, which informed them of the upcoming risk of a well collapse based on the current data provided by the sensor, and this was compared to the historical data that was compiled earlier. This enabled them to operationalize and incorporate necessary measures in their maintenance schedule to reduce downtime. AI systems ingest these data from the components and would feed into the AI algorithms to detect anomalies and provide alerts on time

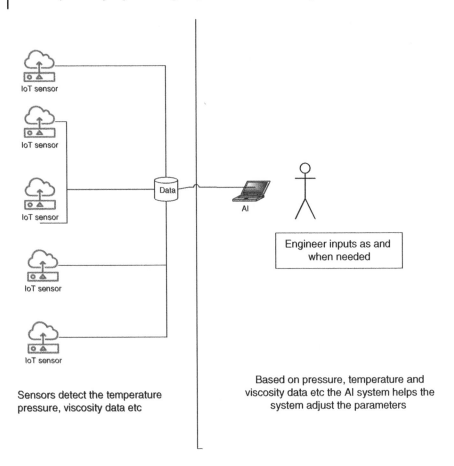

IoT sensor

IoT sensor

IoT sensor

IoT sensor

IoT sensor

Data

AI

Engineer inputs as and when needed

Sensors detect the temperature pressure, viscosity data etc

Based on pressure, temperature and viscosity data etc the AI system helps the system adjust the parameters

10.3   Optimizing Production and Scheduling

In most manufacturing companies, cost and schedule overruns are perennial problems. This can be partially attributed to weather, resource constraints, spare availability, and scheduling issues.

This becomes more complex due to multiple activities involved in oilfield development. It becomes important to find ways to plan projects in a better way to optimize production and scheduling.

In a particular scenario, ML was used to develop a predictive production optimization model that was implemented on an IoT-based cloud platform. Usually, there is a mathematical model which predicts and helps optimize production to a great accuracy. However, the model development is very tedious and is prone to error based on the expert who creates the said model. This is where an unbiased ML model comes into play.

ML helps overcome this problem. ML starts with a programmed neural network model and then based on historical data learns now the particular asset must perform optimally and then predicts when the asset does not perform optimally. Additionally, the AI component of ML can suggest what can be done to bring the asset to green from a red or amber hue.

ML can also compare data it is receiving in real-time to a pattern based on data sets it has processed. For this, there is a requirement of large amounts of data, but once the data have been received, the ML models will start producing results. Based on pattern recognition, ML can detect many anomalies in the system.

One disadvantage can be that in assets which are offshore, getting a huge amount of historical data and regular feed of real-time data can be challenging.

One of the key KPIs for optimizing production scheduling is by improving overall equipment effectiveness. Traditional approaches tended to focus on throughput and utilization rate, but nowadays this is insufficient. The main reason relies on the importance of unconsidered context information, or even small details, which are making a difference. The overall equipment effectiveness indicates how good the equipment is being used. Optimizing equipment effectiveness helps in improving production and scheduling.

Assuming we are monitoring a major set of equipment like pumps, one way to do this was ML-based production maintenance:

1) Pump parameters are monitored using different instrumentation like sensors
2) A Supervisory Control and Data Acquisition (SCADA) system monitors the multiple pumps. The SCADA system feeds processed data to a database. Fiber optic cables were used to transfer data to the ML system
3) The ML-based monitoring system was developed and used to process the large amounts of data
4) However, data quality, signal quality, and delayed responses in change in flow of oil in the pump can change the data received

But once it was decided to install ML-based modeling, there was a need for a large amount of cleaned data. The data that were captured for long time was used by the ML system to learn the patterns. The engineers cleansed the data before feeding it to the ML system, post which the system used the right data to predict the failures that would happen, which in turn would impact production. Also, based on historical data, ML model can predict if there is anything that could fail. So, to overcome the challenges in the data quality and the above reasons, we start implementing data pre-processing steps and needed filters to clean the data. Consequently, extra efforts are needed to cleanse the data to get the data to the desired level that ML algorithms would need.

Different ML-based algorithms were used to test and compare the different ML models for getting optimal results for the ML algorithm result, as shown in the figure below.

In this methodology, a new algorithm predicts the results based on stacking methodology where a new algorithm called blender or learner takes previous predictions as inputs and makes the final prediction usually more precise. The idea of this type of prediction is to combine results from different algorithms based on different approaches, which are good to operate in specific conditions into a new ensemble, which in combination reduces the chances of error.

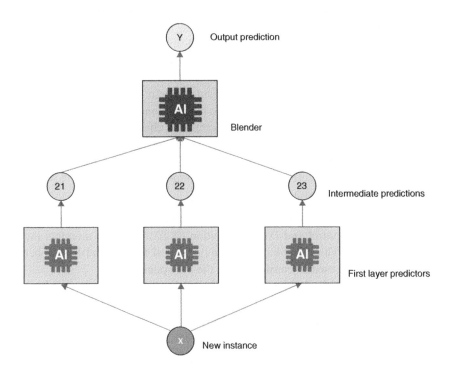

10.4 Detecting Anomalies by Enabling Automation in Assets using Robots

One of the important aspects in the oil and gas industry is ensuring safety of its employees, operators, maintenance personnel, and the assets. There are multiple factors which impact safety in the O&G industry which are in our control, while there are some factors which cannot be controlled like an invisible gas leak can impact the larger society around the asset before being detected and brought under control. For this reason, assets are trying to find innovative ways to detect safety issues and prevent them as much as possible. One of the ways to ensure safety is by conducting preventive safety audits using robots by enabling anomaly detection and control.

Plant operators and engineers visit critical plant locations with gas detectors regularly for inspection rounds. These rounds inspect the setups for any leakage and plan for action accordingly. These rounds are tedious and repetitive and prone to safety incidents, given the explosion risk, when procedures are not being followed correctly. To make the rounds safer and more efficient, companies have deployed unmanned terrestrial robots that can operate in explosion-risk zones for remote inspection and monitoring.

Many times, robots are used in lieu of humans in places where there are safety issues. In this use case, we can explore how robots are used in an oil company to detect safety hazards and use AI to ensure preventive measures are applied.

The most common uses of robots in O&G industry are as follows:

- **Inspection:** Inspection covers activities performed to gain a better insight in the asset integrity and maintenance requirements. Use robotics to perform risk-based inspection (RBI) for identifying (i) the type of damage that may potentially be present, (ii) where such damage could occur, (iii) the rate at which such damage might evolve, and (iv) where failure would give rise to danger.
- **Cleaning:** Using robotics for cleaning ranges from sludge vacuum suction in storage tanks to scale removal in pressure vessels and to spot cleaning for underwater inspection of assets or pipeline cleaning, etc.
- **Fabric maintenance**: Robotics in Fabric Maintenance is organized around blasting and painting. For example, these can be storage tank crawlers or spot repair painting drones.
- **Mechanical repairs:** Mechanical repair robots can be anything to apply force on assets (e.g. cutting and bolting) or additive manufacturing (e.g. welding).

The robots are equipped with optical and gas-detection sensors that are calibrated to detect gas-based leakages. Robots are deployed in assets to do a safety audit and compliance checks on a day-to-day basis. These robots can be used to gather important data and feed it to the engineer to take action to prevent safety-related incidents as part of preventive maintenance and anomaly detection

Robots carry out automated missions based on a particular theme or instruction given. For example, if the robot's mission is to detect pressure changes across the asset, the bot is programmed to run in a specific predefined path and do checks on specific indicators for detecting pressure changes.

Operation can be done in either two modes:

- **ROV (manual) operation**: An operator controls the robot steering and operations remotely, in real-time. The operator can see the robots live feed and take appropriate steps to modify any event inducing changes that can create a significant impact. AI/ML plays a significant role in this activity, where the system gets data from robot, scans it, and alerts the operator to respond. In future, it could be that the system could learn from the data that the robot sent and based on the data can provide specific instructions to reduce any impact.

- **Autonomous mission-mode**: Using the autonomous mode, robot operators can create "missions," where robots will move to a specific point of interest and perform an action. These actions can range from taking high-definition images from gas readings to recognizing a gauge indication. These autonomous missions are programmed to enable the plant operators and engineers to catch any potential anomalies in an earlier stage with minimizing the risk exposure to the engineers.

Operators use fleet management solutions to create robot missions, remotely control the robot and view the robot real-time video and data feed, and take immediate action to prevent safety hazards. The high-level flow of information is shown below.

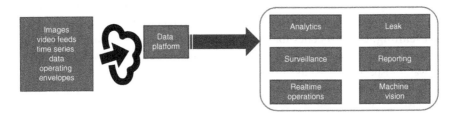

As shown in the figure above, images and video feeds are captured by the robots. This feed is then stored in a data platform based on cloud (AWS, Azure, and GCP) The data are then harmonized and cleansed. These data once cleansed are then stored into a data store which can be accessed by multiple applications for analysis and reporting.

The live feed from the robots can also be combined with data types like timeseries data to give an integrated view of all the important information. This will then alert the engineer for performing preventive maintenance.

Robots are designed in different sizes and shapes to be able to capture information from areas which are inaccessible to humans. These robots have HD cameras to capture images of instrument readings and relay it to the cloud. Then, the operator will be able to view the information in a format that he is able to understand and compare with historical information to make informed decisions. Alerts and events are triggered by the robots based on AI algorithms created using historical data and patterns created from the data. Once the alerts/events are triggered, the operator ensures appropriate steps are taken to prevent the incident from occurring or resolve the issue by making appropriate changes permanently.

10.5 Inspection and Cleanliness of Reactors, Heat Exchangers, and Its Components

Many a times fouling occurs in the pipes of instruments such as heat exchangers and reactors. Fouling occurs due to accumulation of scale, wax, sludge, minerals, etc. Fouling would increase corrosion rate in pipes. Fouling will cause heat resistance and restrict flow in pipes, thereby reducing effectiveness of reactors. To help engineers identify the issues in reactor pipes, any user-friendly dashboard such as PowerBI can be created for viewing of fouling in reactor pipes. Inspection videos of pipes streamed to cloud resources will be processed using AI models and will generate the outputs to be visualized on easy-to-navigate web dashboards using tools like PowerBI.

Real-time inspection videos will be recorded and stored in a cloud storage like Azure Blob storage. Images can be recorded via a borescope to generate real-time depth data during inspection recordings.

These images along with the metadata are stored in a database. The images and video files along with the metadata will be processed by ML algorithms and inferences made. The inferences will be published on a dashboard along with the images and videos. The inferences will be utilized to predict corrosion rate of the pipes, detect early failure, and identify opportunities for optimization.

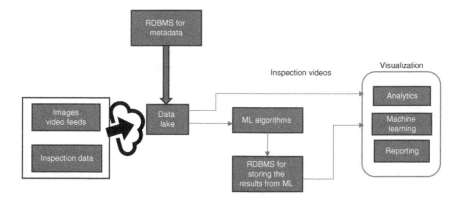

The images, videos, and other data are processed by an ML algorithm to be viewed in a visualization tool by engineers to understand the fouling and take the necessary actions and act on the recommendation.

10.6 AI-Enabled Training and Safety

Many times, when engineers need to be trained, the right infrastructure or content may not be available, the turnaround time may be very high to get the right trainings and be trained, there may be high cost of travel, and we may end up with poor training quality.

To reduce the issues faced by poor training, leading to poor execution of work along with safety hazards, many oil companies have turned to the virtual reality (VR)-enabled systems for training and maintenance. One of the key advantages of using a VR-enabled system is to get the engineers trained without the engineer leaving the office and stepping into hazardous zone.

Users input CAD designs, laser scans etc. An application is used to create 3D models of the site and other asset component models. Once the models are rendered, users can engage in multi-user collaboration in real-time via simulation of the model, ability to drop notes in the model, create procedures for review or instruction, and more. Once the VR models are created, users access the models in the application with a VR interface to create a 360° VR experience. The engineer or operator can get trained on a particular task for perfection. Even when maintenance work has been planned, the engineer can get to the VR system and get trained on a particular course for maintaining the component.

Some of the benefits of AI enabled training are as follows:

- Improved safety with hands on training
- Cost savings in travel and training quality
- Enable the ability to quickly create simulations for learning with minimal resource requirements by leveraging video recordings of environments as assets.
- Improved maintenance by the Asset Operations Team
- Operator training
- Reduced staff on site
- Remote engineering (3D lasers cans available and no need for extensive site visit)
- Reduction of project turnaround time and quicker project delivery

10.7 Summary

Today's drawbacks in AI will be solved in the next few years, and AI will become the game changing technology that many organizations will adopt. It will become more of a do or die situation for organizations with regards to adopting AI in their business processes. Data for AI will be made available for AI systems to learn and make informed decisions. The disparate digital ecosystem comprising IoT, Digital Twin, Sensors, Tags, etc. will soon become more matured and will start working cohesively.

AI has already had a significant impact on the world around us, but as more research and resources are put into the advancement of the field, we will begin to see even more of AI's influence in our day-to-day lives.

In the O&G sector, there will be more response-related AI being deployed. When the AI system feeds data to the engineer, the AI still sends the signal to the engineer. However, in future, the AI system will be able to take decisions on its own and create responses instead of waiting for the engineer to act on the issue/problem.

Safety being one of the key issues in O&G industry, AI will play a great role in the O&G industry. There could be robots with sensors being sent to dangerous areas in a refinery to feedback information about the existing conditions like gas leak to the engineer who can then minimize the safety hazard, there could be maintenance robots which could do preventive maintenance, and there could be robots soon replacing spares in the assets completely independently.

AI-powered drones can start giving more details about an asset such as giving details of visible safety issues like a huge protruding component which cannot be seen usually and can create safety issues.

While AI will grow by leaps and bounds and will also help develop the ecosystem, the key input is clean trustworthy data. No amount of AI will be useful if the data have data quality issues, hence the need to get the data structured, organized, and cleansed for proper use and benefits.

Bibliography

1 Wikipedia (2008–2023). https://en.wikipedia.org/wiki/Artificial_intelligence (accessed 24 September 2022).

2 Madrid, J. and Min, A. (2020). Reducing oil well downtime with a machine learning recommender system. Submitted to the program in supply chain management in partial fulfillment of the requirements for the degree of master of applied science in supply chain management at the Massachusetts Institute of Technology. https://dspace.mit.edu/bitstream/handle/1721.1/126390/scm2020-min-reducing-oil-well-downtime-with-a-machine-learning-recommender-system-capstone.pdf?sequence=1&isAllowed=y (accessed 12 August 2022).

3 Lisitsa, S., Levina, A., and Lepekhin, A. (2019). Supply-chain management in the oil industry. https://www.e3s-conferences.org/articles/e3sconf/pdf/2019/36/e3sconf_spbwosce2019_02061.pdf (accessed 23 August 2022).

4 Bonada, F., Echeverria, L., Domingo, X., and Anzaldi, G. (2020). AI for improving the overall equipment efficiency in manufacturing industry. New Trends in the Use of Artificial Intelligence for the Industry 4.0. https://www.researchgate.net/publication/340170245AI_for_Improving_the_Overall_Equipment_Efficiency_in_Manufacturing_Industry. https://doi.org/10.5772/intechopen.89967.

5 EDP science/https://www.e3s-conferences. org/articles/e3sconf/pdf/2019/36/e3sconf_spbwosce2019_02061.pdf.

6 https://www.researchgate.net/publication/340170245_AI_for_Improving_the_Overall_Equipment_Efficiency_in_Manufacturing_Industry.

11

Illustrating the 5G Communication Capabilities for the Future of the Oil and Gas Industry

11.1 Introduction to 5G Communication

The progression of cellular network generations is primarily influenced by the continuous growth of wireless user device data usage and the desire for a higher quality of experience. For addressing these challenges, cutting-edge solutions are to be developed. In summary, growth of 3D (as given below) encourages the development of 5G networks.

- Device
- Data
- Data transfer rate

The 5G cellular networks highlights and addresses the following three broad perspectives:

- **User-centric:** It provides device connectivity 24/7, uninterrupted communication services, and a pleasant consumer experience.
- **Provider-centric:** It offers connected intelligent transportation system roadside service unit sensors and critical monitoring or tracking services.
- **Network operator-centric:** It offers an energy-efficient scalable low-cost uniformly monitored programmable and secure communication infrastructure.

Hence, 5G networks are conceived to have the three primary characteristics as listed below [1]:

- **Ubiquitous congruence**: In the future, many different types of gadgets will be able to connect and provide a continuous user experience. In fact, pervasive connectivity will make the user-centric viewpoint a reality.

The Power of Artificial Intelligence for the Next-Generation Oil and Gas Industry: Envisaging AI-Inspired Intelligent Energy Systems and Environments, First Edition. Pethuru Raj Chelliah, Venkatraman Jayasankar, Mats Agerstam, B. Sundaravadivazhagan, and Robin Cyriac.
© 2024 The Institute of Electrical and Electronics Engineers, Inc.
Published 2024 by John Wiley & Sons, Inc.

- **Zero latency:** The existing system's real-time applications and facilities will be supported by 5G networks with zero delay tolerance. As a result, it is expected that 5G cellular communication will have zero latency or extremely low latency, on the scale of one millisecond. The zero latency will in fact understand the service provider-centric view in terms of the QoS it provides.
- **High-speed gigabit connection:** The lowest delay property could be achieved by using an increased connection for fast data transfers and reception, which will be in the gigabit per second range to users and systems.

11.1.1 Essentials of 5G Networks

A growing number of UEs as well as an increase in the bandwidth required for massive amounts of data transmission necessitate a novel enhancement to existing technology. This section focuses on the requirements of future 5G networks. The oil and gas industry has a significant impact with the rise in new technology. This gives a cutting-edge advantage to the companies to automate the processes with ease and accessibility.

11.1.1.1 Significant Increase in Device Scalability in the Oil and Gas Industry

The number of devices that can be supported by 5G is exceptional. These can pave new roads in automation with which the oil and gas industry can scale up and explore new ventures. The scalability of devices is pertinent to the industry, as it depends on a lot of intercommunication between the devices for smooth automation.

Soon, the explosive expansion of cellular devices, gaming consoles, elevated TVs, cameras, domestic appliances, laptops, linked transport networks, security cameras, robots, sensors, and wearable technology is expected to continue to increase in volume as a result of strong connectivity offered by 5G networks. As the 5G network can support a high data rate, these devices can be interconnected in large numbers without compromising on the QoS delivered to the end-user.

11.1.1.2 High Data Rate and Massive Data Streaming

A substantial increase in the number of smart routers will result in 100-fold more data trading than that in previous years, which will overwhelm the existing infrastructure. As a result, corresponding data transfer abilities of the network have to be improved, in terms of new framework methods, technologies, and its adaptation capabilities toward distribution density of end-users.

11.1.1.3 Spectrum Consumption

From the point of view of spectrum usage, the two different streams are used, one for an upper layer and another for a lower layer, which appear redundant as significant portions of the presently allocated spectral bands are underutilized [2].

As a result, it is essential to create a connect method of control that can improve spectrum utilization. Furthermore, spectrum utilization and its effectiveness were already maximized through spectrum broadening as well as novel spectrum utilization methodologies that are unquestionably required for improvement.

11.1.1.4 Pervasive Connectivity

Because of the global non-identical operating bands, pervasive connectivity necessitates UEs supporting a wide range of radio rats and bands. Furthermore, the major market is divided into the duration division duplex and frequency division duplex, and as a result, UEs must support both duplex options.

11.1.1.5 Zero Latency

Future mobile cellular networks are expected to support a wide range of real application scenarios. The internet and its supporting infrastructure with different levels of quality of service (QOS) must ensure negligible latency with parameters such as bandwidth, latency, jitter packet loss, and packet delay. The 5G network should ensure quality of experience (QOE) in terms of user and network–provider satisfaction versus feedback, as a result of which 5G networks are expected to provide authentic and lag time services with optimized QOS and QOE experiences.

11.1.2 Characteristic of 5G Networks

Wired connections of the 5G network are less cost-effective, consume low battery capacity, and also have zero bandwidth than 4G wireless technologies. 5G implements Ultra-Wide Band (UWB) systems to improve the amount of data transfer width with lower power thresholds. The frequency range in use for 5G as such is 4000 Mbps, which has the capability to provide 400 times accelerated data transfer rate when compared to 4G portable wireless channels. 5G systems can also endorse thousands of billions of links which can enable enormous machine information exchange, as well as support mega broadband smart device connectivity. Furthermore, 5G provides ultra-low latency of one millisecond, 90% efficiency improvement, 99.9% improvement in response rate, maximum throughput transportation speeds of 10 Gbps, and acellular data quantity of 10 terabyte (TB).

The sections that follow describe the key features of the 5G technology.

11.1.2.1 5G Connectivity Requirements

The 5G standard's fully integrated architecture is divided into two major elements:

1) **Non-standalone (NSA):** The NSA is accountable for the early stages of the 5G technology, as well as infomercial promotional activities that have been started by the end of 2019. Utilizing emerging 4G LTE architecture, the NSA

benchmark controls try to avert load balancing and transmission problems. The 5G network can be viewed as indeed merely a speedy data tubing linked to a conventional 4G LTE infrastructure. The National Security Agency (NSA) initiated the process, which will enable mobile operators to provide subscription business by 2019.

2) **Standalone (SA):** A brand new unique core infrastructure is 5G Standalone (SA). This shifted the network control transformation toward the 5G foundation and introduced significant modifications to system integration. SA would be accessible in 2020, with much more network control sectioning as well as sub-band encrypting. This is aimed to be more effective than 4G LTE and NSA, leading to lower transmission expenses and enhanced usability of the system.

11.1.3 Applications of 5G

5G wired connections are anticipated to support a diverse range of services and facilities which have negligible delay, high data transmission accelerations, as well as robust communication infrastructure. The following are the most influential implementations of 5G technology. The various capabilities of the 5G network are illustrated in Figure 1.

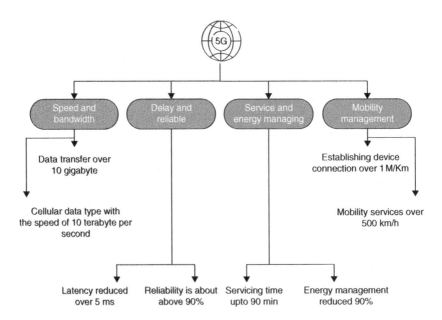

Figure 1 5G network capabilities.

11.1.3.1 Private Applications

One such realm of the 5G infrastructure could accommodate a diverse range of Access Points (AP), from externally powered ones to small embedded systems. On-demand information retrieval should also be managed to meet and sustain expected QoS.

11.1.3.2 Residential Virtualization

Because of C-RAN (Cloud Radio Access Network) frameworks, clients only have user equipment with reduced cost such as set-top boxes for TV and virtualized routers for accessing the network, utilizing physical and data connection layer facilities. Only those implementations within upper layers may indeed be migrated toward the data center for universal coverage or external provider information processing utilities.

11.1.3.3 Intelligent Communities

It really is a porous phrase for interconnected virtualization common in residences, workplaces, as well as in shops. As a result, all technological and electronic solutions, like control of temperature, alert alarm systems, printing presses, LCD screens, air conditioning units, physiological exercise equipment, as well as central locking, would have been interrelated in a way that cooperative activities will indeed enhance the experience for users. Correspondingly, smart shops will indeed aid the buyer with product description, selling adverts, and also item recommendations while on the move.

Challenges of 5G Networks Perception of 5G technology also is not easy to notice. As listed below, a number of issues must be addressed within this perspective:

- **Emergence of available bandwidth and channel strength to improving energy efficiency:** Using a number of base stations are implemented inside a specific region, utilization greater frequency bands, as well as connectivity developments could actively help in upgrading the capacity of the network to millions of UEs, broad bandwidth, high quantity of data, as well as effective fiber optic data access. Even so, placing such alternatives into the activity is indeed a time-consuming but also energy-intensive challenge. As a result, network bandwidth should be substantially increased, whereas preserving energy usage and infrastructure running costs must be kept under strict management.
- **Throughput and versatility:** The following are the most perceptible facets of modern cellular computing. Conventional mobile services and infrastructure as well as research methods should be constructed to a mechanism through heterogeneous networks. Moreover, many prospective clients could demand an

assortment of services at a given moment. As a result, the 5G infrastructure must be able to handle customizable customer requirements throughout its communication range.

- **Uplink and downlink served via a single stream:** A full duplex Wi-Fi broadcasting sends information as well as accepts messages at the same time and recurrence across a single transmission medium. As a result, a full duplex approach performs with the very same efficiency as having various uplink and downlink streams. The 5G network has increasing connection speed but still saving spectral range and funds, though full duplex processes really are not simple to deploy since a broadcasting should now utilize advanced procedures for maintaining effective communication.

- **Interference management:** Tackling inter-device interruption is a common problem throughout Wi-Fi communication technology. Because of the increasing number quantity of UEs, techniques including such as HetNets, CRNs, full duplex, and D2D communication as well as software solutions, interference would then significantly be found more in 5G technology. User equipment along with the 5G network could confront intervention from numerous macro-cell base stations (MBSs), numerous UEs, as well as small-cell base stations (SBSs). As a result, a cost-effective and dependable channel assignment method for spectrum access, voltage regulation, cell affiliation, and bandwidth allocation is needed to sustain QoS in the network.

- **Eco-friendly:** The current radio access network (RAN) consumes 70–80% of total power. Wireless technologies use a lot of energy, which causes a lot of CO_2 emissions. This presents a significant potential damage. As a result, it is indeed required to create energy-efficient communication infrastructure, hardware, as well as techniques in order to attain an equal and fair proportion of network bandwidth to power consumption.

- **Durability and reduced latency:** Data transportation in robotic systems, patient monitoring, existence safety devices, cloud-based online games, nuclear reactors, detectors, unmanned aerial vehicles, as well as linked transportation infrastructure always demands reduced delay and increased consistency. Even though having incredibly low propagation delay and accurate data distribution to a large number of nodes without rising communication infrastructure prices is challenging, it demands same growth methodologies that would provide better connectivity, fast deployments, and high data transmission rate.

- **Enhancement of network connectivity:** Maximum throughput in terms of geographical coverage, bandwidth efficiency, quality of service, quality of experience, convenience of interconnections, energy efficiency, response time, consistency, fault tolerance, and computational cost all seem to be essential performance metrics for a next-generation mobile network. As a result, a general infrastructure for the 5G technology must substantially enhance on all such metrics. Even though a few trade-offs between the acceptable threshold for abovementioned variables are tolerated, it is highly significant to provide an optimized and efficient network as the end goal.

- **Economic implications:** The great revolution through generation of cellular communication techniques may have significant economic consequences in terms of continuous integration and a personalized patronage with customer determination. Because of financial constraints, it has become essential to construct a completely new transportation system. As a result, the cost of infrastructure deployment, preservation, strategic planning, and procedure has to be cost-effective as from the viewpoint of government agencies, regulatory agencies, as well as internet providers. Furthermore, the expense of just using wireless links ought to be viable, such that gadgets implicated in device-to-device transmission really should not cost higher than the expense of employing a D2D facility. Furthermore, revenue growth is expected to be much lower than traffic growth. As a result, 5G networks must be designed in such a way that both network administrators and users benefit from it.

11.2 5G Architecture

The important objective of preceding cellular communication peers was to provide users through effective dependable voice and information services. 5G has expanded this possibility to provide a diverse series of Wi-fi communications to target users throughout multi-carrier portals and cross networks. 5G establishes a vibrant consistent and modeling approach of progressive technologies that can facilitate a broad range of applications. 5G employs intelligent architectural style of the radio access networks, no longer constricted by propinquity to cell towers, with unique integrations providing new data access points. 5G paves the way for heterogeneous versatile and virtual RAN [3, 4].

11.2.1 5G E2E Network Architecture

The 5G E2E network topology is depicted in Figure 2. It describes the 5G E2E system. Due to the bottlenecked Wi-fi connection, base stations are primarily utilized for the transition from 3G to 4G communication between base stations. Although while dealing from 4G to 5G, and the 5G E2E layout, because of the improvisations imparted to channels in 5G, the central server does not form a main bottleneck in the network [5].

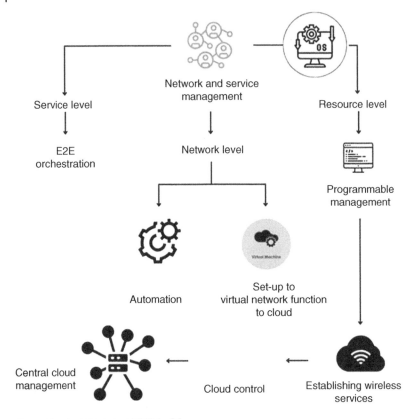

Figure 2 Architecture of E2E in 5G.

11.2.2 Network Slicing Architecture

In the 5G vertical equities paradigm, different operators can build their own network personalized for an intended task. Different data commitment models are used to provide the required infrastructure, varying from quite static requirements to some dynamical changes while procurement [6, 8] (Figure 3).

11.2.3 5G Mobile Network Architecture Design

The architecture includes a computer terminal and a number of self-contained radio system innovations. Any one of the broadcast methodologies is allowable in each port as a result of an IP link to the outside of the internet's environment. However, within the mobile unit, there is RAT, which requires a unique network interface. For example, we need to provide four separate RATs in order to get four distinct functions to related connections within the mobile terminal and activate all of these interfaces simultaneously at the same time in order for the architecture to function properly and deliver its intended purpose.

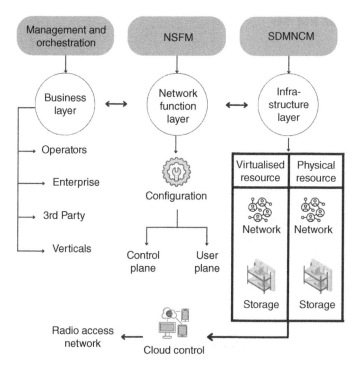

Figure 3 Network sliced architecture.

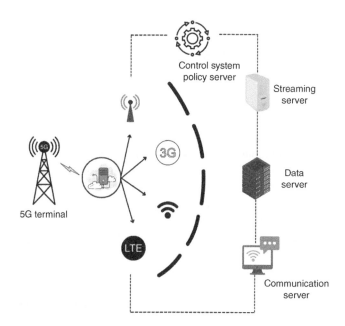

Functional Architecture of 5G

11.3 Antennas For 5G

The 5G digital cellular technique would provide higher data transfer, increased protection, reduced latency, and the latest previous experience within the sector and the environment. But besides a most recent large data analysis on 5G antenna elements, there remain numerous hurdles that involve several alternatives. Common antenna used during mobile telecommunication system applications utilize capacitor banks, underground cables, as well as steel rods and are therefore categorized as passive antenna elements.

While the most recent antennas suggested for 5G channels have been recognized as active antennas, this is a powerful technology which differentiates 5G networks from earlier networks throughout with regard to its speed, end-to-end delay, as well as protection of data.

5G networks necessitate more complicated antenna implementation and layout achievements which should ensure accelerated speeds and reduced delay. By carrying out a thorough examination of the literature, a few of the most effective antenna elements for such forthcoming 5G networks have been mentioned, together with their technical specifications and efficiency improvements.

11.3.1 Circular Patch Antenna with Three Notches

The three-notch circular patch antenna was intended as well as evaluated by millimeter-wave (mm-wave) pin feeding. This method utilizes within 58.5–60.5 frequency band. A mature antenna design does have a radiation efficiency of over 88% (percent) within the reverberance band and even a rational heading back loss that should be less than −10db. A three-notch circular intended antenna does have a highest proportional gain of 7.839 db at the 60 Ghz band and a form factor of $5 \times 5\,mm^2$ (Figure 4).

11.3.2 MIMO Dual-Band Eight-Antenna Cluster Layout

This type of antenna technique relies on the Stepped Impedance Resonator (SIR), and it comprises four combinations of L-shaped slot machines. For prospective 5G cellular modems, a dual-band eight-slot antenna of multiple-input–multiple-output (MIMO) has been used. A resistive proportion of the SIR is used in such directional antennas, which works as a double resonant frequency, allowing us to acquire signals by trying to adjust input resistance.

A simulation model project demonstrates that inter-component insulation higher than 11 db and also return loss more than 10 db were acquired. An overall efficiency of the proposed antennas seemed to be as in almost 51% from across long-term evolution (LTE) spectroscopy 42 and LTE spectrum 46. For both operating bands, the proposed MIMO antenna design accomplished a simulation model fully equipped with much more than 36.9 bps/Hz. Moreover, the premeditated envelope correlation coefficient (ECC) in between two antenna arrays is less than 0.1.

11.3.3 Leaky-Wave Antenna

Leaky-wave antennas (LWAs) eventually show a stunning response for reduced price for scanning and elevation, while preserving the qualities of high gain, recursive structure, as well as simple feeding. This proposed antenna performs within 37–43 GHz mm-wave spectrum. The effectiveness of this antenna comes to an estimated 85%.

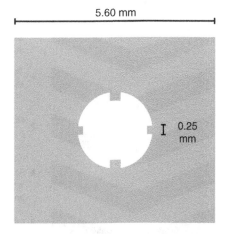

5.60 mm

0.25 mm

Figure 4 Circular patch antenna with millimeter wave Pin-fed.

11.3.4 Magnetoelectric Dipole Antenna with Circular Polarization

A high-efficiency circularly polarized magneto-electric (ME) dipole antenna is used based on the Printed Ridge Gap Waveguide (PRGW) technique. Its complete analysis produces a gain of more than 10 dBi on a frequency band of 31–35 GHz, with estimated 94% coefficient of performance reported on 34 GHz. A wide-band lens was utilized to enhance antenna performance. A lens used in this is composed of three layers, each containing a 3×4 mu-near zero (MNZ) atom. The printed ridge gap waveguide (PRGW) is an established millimeter-wave (mm-Wave) electromagnetic band technology. A participant for 5G broadcast mobile networks is indeed the Mm-Wave band from the 30–300 Ghz frequency range.

The following Figure 5 demonstrates a three-dimensional representation of a dual-polarized split-ring resonator (SRR)-aided magneto-electric antenna array.

11.4 5G Use Cases

It is vital for companies to identify areas at which 5G can add value to its business. Figure 6 illustrates some example use cases in the 5G realm that can deliver much more than the previous version network, but there remain few situations in which reliability, high accessibility, and reduced inactivity of 5G has a significant impact [9].

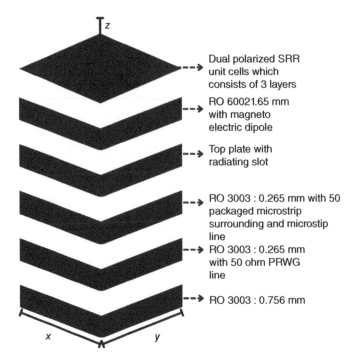

Figure 5 Magneto-electric dipole antenna loaded with a dual-polarized split-ring resonator. Three-dimensional representation of a dual-polarized split-ring resonator (SRR).

Figure 6 Use cases of 5G.

There are three kinds of users and the system in the context of 5G.

- Enhanced Mobile Broadband (eMBB)
- Massive Machine-type Connectivity (mMTC)
- Ultra-Reliable and Low Latency Connectivity (URLLC)

11.4.1 Enhanced Mobile Broadband (EMBB)

The EMBB is a relatively simple progression of enhanced client knowledge in device bandwidth. When millimeter-wave (mmW) frequencies are available, throughput speeds for eMBB could eventually reach 20 Gbps, resulting in a quicker and better user experience. eMBB will enable new data-driven experiences that demand high data rates. However, eMBB offers new potential and a more seamless user experience, in addition to quicker download rates.

11.4.2 Massive Machine-Type Connectivity (MMTC)

MMTC establishments are characterized by a variety of devices, including remote controls, sensors, and tracking of multiple components. The primary considerations for such systems involve relatively low costs and comparatively less computational power, allowing for extremely long battery life, enduring up to at least a couple of years. It is designed for working with a large amount of devices and working closely with the capability of sending short messages among themselves for coordinating the activities and completing their common goal.

11.4.3 Ultra-Reliable Low-Latency Communication (URLLC)

This type of communication is necessary for devices which cannot tolerate delay in functioning like automation in the industrial sector, advanced driver assistance systems, and augmented reality surgery. These devices require sub-millisecond delay with a successful data transmission rate of below one packet loss for every 10^5 packets.

11.4.4 Increasing Agricultural Productivity

Precision agriculture is already going to take place, with modern computing functionalities and IoT enabling information gathering, predictive analysis, and decision-making to reduce costs, reduce resource utilization, and increase yields. According to industry reports, 5G can expand the global footprint of smart farming and reduce prices by attempting to bring high-capacity connectivity to remote agriculture sectors.

11.4.5 Optimized Remote Knowledge Acquisition

The global pandemic restrictions that pushed remote knowledge acquisition highlighted the flaws in the current connectivity framework. Educators and students were made to rely on erratic and deceitful channels. As broadband providers expand their 5G networks, more groups must be able to benefit from the high speeds, improved efficiency, and durability of 5G connectivity.

11.4.6 Improved Supply Chain

The supply chain, including mobility, is increasingly utilizing IoT to track shipments as they travel throughout international borders and around the world. The sector has also been making big strides in the use of self-driving vehicles in warehouses and on the road.

However, the number of sensors necessary to transport and process all that data strains the 4G and LTE systems. As a result, in the current scenario, an industry's ability to capitalize on innovative logistics is limited. 5G eliminates that cellular capacity constraint, allowing the industry to increase its use of digital sensors and improve on its logistics and supply chain.

11.4.7 Advanced Medical Services

The health service system is yet another sector that is utilizing 5G to promote and enhance its operations. 5G can enable the vital and existence use cases that are common in the diagnostic space. Care providers use the 5G in a variety of ways, including data analysis, health monitoring, telemedicine, and technology surgery. 5G technology can provide same reliability and strength of a fixed-line network, which is much essential in medical service delivery.

11.4.8 Advanced Production Methods

5G has a great deal of potential for the industrial sector, which needs higher mobility than wired networks while maintaining the high-capacity, high-reliability, and low latency requirements of the industry. 5G enables automated manufacturing operations to be transformed more quickly to meet changing business needs.

11.4.9 Revamped Mining, Oil, and Gas Operations

5G can aid in the modernization of processes in the resource extraction in the oil and gas industry, which are frequently situated in rural and extremely remote locations. Because of costs and procurement, many facilities are unable to install wired networks, and they cannot rely on 4G/LTE channels to disseminate the massive amount of data generated with acceptable performance. More companies are switching to 5G because it can endorse the large-scale industrial IoT (IIoT) rollout required to track workplace conditions and guide automated machinery. In addition, 5G, in conjunction with edge devices, may assist the oil and gas sector in capitalizing on the massive amounts of data generated by machines.

11.4.10 Customized and Productive Retail

Vendors are attempting to improve customer engagement and create more personalized experiences by combining digital and in-person services. Many individuals are resorting to next-generation innovations and 5G interconnection to enable this. Wearable technology helps customers control home furnishings and devices in their own residences and sensor systems supplying data to business intelligence systems to more effectively streamline the supply chain, so customers do not find excess inventory, and personalized advertisements geared to each customer's special circumstances are examples of this.

11.4.11 Smarter Administration Presidency and Systems

Communities around the globe are utilizing a variety of tools to create "digital framework," in which buildings, infrastructure, and users are all linked to ensure that everybody and everything connects securely and seamlessly as much as possible. A city, for example, can collect and analyze end-user data regarding pedestrian traffic and then utilize the knowledge to public transport operators around densely populated areas as need arises.

11.4.12 Efficient and Productive Infrastructure

Advanced technologies depend on a massive IOT infrastructure, with endpoint sensors, end devices, and analytic tools spread across large geographic areas. This necessitates fast and robust interconnection provided by 5G.

According to the experts in the 5G network, a super-intelligent grid enables tools to decrease operational expenses and enhance the strength and resilience of power generation. A utility, for example, can use sensors to analyze the potential for forest fires instantaneously, improving preventive action and accelerating reaction times in the case of an actual crisis. This can then assist utilities in avoiding service interruptions and preventing major harm to their infrastructure [9].

11.4.13 Massive Support for the Labour Force

5G has the chance to have a major effect on various organizations across the industry segments. It can permit transmission over longer distances by providing a much more rapid and secure link with ample perseverance to support even expedition telecommuting to remote locations. It also enables direct collaborative efforts through the use of virtual reality and other innovative technologies. It also supports training that makes use of virtual reality and other similar simulation tools.

5G also provides enhanced worker protection measures. For instance, a industry can use a data analytics system powered by 5G communication to analyze and respond to crucial safety hazards, such as attempting to prevent equipment from starting, if the sensor recognizes a worker who is missing safety gear [5, 7].

11.5 5G and Digitalization in Oil and Gas

11.5.1 Intelligent Automation and Economic Contributions of 5G Networks

Production industries are shifting toward digital revolution for a wide range of reasons, including additional profits through better customer service, increasing supply, beating the competitive market, decreased production costs through enhanced efficiency, as well as reduce the potential loss through increased safety and security. A new study identified a number of issues that need to be addressed toward digitizing industries.

Some of the use cases for in 5G digitized industry are as follows:

- Interconnection for thousands of gadgets that is ultra-reliable, resilient, and instantaneous.
- Low-cost equipment with increased battery lives.
- Asset tracking across dynamic production processes.
- Remote health and safety operations are monitored.
- Using AR/VR to optimize customer interactions.
- Using AI to improve processes in numerous regions or across the organization.

In summary, 5G platforms allow network operators to generate virtualization that is customized to the requirements of specific applications, such as

- Internet communication such as communication via mobile broadband, multimedia and entertainment, and the world wide web.
- Machine-to-machine (M2M) communication such as online shopping stores.
- Response time that is predictable in domains such as commercial vehicles, diagnostic implants, and building automation.

11.5.2 5G for the Internet of Things (IOT)

11.5.2.1 Structured Internet of Things

The IoT is a communication system that broadens the connected environment into all kinds of sensor devices. This includes it all from automobile engines, jet aircraft, offshore oil drills, washers and dryers, coffee machines, cellular telephones, fitness trackers, and much more. IoT gives even separate devices greater computing and methodical capabilities. IoT is a fast-changing innovation that several sectors are eager to adopt in order to increase their productivity. Smart connectors, wireless data, and data storage facilitate ubiquitous interconnection, altering our perceptions of the world around us.

11.5.2.2 IOT Architecture and Working Principle

- **Wireless sensors and network layer:** Motion detectors, hydraulics, and core network, radio frequency identification tags and interconnection channel are all part of this model. They are the essential "items" of an IoT system that collects real-time data. Sensors easily record data, and then it is passed as information to be analyzed. Actuators intervene in physical reality by turning off lights and adjusting room temperatures. Controllers and actuators encompass and adapt every component of the physical realm that is required to gather the essential data for further analysis. The detailed step of this process is shown in Figure 7.
- **Internet and data acquisition system:** At this phase, data are both stored data-based and segregated as clustered. Data acquisition systems (DAS) communicate to sensing devices and gather information. This stage reduces the massive amount of data collected by sensors to a reasonable amount for further assessment and analysis. Access points are accountable for routing information from the sensor, provide interoperability, and reliable data link connection and passing it onto the following layer for further processing.
- **Data center and application layer:** The principal research methodology, strategic planning, and memory procedures occur in the central cloud. Such a process enables for in-depth processing as well as analysis, and based on the findings it can even modify responses.

11.5.3 IoT and 5G in Retail Applications

11.5.3.1 Tracking of Foot Traffic

Inside a smart store, video-based pedestrian traffic surveillance may be used to evaluate whether consumers are spending more time for one product area than the others. Next, in real-time basis, individuals could indeed instruct an affiliation

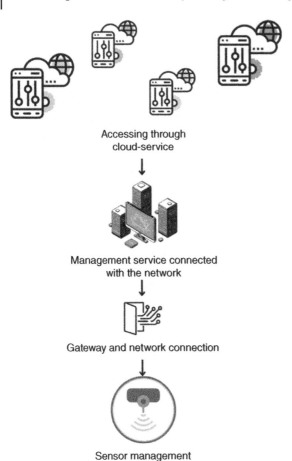

Accessing through
cloud-service

Management service connected
with the network

Gateway and network connection

Sensor management

Figure 7 IOT architecture layers.

to support the client or analyze the information subsequently to enhance window displays to facilitate even more effective customer visits. Moreover, individuals can personalize in-store buying experience through surveilling store data as well as customer needs through real-time.

11.5.3.2 Automated-Driven Depository

In online shopping, demand–alert warehouse fulfillment is about automation technologies powered through high-speed network-connected online services. The IoT enables you to monitor selling and inventory management throughout real-time basis. It indeed allows you to keep record of current in-store revenues.

RFID is an established and proven element of IoT, which may be used for further precise level of service enhancement as well as stock management.

11.5.3.3 Prescriptive Device Management

Inferential device management aims to manage energy, prediction equipment malfunctions, as well as preclude numerous different problems before they happen. For example, a grocery item contains a significant quantity of sophisticated equipment, including refrigeration systems as well as ventilation systems. When these components are equipped with sensors, the staff could indeed understand future servicing problems that could reduce energy usage to save money and even control temperature to maintain food safety.

11.5.3.4 Intelligent Enlightenment

Relocating commodities, effectively based on changing customer demands, is indeed a key objective in retail. IoT can assist in mass transit servicing, monitoring, as well as network optimization to achieve this goal. Furthermore, numerous retail stores have used the global positioning system (GPS) to monitor and route goods transport for decades, but now with IoT, people could indeed recognize how similar a pallet of commodities is provided to a particular shop with high precision.

11.5.4 IOT and 5G in Manufacturing Industries

11.5.4.1 Increasing Manufacturing Performance

5G would lead to increased efficiency inside this industry by improving production output by automating processes with technologies like IoT, as well as AI deployed on production lines. This will improve productivity even more by equipping AR layouts throughout workloads and making sure safety procedures via virtual reality (VR)-based strength and conditioning. Through the prediction as well as critical damage diagnosis, a number of problems will be reduced, which will lower the output. Digitalization as well as latest technologies will have a comprehensive influence on the manufacturing sector.

11.5.4.2 Obtaining Real-Time Utility Insights

Detectors supported by 5G might provide real-time insightful information on power failures and also power consumption in the oil and gas industry. Tracking systems linked to 5G will also provide building owners with such a wide visualization of their own power consumption, enabling them to utilize it more effectively. According to data analysis, 5G networks seem to use approximately 90% less resources for data delivery than 4G technology, although some variables including allowing massive network activity could indeed impact overall power savings of 5G. The utilization of 5G technology in drones to track, preserve, as well as investigate electricity systems would then lead to increased efficiency along with a decreasing impact on the environment.

11.5.4.3 Improving Agriculture Production via Data Integrity

A conventional employment sector like farming is increasingly dependent on technology even as the working population shifts toward the service industry. Emergence of 5G would provide farmers with accurate information regarding their crop production, resulting in an increased production of high standard. Devices and sensors, for instance, would provide real-time notifications on industrial equipment, infrastructure facilities, and much more. The time required to aggregate smart agriculture procedures and functions has reduced to a fraction of a second, thanks to 5G. Agricultural production will be more effective and productive than it has ever been.

11.5.4.4 Smarter Transportation

Smart traffic control, enabled through 5G technologies, could save drivers time and decrease fuel emissions. Sensor-based traffic control will assist operators through deciding most effective routes, cutting costs, fuel, as well as duration. 5G is also essential for automated vehicle processes, which include handling the large amount of data, for example, in vehicle-to-everything (C-V2X) communication systems. Via real-time route planning and high-speed information exchange, the public transportation system can be made more versatile and efficient.

11.5.5 5G in the Oil and Gas Sector

The oil and gas industry seems to be well-known for laying emphasis on accurate, safe, and adaptable networks to enable their operational functions. A firm's innovative business practices and its critical applications must be developed on rigorous operating systems. They offer comprehensive data to the control center point for actionable decision-making. Network administrators had to constantly evolve to provide a cost-effective method to meet application-specific services. Including its service-oriented virtual private network (VPN) functionality, Multiprotocol Label Switching (MPLS) has already been meeting that kind of prerequisites, besides integrating implementations.

Furthermore, 5G standalone mobile-virtual network protection workarounds as well as network slicing characteristics enhance network performance standards, in addition to data confidentiality, real-time regulation, network capacity, and fast broadband capabilities. This would be available due to 5G networks' autonomy from open networks, better control from the application layer, and versatility in developing connections for optimum as well as huge performance improvement.

5G systems have emerged when the oil and gas sector has been enduring massive change by implementing higher robotization via the use of technologies like IoT, machine learning, and robotics. Numerous manufacturing processes already are streamlined, and oil and gas businesses continue to push for higher efficiency as well as delivery performance through utilizing existing and developing communication infrastructure including 5G due to a highly competitive market.

11.5.5.1 Data Analysis and Video Monitoring

Physical securities in the oil and gas industry are just as essential as securing the manufacturing process. The Wi-fi monitoring system is indeed an efficient and tested method for overseeing the protection of oil and gas corporation campuses as well as drilling portals. According to recent market forecast, a worldwide surveillance video economy would surpass $60 billion by 2025. Wi-fi bandwidth empowers a setup of camera systems on unmanned aerial vehicles, which gives it access to areas that seem to be challenging to reach with the help of fixed connectors.

A video camera can also provide recordings that can be utilized to enhance consciousness as well as tactical decisions inside a variety of circumstances. Unmanned aerial vehicle (UAV) cameras, for instance, could effectively report seismographic changes, catching fire, as well as environmental catastrophes quite efficiently. Similarly, operations for search and rescue could just use sensor arrays as well as surveillance systems to review remote regions without the requirement for human involvement.

Wi-Fi surveillance footage requires a huge quantity of bandwidth, and as a result, oil and gas firms require a robust network connection as well as high bandwidth offered by 5G technology in order to implement high-resolution camera systems. 5G systems will also allow organizations to use 4K camera systems for video surveillance as well as assistance in accurate object recognition on production lines.

11.5.5.2 Governance of a Large Campus

Virtual Reality (VR), Augmented Reality (AR), and Mixed Reality (MR) are improvising customer perspectives and also are extremely relevant within the industrial world. AR and MR both are enabling collaboration for diagnostic tests as well as preservation of vital resources, ensuring its high level of utilization. The oil and gas industry can also employ AR and MR techniques to enhance employee instruction, which can increase employee performance. The VR technique, for instance, is now being utilized to recreate case studies, such that employees could indeed securely start practicing their abilities. It may even permit data transmission over longer distances, allowing to collaborate between employees at two different locations.

5G would therefore assist in attracting industries through providing the necessary bandwidth as well as capabilities for high-resolution and CPU utilization. Furthermore, 5G network slicing would then allow the application to have specified quality of experience (QoE) to maintain continuity. 5G would provide coherent transfer of information for AR technologies while retaining ultra-low transmission delay and good stability. Edge computing, on the other side, would then confront transmission delay even while maintaining the security of confidential documents.

11.5.5.3 Machine Administration and Navigation System

Oil and gas companies, like other industries, would prefer to employ remotely controlled machines such as drones, cranes, and robot arms. Among the top objectives of remote devices as well as security cameras would be to evolve toward unsupervised manufacturing environment through AI to reduce costs. This is not just to improve efficiency in demanding environments, but it is also to enhance safety at work and protecting employees from potentially dangerous surroundings.

This industry is gradually moving toward completely automated remote locations, which perform its operations on a wide variety of machines. Portable machineries which work on remote administration want reliable, ultra-high bandwidth Wi-Fi connection over large distances. Current control strategy specifications could be managed by the LTE network, but future requirement can be handled only by 5G. A residential LTE network, for instance, might endorse video in high-definition channels, but 4K recordings are possible only on the 5G network. Quite pertinently, data-intensive control techniques such as tactile feedback, along with the status of remote-controlled equipment, would definitely require the utilization of 5G technology.

11.5.5.4 Automation and Robotic Systems in the Cloud

Digitization and industrial automation have been growing rapidly throughout enterprises, such as oil and gas, in such an effort to streamline mechanisms, make it more efficient, as well as completely avert risk of human error. But somehow it necessarily requires constant monitoring for low latency and a substantial frequency band, along with an ecosphere of machineries which endorse LTE and 5G in their components. Wi-Fi eliminates the restrictions of static factory manufacturing lines, enabling machineries to be become portable, and therefore reducing the time it takes to completely rebuild the development environment as well as modify production lines as requirement arises. 5G networks are required to enhance effectiveness and provide more opportunities for sophisticated automated processes. Moreover, the distributed system would then facilitate oil and gas producers to adopt robotic controllers both in manufacturing and non-manufacturing configurations. It thus eradicates the requirement for local controllers with programmable logic as well as provides new possibilities for cost savings and device co-ordination.

11.6 5G Smart Monitoring Instruments

5G does not quite inherently describe manufacturing processes; instead, it functions as a catalyst for novel scenarios as well as to improve upon operational processes. The following are few of the most pertinent 5G scenarios related to the oil and gas industry.

11.6.1 Multi-Vision Smart Surveillance (MVSS)

MVSS is indeed a 24-hour emergency live tracking webcam along with AI and adaptive zooming functionality that really can analyze video and identify behaviors over 5G. It has the ability to identify traffic congestion as well as undertake behavioral analysis including such tripwire crossing, quick motion recognition, unauthorized parking, and neglected object recognition. Such scanners are typically fixed and assembled around industrial sites or public places to enhance field protection and safety.

Important aspects of MVSS are listed as follows:

- Multimedia validation of an alert system.
- Batch processing for Incident Handling.
- The centralized command core helps control the camera systems.
- Each and every location can be tracked in real-time.
- Virtualized capturing of alarm systems or activity video files.
- Geographic location confirmation.
- Endorse for full-featured virtual servers, as well as web and mobile consumers.
- Enable for third-party Digital Video Recorder (DVR), Network Video Recorder (NVR), and Video Management System (VMS) integration.
- Emergency alert messages via SMS or email.
- Audio- and video-based predictive analysis assistance.

11.6.2 Wearable Webcam

A wearable camera transfers video, audio, as well as information to an emergency customer service center like a police control room. One such webcam could provide real-time video streaming throughout foot patrolling for accident investigations, proof recording, and live control interaction, utilizing the 5G high-speed information dissemination system.

11.6.3 3D Augmented/Virtual Reality (AR/VR) for Diagnostics and Preservation

The Digital Helmet will be employed to continue providing 3D augmented/ Virtual Reality features across a 5G access network in order to enhance workforce productivity and efficiency. AR/VR would be capable of obtaining remote location help and support for following scenarios:

- Allow for hand-free document analysis.
- Video surveillance guidance and specialist collaborative effort.
- Enable potential workplace strength and conditioning and field procedure.
- Resolving technical and visual predictive management.
- Recognizing objects.
- Real-time monitoring for healthcare devices.

11.6.4 Unmanned Aerial Vehicle Investigation and Video Monitoring

Unmanned Aerial Vehicle (UAV), further known as drones, can be used for real-time video broadcasting for scenarios like wild fire investigation over 5G to enhance emergency management. Within the scenario of an emergency, it is highly necessary to focus on providing real-time surveillance footage to gather field information and ensure accurate investigation.

11.7 Conclusion

Through its incredible data transfer speed, superior reliability, and high throughput, 5G might very well flourish as a fresh era inside the digital environment. It will have a major effect on our society since it offers higher connections, improved throughput, as well as ultra -low latency than existing communication devices.

Informing the end-users well about IoT devices and information security seems to be vital since it will give them a clear knowledge of a device's security considerations. 5G will have an impact on every industry, making safe and sustainable means of transport, virtual medical services, and computerized supply chain. It has the potential to construct the infrastructure of energy firms, including cloud-based solutions, Big Data analytics, automation evaluation, and unmanned aerial system monitoring, as well as enhance on virtual worlds.

The oil and gas industry can leverage the power of just what one's contemporaries are undertaking in international markets and simulate such a situation to check its feasibility with our business model. Collaborations among businesses, distributors, as well as shareowners may also be established to expedite and streamline 5G implementation throughout the company to obtain a balanced, secure, and stable enterprise.

References

1 Panwar, N., Sharma, S., and Singh, A.K. (2015). A survey on 5G: the next generation of mobile communication. https://doi.org/10.48550/arXiv.1511.01643 (accessed 13 September 2022).

2 Rogalski, M. (2021). Security assessment of suppliers of telecommunications infrastructure for the provision of services in 5G technology. *Computer Law & Security Review* 105556. https://doi.org/10.1016/j.clsr.2021.105556.

3 Roy, S. and Pani, C. (2021). SDN integrated 5G network architecture: a review, benefits and challenges. *Advances in Wireless and Mobile Communications* 1: 1. https://doi.org/10.37622/awmc/14.1.2021.1-13.

4 Sengupta, R., Sengupta, D., and Pandey, D. (2021). A systematic review of 5G opportunities, architecture and challenges. In: *Future Trends in 5G and 6G* (ed. M.M. Ghonge, R.S. Mangrulkar, P.M. Jawandhiya, and N. Goje), 247–269. CRC Press.

5 Noohani, M.Z. and Magsi, K.U. (2020). A review Of 5G technology: architecture, security and wide applications. *International Research Journal of Engineering and Technology (IRJET)* 07 (05): 3440–3471.

6 Gramaglia, M. and Kaloxylos, A. (2018). 5E2E architecture. (accessed 13 September 2022). https://doi.org/10.1002/9781119425144.ch5.

7 Singh, R.K. and Bisht, D. (2017). Development of 5G mobile network technology and its architecture. *International Journal of Recent Trends in Engineering and Research* (10): 196–201. https://doi.org/10.23883/ijrter.2017.3475.vmof0.

8 Abdel Hakeem, S.A., Hady, A.A., and Kim, H. (2020). 5G-V2X: standardization, architecture, use cases, network-slicing, and edge-computing. *Wireless Networks* (8): 6015–6041. https://doi.org/10.1007/s11276-020-02419-8.

9 Panek, G., Fajjari, I., Tarasiuk, H. et al. (2022). Application relocation in an edge-enabled 5G system: use cases, architecture, and challenges. *IEEE Communications Magazine* (8): 28–34. https://doi.org/10.1109/mcom.001.2100623.

12

Delineating the Cloud and Edge-Native Technologies for Intelligent Oil and Gas Systems

12.1 Introduction

There has been a dramatic transformation in software methodologies and paradigms over the last 10–15 years. What has been driving all this? The consumer web service was one of the early markets with scalable web services that provided more new, convenient, or automated ways to do things we did differently in the past (ride sharing, online retail shopping, smart home, content streaming, etc.). Many other markets in the Internet of Things (IoT) field are following the footsteps of this and increasing digitalization of the architectural patterns and paradigms, that paved the way here. As end consumers today, we expect our digital services to work and are available all the time. If a popular retail website would become unavailable due to a sudden increase in traffic, or our favorite streaming service suddenly could not stream our favorite TV show – it would clearly be a nuisance, but in the IoT production and manufacturing space, it could have a devastating impact such as hazards, accidents, and economic loss. In the IoT and manufacturing and production, availability is key. Many scenarios in this market are to ensure optimized uptime of production, while retaining safety with high constraints on monetary spending to accomplish this objective. There are however many additional reasons for digitalization in this space, which we will also touch on in the subsequent chapters.

The cost of computing in general has decreased dramatically, which has opened and scaled the compute space in an enormous rate, with a vast number of compute nodes with a remarkably diverse set of compute capabilities to serve different workflows, usages, and scenarios spanning from very constrained leaf or end devices to scalable server platforms. This has led to a massive transformation of

The Power of Artificial Intelligence for the Next-Generation Oil and Gas Industry: Envisaging AI-Inspired Intelligent Energy Systems and Environments, First Edition. Pethuru Raj Chelliah, Venkatraman Jayasankar, Mats Agerstam, B. Sundaravadivazhagan, and Robin Cyriac.
© 2024 The Institute of Electrical and Electronics Engineers, Inc.
Published 2024 by John Wiley & Sons, Inc.

traditional usages or scenarios and jobs becoming digitalized from previously being manual, error-prone, and potentially more prone to increased predictability, lower cost, and enabling a high degree of insights and observability.

It has furthermore helped identify new areas to monitor and digitalize to overall address specific business challenges that were previously not conducive with manual labor or truck-rolls (e.g. predictive maintenance). The affordability of computational resources in general from sensor devices, wireless technologies, free open-source software, to more capable standard computer devices and servers has also enabled a broader market to adopt and digitalize solutions that previously were not economically feasible.

With the increased number of computing devices that must be connected and available 24/7 to support the different markets and the increased number of users – there is an ever increased need to ensure a modern and efficient methodology and paradigm that can ensure reusability, scalability, efficient development, and delivery efficiency of the underlying software that enables the different usages. In this chapter, we will discuss what cloud and edge native technologies are and how they are used and applicable from cloud to edge today in the oil and gas (O&G) and IoT market.

We will explore and discuss the different tiers in which computing can be done. There are many reasons that may drive the need to do computing closer to where the data are emanating from – which is many times the edge, therefore the term edge computing. The processing of data may have to be done closer to the edge for latency reasons, privacy / security, or others. The second aspect of this chapter, i.e. edge computing, that will be discussed in more detail.

12.2 Cloud Native Technologies – Motivation

In cloud native architectures, we strive to develop, build, deploy, and maintain distributed applications like the big FANG (Face, Amazon, Netflix, and Google). We will explore some of the key pillars or foundational elements when it comes to cloud native architecture.

There are four main elements here that make up any cloud native architecture.

- Containers
- Microservices
- DevOps
- CI/CD (Continuous Integration, Continuous Delivery)

Our objective and motivation here is to pave the way for scalability, and in particular on-demand scale and elasticity (i.e. bring up and down computing resources and associated services when needed), enable automated and quick delivery of software focused on what needs to be updated, and software packaging with decoupled dependencies on platform and OS software. "CNCF is an open source vendor neutral hub for cloud native computing [1], [2]". All of the details including code of conduct and details can also be found in their github location [3].

12.3 Containers

Container technology plays a pivotal role in the cloud native ecosystem in general. It provides the portability of software and eliminates a lot of the underlying platform and operating system dependencies that traditional software architectures and methodologies have used. Think of a container as a complete bundle of the software, component, or service. A container provides the facility for the software to include all its dependencies, custom libraries, frameworks, etc. into one bundle that can be ported from one system to another.

All cloud native software are packaged and bundled together as containers and are essential when we discuss resource orchestration in subsequent sections. Containers really became prevalent in 2014 with the introduction of docker containers, although there are plenty of other container technologies out there, each providing its own benefits to the ecosystem. Docker quickly became the defacto container technology in the market, although there are other popular choices such as BuildKit, containerd, LXD, and Buildah.

12.3.1 Containers and the Operating System

We have already mentioned that containers provide a mechanism to package and deploy them in different environments and infrastructures, ensuring that all runtime dependencies of the container are included in the container image. Containers need an infrastructure to manage the lifecycle that supports starting and stopping them in addition to control permission and resource management of containers running on the system. The facility providing this is often called the container runtime. The container runtime provides an isolated execution environment for each container – like that of a virtual machine but more lightweight and with more limitations.

Figure 1 shows a representation of how containers can be viewed from a platform perspective.

The container engine is installed on the operating system and provides a CLI (Command Line Utility) to control and manage the containers. Containers are provided in an isolated environment where they by default see themselves as the only application running on the platform, oblivious to other containers that may

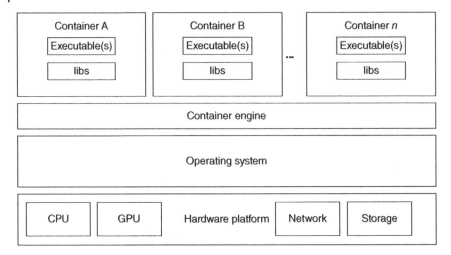

Figure 1 OS and container runtime.

be running side by side. The container engine provides this environment for the container, and the way it is enabled can vary from container engine implementation. On Linux, all the platform resources are managed through the operating system and governed by the OS kernel. Through the file system you can see attached peripherals, information about the CPU cores, memory, etc. There were two popular technologies that really enabled the lightweight isolation environment that containers provide: namespaces and cgroups.

12.3.2 Namespaces

Namespaces in Linux provide a way through which hardware resources managed by the kernel can be abstracted to allow for different processes to see one set of resources, while another process sees another set of resources. This provides a basic mechanism to isolate processes from each other. Examples of resources provided within the namespace feature are process IDs (PIDs), network, and IPC (Inter-Process Communication). This mechanism also provides a guarantee that there cannot be a resource conflict due to this isolation. Let us take the PID namespace as an example. When a process is launched in a namespace and spawns child processes, that process or child processes could have been assigned the same PID in parallel branches of the PID namespace tree.

12.3.3 cgroups or Control Groups

Control groups provide the facility to limit resources within different namespaces on Linux, which is a particularly important part of the runtime aspect of containers running on the system. This provides the system administrator with powerful and fine-grained control when it comes to, for example:

- CPU quota. Control how much CPU time is allotted to the container and control CPU core affinity and exclusivity to a particular container that may have real-time requirements.
- Memory allocation provided to the container.
- Network control such as policies that control how the network is set up, if the network is isolated to the container itself, is authorized to access external services, communicate with other containers, etc. There could even be possibilities to control things like quality of service (QoS) such as rate control and TOS (Type of Service) at the IP layer through traffic control facilities provided by the operating system.

12.3.4 Container Interoperability

There has been standardization efforts in the container technology space to help provide compatibility between one container runtime and that of another. The OCI (Open Container Initiative) provides interoperability of the container runtime through two specifications. The image specification ensures that a container image is represented and stored on disk or other non-volatile memory in a consistent format to allow for vendor interoperability (e.g. a container generated and stored by one container runtime is compatible with another OCI-compliant container runtime). The runtime specification defines a set of primitives that every container runtime should be implementing, required to support the lifecycle of the container such as starting, stopping, pausing, and deleting.

12.3.5 Container Images

Containers are stored and represented on the file system as layers which have some extremely useful practical aspects. The specifics in terms of formatting and representation are covered in the image part of the OCI specification.

Figure 2: Container representation illustrates conceptually how a container is organized and stored on the file system. The container consists of one or more layers, where the top layer is referred to as the container layer and the others are referred to as image layers. Each layer defines a set of files, libraries, and other files, with each additional layer adding to what was defined in the previous layer. The joint set, or union of the content across all these layers exposed, defines the entire image of the container. The operating system mounts the union set across all the layers to one union view across all the artifacts that defines the overall container image.

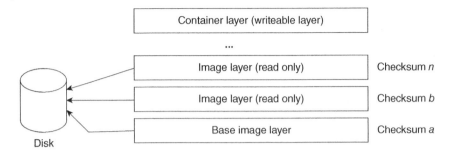

Figure 2 Container representation.

Each container is stored exactly once on the filesystem and has an integrity checksum and unique identifier associated with it that is used to ensure that the image is intact. The fact that each layer is stored exactly once is important as it allows for layers to be reused and referenced to the disk without storing another copy of the same helps reduce filesystem usage. Another important consequence of this is that when a newer version of a container must be downloaded from a remote storage location, it only needs to download the layers that are different from those of the previous version. This also helps reduce time and overall network usage. Conversely, if a container image is modified and pushed to a remote registry or storage location, only the affected layer of the new revision is uploaded.

12.3.6 Container Registries

Containers are commonly stored and managed through container registries that can be deployed and managed on-premises or as a storage service from a cloud service provider. They commonly comply with the Docker Container HTTP v2 API specification, which provides the core set of primitives needed to pull, push, and manage containers from the hosted registry. There are several vendors that have built their service around the Docker Container HTTP v2 API and then extended and provided additional functionality and capabilities such as security hardening and more granular access control.

12.4 Microservices

The microservice design pattern provides a scalable foundational loosely coupled architecture that breaks some of the issues mentioned previously. There exist many different definitions around microservices and the direct properties and benefits that come with such architecture. In this chapter, we will explore the architectural design pattern and give a few examples to that end. There are several adjacent technologies that have helped the adoption of this design pattern such as container technologies and resource orchestration. We will discuss those too in the upcoming sections of this chapter.

12.4.1 From Monolith to Microservices

One of the foundational technologies and design patterns is around containers and microservices. It has been a long journey from the software monolith, which was composed of a set of static or dynamic libraries, delivered in its entirety many times with underlying OS and system-level dependencies that could cause interoperability and back-forward compatibility issues with other software running on the same system. It made update processes more brittle and cumbersome, requiring more network and storage resources to manage, but it was a reality for an exceedingly long time – from user space software all the way down to firmware.

The Service-Oriented Architecture became popular in the hosted service space, which provides a domain functional separation of the services through dedicated fronted services that provide more flexibility, governance, and operational efficiencies for development teams but would still depend on common backend infrastructure services, where there is a service function dependency on the common infrastructure components, which requires that backwards compatibility and interoperability is retained across the horizontal set of services exposed by the SOA.

The microservice design pattern plays a pivotal role in cloud native computing and provides the necessary packaging and encapsulation to eliminate that cross-software layer dependency which allows for it to be deployed and operates more autonomously without those challenging dependencies. Figure 3 illustrates that change in thinking.

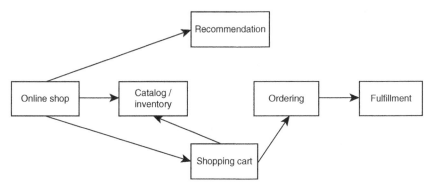

Figure 3 Online shopping.

12.4.2 When Did Microservices Start Emerging?

Around 2011, the well-known 12-factor app principles emerged, which describes twelve different principles and associated patterns for web applications and more generally for SaaS (Software as a Service) offerings, which has had a monumental influence and been adopted very broadly since it came out and is still relevant. Docker as a container technology came out around 2013 and is a key technology when it comes to the packaging- and platform- independent nature of any microservice, and finally in 2014 Kubernetes from Google was open source software, which originally came from the previous internal framework at Google called Borg.

12.4.3 Microservices as a Design Pattern

The reader is highly recommended to read up on the 12-factor application principles mentioned earlier as that provides an exhaustive set of principles to keep in mind and adopt when it comes to the microservice design patterns.

Let us start with a simple example to illustrate how one could enable a distributed web application realized over a set of microservices. Figure 3 outlines an elevated level, commonly use example for this purpose – a small online shop.

In this example, let us not worry about how these different services are communicating with one another or where and how they are deployed in the infrastructure. We will cover those details later in this chapter.

We have a fictional online shop, let us say selling books, where the user can browse and purchase books. In our example, we have a front-end web application that provides the overall experience to the end user and communicates with the different microservices. We have a microservice that handles the online catalog and inventory, which has a database encapsulated that contains information about what is in stock and general inventory management information.

We have a shopping cart that keeps information about everything the customer has placed in his shopping cart. The shopping cart may need to communicate with the catalog/inventory system and the ordering system.

The ordering system collects customer information and payment data for the items purchased and ensures that the payment or any discounts, etc. are applied and verified correctly. Finally, there is a fulfillment service which may be interacting with other backend, legacy, or third-party systems to realize and get the item out to the customer through the partner supply chain.

12.4.4 Characteristics of Microservices

A microservice-based architecture adheres to a set of common properties that characterizes the microservice and lends itself to the benefits they provide for an organization.

- **Specialized**: designed to perform a specific task or related set of tasks.
- **Contract**: exposes an interface to its consumers through an API (Application Programming Interface), commonly HTTP, gRPC, or a similar protocol.
- **Autonomous**: packaged and distributed in its entirety to the infrastructure, which means separate code base from other business logic.
- **Scaling**: allows for independent scaling of the service when demand increases or decreases to help conserve infrastructure resources
- **Discoverable**: provides means for discoverability by its consumers in the infrastructure through DNS, mDNS, or some other mechanism.
- **Resilient**: provides a higher degree of resilience by limiting the impact of failure to other parts (e.g. microservices) in the system.

Ref. [4] provides additional insightful detail on microservice architectures.

12.4.5 Advantages

An architecture like this comes with a lot of genuinely nice advantages. First, all the microservices are deployed and managed independently in the infrastructure. This means that the only dependency between them is the contract or interface a microservice is exposing to other microservices that are consumers of the interface. If the definition and semantics of that interface do not change the underlying implementation of the microservice, it is irrelevant to other parts of the system. This is a substantial difference between a microservice-based architecture and an SOA-based architecture, for example.

A microservice can be taken and reused in a different project or product all together much more easily than other architectural paradigms would allow for. This can be a huge advantage for companies that are building out multiple products with reusable elements as services across their portfolio.

The abovementioned model furthermore means that development teams may be able to work more effectively and efficiently, particularly if distributed globally as we commonly see in this industry. Each microservice is independently verified and rolled out in the infrastructure, which means that different teams may choose to implement the deliverables in a language that best suits the skillset of the team or the type of problem the microservice is solving.

Vulnerabilities and issues discovered in the system can be subjected to root cause analysis and addressed to a particular microservice and managed more in isolation than traditional architectures, which may have a much higher impact and blast radius when those issues occur. Because they do.

Finally, this architecture leads to the various parts of the system (i.e. microservices) to be scaled up and down as needed depending on load and users. In a more monolithic system scaling, you would be forced to scale the entire system based on the resource needs of the most demanding part. This enables economy at scale to a significantly different degree than traditional architectures, where you can allow for replicas of this set of microservices to be spun up and down depending on load and demand during the day. We will discuss in later chapters how that is realized and how it works in practice (i.e. how does an orchestrator help with service scaling and how does the system know when to scale up or down an instance of a microservice).

12.4.6 Disadvantages

While there are a lot of nice properties and benefits with microservices, one should make sure to understand what some of the drawbacks are with such architecture and the fact that it may not be the best option for everyone.

Microservice architecture comes with increased complexity and requires initial development teams to ramp up and understand a slew of modern technologies. The operational side of things must be taken into consideration as well, including the infrastructure needed to realize a microservice deployment, use of an orchestrator, scaling strategies, how upgrades and rollouts are managed, how the delivery system is integrated with the infrastructure, and need for pre-production / production separated environments.

As each microservice is containerized and integrates additional software, replicated across other parts of the distributed system, an amount of CPU, memory, storage, and network resources is lost due to the distributed nature of a microservice architecture of a solution compared to its monolithic counterpart.

Finally, debugging a distributed application across a set of microservices can be challenging, although there are many helpful tools and frameworks that can help in this space. That said, this aspect should not be neglected.

12.4.7 The Service Interface HTTP and gRPC

A microservice should focus on delivering a well-defined function or service to its consumers. It is commonly deployed and exposed through a RESTful (e.g. HTTP and HTTPv/2) interface or gRPC. There are no hard-set rules when it comes to the specific type of interface the microservice exposes, but HTTP and gRPC would be the most seen today.

HTTP and newer interfaces are of course the protocol that helped power webservices and standardized through W3C, which provides the RESTful paradigm the entire web is built on today. gRPC may be new to some readers, although it has been around for a long time. gRPC is a powerful Remote Procedure Call (RPC) communication paradigm open sourced by Google in 2016. gRPC is highly efficient in part due to its efficient on-wire representation of data through protobuf (Protocol Buffers) and supports a variety of different messaging paradigms such as

- Remote procedure calling
- Unidirectional streaming
- Bi-directional streaming

The gRPC consists of a client and server entity. The server exposes its functionality by defining the interface in an IDL (Interface Description Language) that provides the definition of the supported messages and related data types. gRPC implementation exists for most common languages today such as Go, C, C++, Java, Python, and Rust. The IDL is later compiled down to the client and server stub code that can directly be used within the software, allowing for the developer to focus on the task at hand, serving or consuming a specific service interface or streaming capability over worrying about efficient serialization and transmission of the data over the network.

12.4.8 Persistent Storage and Databases

We have already mentioned that a microservice should include all its dependencies required for operations and not share direct resources with other entities. This includes data as well. In other architectural patterns, shared databases, etc. would be used by multiple consumers, but when designing microservices, any need for a database should be contained within the microservice itself. This brings up the question of how to deal with consistency across a set of distributed services. There are several approaches to this problem, but a common design pattern here is the Saga design pattern which helps in getting consistency across a set of transactions. Each transaction should be following the ACID (Atomic, Consistent, Isolated, and Durable) properties. A microservice would perform its local transactions and publish a message to other microservices related in the Saga. If any microservice in the Saga fails to execute a transaction internally (e.g. it violates some business rule or aspects), other microservices involved in prior steps will undo their internal steps, leading up to the point in the Saga. Figure 4 illustrates this concept:

There are two primary ways in which Sagas are realized. These fall into **Choreography** and **Orchestration**. The main difference here is that the former is a more distributed model, where the individual microservices rely on a messaging system and are themselves responsible to determine what to do based on events from other parts of the system.

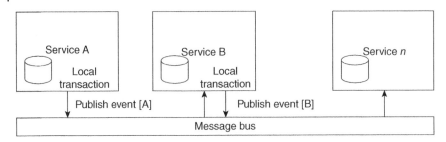

Figure 4 Saga pattern.

The orchestrated mechanism involves a centralized entity that helps inform the different microservices about local transactions to perform. There are several especially important aspects and considerations to understand based on the system, complexity, and anticipated new workflow and scenarios.

The reader should more carefully read up on these patterns if multiple distributed microservices involving databases and transactions come into the picture.

12.5 Continuous Integration, Continuous Deployment (CI/CD)

It is important for any business that is looking to embrace cloud native technologies and paradigms to be able to quickly release and deploy updates to the software in the infrastructure. CI/CD is a methodology that helps provide an automated faster path to getting new functionality and contributions integrated and deployed to the software and is a cornerstone in DevOps.

Figure 5 provides a high-level diagram outlining a general CI/CD pipeline. We will take a closer look at what this is and how it works.

A CI/CD pipeline is normally integrated with the SCM (Source Control Management) system like a git repository such as GitHub, GitLab, or similar

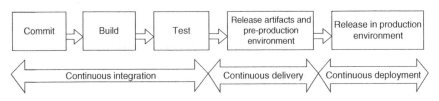

Figure 5 CI/CD pipeline.

offering. These git platforms provide facilities to define pipeline steps that should be executed in the two distinct steps of

- Continuous Integration
- Continuous Deployment

Figure 5: illustrates a simplified view of a CI/CD pipeline in terms of the stages. Continuous Integration (CI) is involved with integrating parts of the software from multiple contributors and performing various levels of automated validation of the code base such as unit-level testing, integration testing, and API level testing. A CI pipeline can also be used to automatically generate boilerplate client- and server stub code defined from a specification such as OpenAPI.

Continuous deployment is concerned with automating the delivery of the software to a production system once approved by quality assurance. The approval process takes place once built artifacts are provided through the CI pipeline and deployed and tested in a pre-production or staging area.

The CD process delivers the approved software to the production system, further decreases the feedback loop, provides the automation of releases, and hence increases developer productivity, such as being able to provide updates, address bugs, and fix vulnerabilities, providing the value to customers more efficiently.

12.5.1 Interface Specification

OpenAPI, formerly called Swagger, provides a YML-based specification tool that enables to specify RESTful HTTP-based services. OpenAPI is funded by the Linux Foundation and has a lot of active contributors and adoptions throughout the industry and founded by Google and Apigee. The user can define entire RESTful services that comprehensively define supported endpoints, operators, return codes, content-types etc. OpenAPI is commonly also used to translate the formal specification to user-friendly documentation (e.g. HTML and PDF) which can be published and can also be used as a tool integrated with the CI/CD pipeline to get the boilerplate client and server code generated. This allows for specification, documentation, and code to be in lockstep with each other. OpenAPI can be used to generate the code for many common development languages such as Go, Python, C, C++, Java, and Rust.

Figure 6: OpenAPI and CI Pipeline show a simple CI pipeline with a specification tool like OpenAPI and what that could look like in a continuous integration pipeline using GitLab or GitHub.

It should be mentioned that there are other API documentation tools other than OpenAPI that may be a better fit if the interface specification is something other than RESTful.

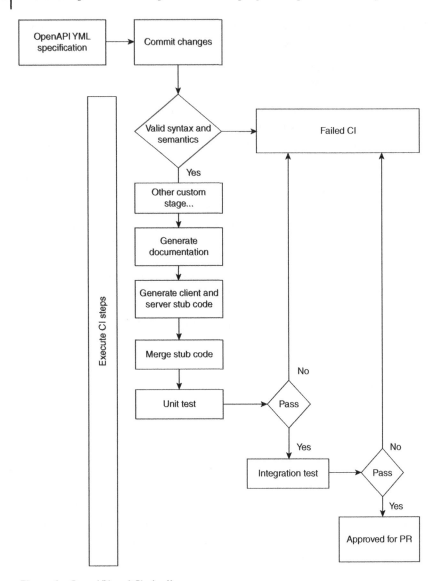

Figure 6 OpenAPI and CI pipeline.

12.5.2 DevOps

DevOps is the operational aspect of the previously discussed CI/CD pipeline, which is short for Development and Operations. The entire process of DevOps governs the key aspects of the steps outlined in the above CI/CD pipeline, with the goal of providing a fast release process of software with rapid feedback.

12.6 Edge Computing

We have discussed the key principles and pillars around cloud native development and will in this section shift our focus on how these principles can and are being applied in hybrid cloud to edge architectures and what considerations and benefits come with those distributed deployment models.

Cloud native software originally designed to run and scale well for the cloud could in theory directly be applied to the industrial IoT and particularly the oil and gas industry. The elasticity of the cloud when it comes to computing, storage, and other vital resources is important, but there are aspects of production and manufacturing systems that may not be well-suited to exclusively run on the cloud. We will explore those next, but first let us look at what edge computing is in general and establish a taxonomy.

The main notion of edge computing is to move the processing of data closer to the source of the data. If we look at a general diagram outlining the various parts of a hybrid edge deployment, there are multiple edges here where the processing of data could take place.

Figure 7: Network Edges illustrates a few of the network edges between the cloud service provider and regional data center to the edge devices. The distance between the edge and end devices increases as we move toward the cloud and regional datacenter, and the network latency reduces as we perform the data processing closer to the edge devices.

12.6.1 Motivation

There are several varied reasons why enterprises find it desirable to embrace edge computing and move more of the data processing and analytics closer to the

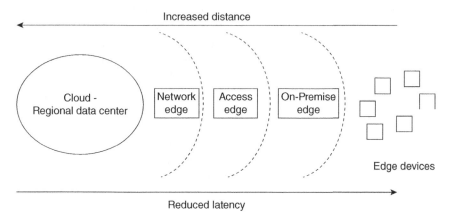

Figure 7 Network edges.

source of the data. The cost of processing data increases as you move further away from the edge devices, in addition to the latency. Cloud instances charge by the type of the computer required for processing in addition to storage (non-volatile memory) and network traffic (e.g. ingress network traffic to their data centers). Investing in infrastructure equipment closer to the devices (for instance, on-premise infrastructure) would reduce the operational cost over time and additionally provide several other benefits:

- **Latency**: Latency-sensitive workloads such as closed-loop control and automation may have tight latency requirements well under 1 ms. Certain deployments may require even more optimized computer devices and network fabric utilizing TSN (Time-Sensitive Networking) and TCC (Time-Coordinated Computing) to meet real-time performance requirements and provide a shorter, optimized response time.
- **Cost**: Analytics-based workloads processing computer vision/AI (Artificial Intelligence) and media workloads may consume a significant amount of network resources (e.g. consider a surveillance camera ingesting HD or 4K video frames at 30 or 60 FPS (Frames Per Second). Performing the analytics at the edge may be cost-effective as a result of doing the same analytics in the cloud. At the very least, ingestion and pre-processing of data at the edge and final analytical step done further away from the source would be beneficial. A video stream that includes a pipeline consisting of multiple steps could perform pre-processing steps at the edge, hence reducing the amount of traffic that has to be sent to the cloud (e.g. object detection, cropping the stream at the edge, and classification done in the cloud). It should be noted that an analytics workload may have its own latency constraints for various parts of the pipeline or pipeline that may drive the deployment of the analytics down to the on-premise edge even if cost was not a factor.
- **Availability**: Certain classes of deployments may not always have consistent, permanent backhaul connectivity to other network edges. In these situations where the connectivity may be intermittent and availability requirements are high, it may be critical to deploy important workloads that guarantee increased uptime at the edge for autonomous and independent operations if network failures occur.
- **Security/privacy**: There are reasons why sensitive data and information must be kept closer to the data source, or on-premises entirely. The type of assets that may be required to keep data undisclosed and minimize the threat surface and exposure to malicious adversaries may include sensor data, video, images, or even intellectual IP found in the implementation of services performing the processing or AI models. The cost of model training and general cost when it

comes to MLOps is substantial, and the protection of the resulting models can hence require huge investments that the enterprise needs to protect from disclosure. There may also be other compliance reasons enforced by IT/OT.

The other network edges, regional datacenters, and cloud may still be critical elements in the edge computing strategy as they offer benefits that the far network or on-premises edge cannot meet, such as compute and storage elasticity, external service exposure and connectivity, and scaling.

12.6.2 Edge Architecture

Let us take a closer look at a few different options when it comes to how the edge node (end device or edge server) is enabling workloads to be deployed and managed [5–7].

Figure 8: Edge Architecture Management illustrates an edge node connected to an orchestration – and DMS (Device Manageability Service) control plane nodes. We will discuss how these are used and how workloads can be deployed using both mechanisms depending on the nature and requirements of the workload and the rest of the edge infrastructure.

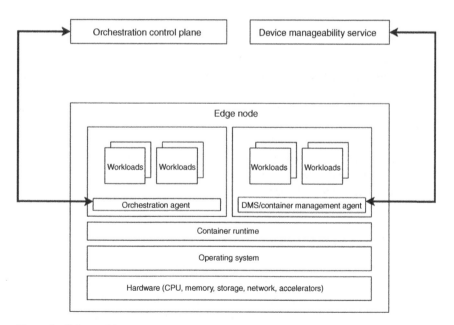

Figure 8 Edge architecture management.

12.6.3 Orchestration vs. Container Management

There are two common deployment vehicles for workloads or applications to an edge infrastructure. Edge devices are commonly enrolled to a Device Manageability Service (DMS) where each edge device has an agent running, which provides functionality to a centralized DMS through which the entire fleet of devices can be managed. One of the functions commonly exposed by a DMS is to deploy jobs, applications, or workloads down to the edge devices. These could be deployed in any manner the edge device is compatible for but commonly today is as containers. The DMS provides a single pane of glass down to manage every aspect of end devices.

This provides an efficient way for an IT administrator to push various applications down to edge devices, which is particularly useful when there must be an affinity between the application and the targeted edge device. You can think of this as a hub and spoke model where a DMS is the hub with spokes (established secure tunnels) down to each enrolled edge device.

Orchestration is another mechanism where an edge cluster consisting of an orchestration control plane is responsible for scheduling and deploying workloads across the nodes in the infrastructure. The scheduler in the orchestrator primarily looks at the set of resource types it is aware of such as:

- Compute/CPU
- Memory
- Storage

Many orchestrators, such as Kubernetes (K8s) and derivatives, introduce custom resources (by default unknown resource types) which the scheduler can also consider when making placement decisions.

Nodes may be heterogeneous or homogeneous where the scheduler can make placement decisions not only on available resources but may look at unique properties and characteristics of nodes in the infrastructure.

It is common to see deployments where an IT/OT persona has both manageability and orchestration to their disposal as mechanism through which workloads can be deployed to the end nodes.

12.6.4 Device Manageability

We introduced the basic elements of a DMS in the previous section and explained that the DMS could be used to push software down to edge devices. Device manageability in general deserves more detail as this is an important topic of edge computing and manageability of devices on-premises and out in the field. The importance of being able to access and manage the devices remotely reduces truck rolls or to otherwise dispatch personnel to local or remote locations to perform maintenance on devices.

There are fundamentally two classes of device manageability which we will cover here from a definition standpoint.

- In-band manageability provides a management channel to the device, like the previous diagram, which requires the operating system and DMS agent to be running and responsive. This could be considered the normal operational mode of a DMS where the system is working as expected. However, if one would have to remotely access the BIOS of the device or reboot the device if it stops responding – in-band manageability falls short.
- Out of band (OOB) manageability provides an exclusive path to the device, sometimes through a dedicated network. The OOB manageability on the platform is managed by dedicated HW and firmware on the platform so that the system can be accessed and managed even when the machine is turned off or the host has crashed or is otherwise not available. The OOB manageability is standard on server class platforms and available on certain standard CPU platforms, such as Intel vPro with AMT (Active Manageability Technology).

One obvious question may be how to reconcile the in-band and out-of-band manageability from an IT/OT perspective. There are software solutions that provide a comprehensive single pane of glass for the user as an abstraction and uses one of the two control paths down to the device.

12.6.5 Device Life Cycle

An edge device goes through a well-defined set of phases throughout its lifecycle, which maps down to day 0, day 1, and day 2 operations. Day 0 operations is referred to as the preparation of the system with respect to initial software, configuration etc. Day 1 is the first day in the production environment and is associated with a few unique workflows. Day 2 is ongoing days in the production environment and associated with its own set of workflows and required tasks.

Figure 9: Device Lifecycle illustrates the lifecycle of a device.

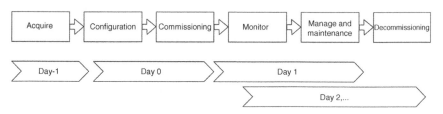

Figure 9 Device lifecycle.

- Device acquisition involved getting the hardware perform any inventory management related tasks, identifying deployment location, software, and configuration requirements.
- Configuration involves getting the initial set of software installed on the device and associated configuration (e.g. network, storage, and other general platform settings). This step would include getting a DMS agent installed on the device with details about the DMS end point and credentials needed for first time connection and registration.
- During the commissioning phase, the device is deployed in infrastructure and connects to the DMS for the first time. The DMS may have a specific onboarding step where the device and backend mutually authenticate each other for purposes of a secure device onboarding. Once onboarded, the device has necessary credentials for the in-band connection to the DMS. This phase may also configure telemetry collection details required from the device and ensure that initial software is properly configured.
- The monitor and maintenance are the normal operations of the deployed device, where DMS may perform a variety of tasks including monitoring telemetry data of the device, issuing alerts, or taking actions based on that data. Any software, firmware, or deployment of workloads and software is also part of this phase.
- Finally, the decommissioning stage is where the device is retired and may include things such as remotely wiping the device of sensitive data, invalidating credentials, disconnecting the device, and disposing of the device.

12.6.6 Kubernetes

There are several choices when it comes to orchestration, but Kubernetes or K8s is the most adopted orchestrator today. The abbreviation K8s comes from the name Kubernetes with 8 letters between the K and the s, hence K8s. In a previous section, we provided a gentle introduction to orchestration, unlike deployment of workloads or containers through a centralized container management system. Here, we will introduce K8s a bit more and introduce the reader to some of the basic concepts and constructs in K8s.

Figure 10: Kubernetes Overview shows a general overview of a K8s cluster. The control plane manages all aspects of scheduling and lifecycle management of the workloads. The cluster consists of one or more control plane nodes and worker nodes. As the name suggests, the control plane has software which is responsible for managing the worker nodes and the deployed workloads. Conversely, the worker nodes provide the different resources (computer, storage, memory, and custom accelerators) for the workloads to use. The control plane nodes and worker nodes are connected to each other over a flat, NATless network.

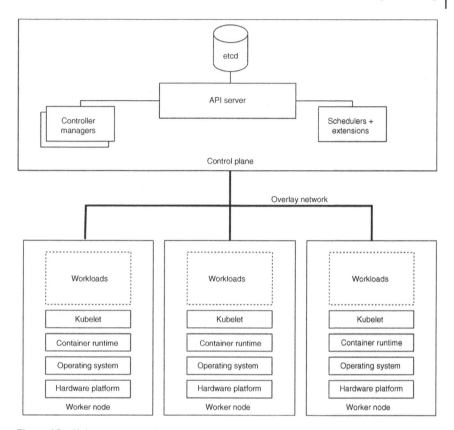

Figure 10 Kubernetes overview.

The key capabilities that K8s provide to DevOps and its users are as follows:

- Services through which discovery and load balancing are the key aspects. K8s provide a service construct that allows for microservices to be discovered over standard DNS and provides several load balancing functions, where traffic can be balanced and routed to the different backed instances of the service.
- Storage orchestration which enables storage to be separated from the actual workload itself and is a notable characteristic. The lifecycle of workloads and their associated storage are not the same, i.e. a workload could be started and stopped many times while the storage being available and not deleted as ephemeral data with the workload.
- Rollout and upgrades to workloads simplifies the workflow of DevOps when it comes to providing upgrades to workloads in the cluster as new versions are becoming available. It provides different policies which define how the rollout

strategy of the latest version of a workload should be handled and how much of the residual computer in the cluster can be used for the upgrade process. On deployment failures of those upgrades, it also provides rollback functionality to go back to the previous well-known good state.

- Monitoring and healing capabilities by K8s come in the form of different probes that workloads can instrument. K8s will periodically check the health of workloads by invoking the appropriate probe and restart or move the workload to a different node on failure.
- Workload configuration and secret management is provided by K8s, that allows for developers to separate the workload packaging from tangible configuration and secrets/assets (such as passwords and tokens) and allows for those to be securely stored by K8s and provided into the workload runtime when instantiated.
- Resource management and scheduling that provides for workloads to annotate what type of resources it needs from the underlying node, quality of service (QoS) settings, and other policies which are used by the native scheduler to make informed placement decisions of the workload when it is to be deployed.

Internally, the control plane consists of an API server which can be accessed locally or remotely using the standard command line interface kubectl, or via third-party and open-source software providing web management consoles for a more user-friendly access to the cluster. The API server has a key-value stored based on the open source etcd key-value store. The key-value store helps the API server to retain the current state of the cluster when it comes to the different resources.

The worker nodes have a software agent running on them called kubelet. The kubelet is an agent which has a continuous connection to the API server and performs requests from the API on the node such as facilitating the deployment of a workload when the API server scheduled.

There are many built-in primitives in K8s which are essential to understand as a user of K8s. We will look at a few basic ones here, but the reader should refer to the latest online documentation for K8s which can be found at Kubernetes.io. All the resources are defined in yml and deployed through the API server.

- **Pod:** this is the smallest deployable unit in K8s. A pod consists of one or more container images. The pod specification contains all the necessary information the API server needs to instantiate the pod such as the location of the container image and related configuration information, what resources the Pod needs, any network ports it needs open, and type of storage it needs and how it will be mounted to name a few.
- **Service:** the service construct allows for Pods to be fronted with a DNS, a service discoverable name via DNS, along with load balancing functionality using

iptables or IPVS. The service specification contains things like the desired service name, the pod that should be backing the service, ports and how it should be exposed in the cluster. A service can be exposed in a few diverse ways such as accessible locally to the cluster or accessible externally to the cluster.

- **Secrets:** Allows for a secret to be stored in the etcd key-value store. A pod specification can refer to the secret it depends on, and that secret will be propagated into the pod when instantiated.
- **Config-map:** Allows for general configuration data to be stored and provided into the Pod when instantiated and again eliminates the need to hardcode anything inside of the workload itself.
- **Daemonset:** Is a singleton workload by name which gets deployed on each worker node. This type of workload is useful when there must be exactly one instance of a workload type running on each worker node and automatically get deployed to new worker nodes that are introduced into the environment.
- **Replicaset:** Enables multiple replicas to be created of a particular workload/ Pod type in the infrastructure. It should be noted that K8s also provides an auto-scaling function which allows for the number of replicas to be automatically managed by K8s based on threshold/conditions that can be defined at the Pod level.

These are just a few of the built-in constructs but there are many more. The main idea here is to give a high-level overview to the reader of what types exist and what they do. Custom constructs can be added to K8s through CRD (Custom Resource Definition).

Custom resources in K8s are essential to edge deployments as these provide the opportunity for the orchestrator to be aware of custom platform accelerators and fixed functions that can have a performance or power impact to the workload. Kubernetes by default is only aware of three types of internal node resources

- CPU
- Memory
- Storage

Custom resources could allow for network accelerators, Vision Processing Units (VPU) and other workload special purpose hardware to be exposed to K8s and their scheduler. Without that knowledge, the scheduler would make scheduling decisions without those in mind which may end up a workload on an infrastructure worker node not having the right type of GPU or lacking some other platform accelerator it needs to efficiently get its job done.

Edge infrastructure deployments are many times heterogeneous in nature which emphasizes the need to have the platform awareness functions exposed to the K8s control plane.

The built-in resources that come with K8s are managed with a controller-manager. The controller-manager is running a control loop which objective is to manage the state of the cluster and move it towards the "desired" state of the cluster. This is a key backbone to the declarative paradigm used in K8s.

12.6.7 Declarative Paradigm

Means that the paradigm is more centric towards "what" should be done, and the infrastructure tries to accomplish that (the "how") internally without explicit instruction from the user. This is a core concept in K8s as deployments are typically stated in a declarative fashion and the controller-managers are continuously trying to move the internal state of the cluster from the *current state* to the *desired state.*

12.6.8 Packaging and Helm

We discussed container registries previously and the Pod specification or Deployment specification would refer to the container registry and image(s) which a Pod should be pulling. When the distributed application starts to grow the number of pods, daemonsets, services etc. it can quickly become hard to manage because DevOps is dealing with an increasing amount of yml specification files. Packaging of these distributed applications is what Helm provides. It can be viewed as a package manager (e.g. apt, yum ...) for K8s. All the resources required can be packaged into a Helm chart which is just a compressed archive of all the yml files with additional configuration information in a structured way. Helm is the defacto standard for packaging anything that should be deployed in K8s and comes with a powerful, easy to use command line interface in addition to several web based interactive solutions. As with container registries, Helm charts can also be stored and managed from central repositories, called Chart Museum.

12.6.9 Deployment of a Service

We have introduced some of the fundamental capabilities of Kubernetes and the associated constructs that exists in K8s, now we will look at the specific steps involved to deploy and application or more specifically a service in K8s. Figure 11 illustrates a service deployment in Kubernetes.

In this example we have a pod containing two containers implementing the business logic of the service. A pod alone is not a good mechanism to offer as a service because the lifecycle of a pod is ephemeral, it may be rescheduled and spun up on a different node, which causes the underlying node's IP address to be different and hence also ephemeral, which makes it impractical for an application to use. The pod must expose the relevant port or ports on which it is providing its services.

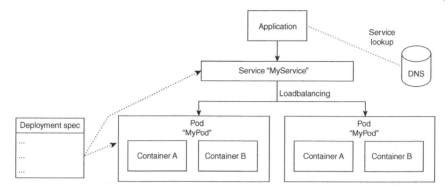

Figure 11 Service deployment.

The service specification in the deployment specification helps address this specific issue along with another one – load balancing (i.e. allowing for requests to the pods to be load balanced across a set of identical pod instances of type "MyPod."

The deployment specification holds information about the definition of the pod, i.e. the container images it is composed of, where they can be fetched from, and the port(s) that should be opened at the pod level for incoming traffic. The service section contains the port applications/consumers would be connecting to and are mapped to the pod port. The service also allows for the service to be provided with a fully qualified domain name and managed in the local DNS server the cluster provides. This eliminates the need for an application to have to know a specific IP address to connect to and provides the necessary façade abstraction to refer to the service as "MyService."

When the application is connecting to MyService, K8S looks up and returns the service IP and handles the load balancing of the traffic across different backed pods of the service. There have been multiple underlying techniques used by K8s to support this load balancing, from user mode proxy to the use of iptables to most recently IPVS. The typical and default policy has been a round-robin load balancing pattern.

12.6.10 Multicluster Orchestration

In IoT edge infrastructures, it is not always possible to set up one infrastructure providing a single K8s cluster. There are multiple reasons why organizations must segment their infrastructure into multiple K8s clusters based on

- Geography
- Security
- Assurance and Availability

- IT/OT compliance
- ISA 95/Purdue model

12.6.11 Serverless

Serverless is another important technological advancement that has been around in the cloud environment for many years. Serverless can be thought of as FaaS (Function as a Service). That means instead of hosting an entire VM for an extended period, a piece of software function is deployed and executed to perform a particular task. This provides an opportunity for cost reduction by the enterprise – instead of hosting a VM and paying continuous cost for it, including maintenance by the subscriber and the running cost of the provider based on the capacity of the VM and the traffic incurred cost to the infrastructure the user pays for the computing and network used.

Figure 12: Serverless Lifecycle illustrates the lifecycle of a typical FaaS flow. The subscriber must configure conditions or triggers that should result in the function to be executed by the hosted service provider. The software that constitutes the function has been uploaded and configured in that workflow.

When the condition has been met, e.g. an incoming call to a particular URI endpoint is invoked. The provider deploys the VM which provides the runtime environment in which the function should be executed. The function is deployed to the environment, and the function is invoked along with any additional contextual data required to perform its work.

Once the function has been finished executing, it has performed some tasks and generated some data that must be returned or persisted. The VM is finally shut down. In this model, the subscriber is only paying for the use of the hosted infrastructure during the time in which the function or functions are executing, which can yield substantial amount of cost savings.

There are several advantages to this model, but it should be noted that it does not fit every type of software or workload. The more frequently you need to call the FaaS, the cost benefits diminish.

Advantages:

- Faster development time allows the business and developer to focus on the specific business problem at hand.

Figure 12 Serverless lifecycle.

- Reduced maintenance cost of the infrastructure compared to what a traditional IaaS or PaaS approach would have been.
- Reduced cost since you are only paying for what you are using in terms of provided resources.
- Scalability provided by the service provider as the traffic and demand increase.

Disadvantages:

- Limitations to workflows and scenarios depend on what order of flexibility the vendor provides. Not every workload may be possible to integrate as a FaaS, and some may end up **costing more**, such as long-running workloads.
- Latency can become an issue as instances may have to cold start, which can take time, although many vendors have improved this significantly from when the technology was first incepted.
- Testing and validation can become a challenge in the actual deployment of the FaaS.

12.6.12 Serverless at the Edge

The whole paradigm of serverless computing at the edge is something we start to see and will increase in adoption and popularity over time. We have already seen this trend from some service providers, which provides FaaS at the edge, which would be at a location reasonably close to the end user or data source, typically using the CDN (Content Delivery Network) residual compute resources to allow for the FaaS to run at the edge, which helps reduce the latency, which is one of the cornerstones in edge computing.

However, with serverless technologies, the user is locked in by the vendor, and the provider has limited functionality and flexibility when it comes to its FaaS, overall cost structure, geographical deployments, availability guarantees, etc.

This entire paradigm has also started to see a shift to the on-premises edge where FaaS can be enabled on infrastructure owned and maintained by the enterprise. Why would someone do that if you get all these benefits going with a cloud service provider?

- Improving utilization of the overall on-premises edge infrastructure better. For traditional workloads, you may need dedicated VMs available for the workload when it needs to run.
- On-premises security reasons where you may have sensitive data or code that should not go outside the corporation and therefore prefer running that locally.
- Workload optimizations for the underlying platforms such as specific instruction sets that provide optimization for the workload or use of accelerators that are not provided by the service provider – such as offloading Media/AI workloads to GPU and optimizing network data plane traffic.

There are several solutions to on-premises serverless computing such as Apache OpenWhisk and extensions to Kubernetes such as Knative.

12.6.13 Apache OpenWhisk

OpenWhisk is an open-source framework that enables serverless function execution at on-premises infrastructure. It supports a variety of deployment models and underlying programming languages including Node.JS, Rust, Go, and Python.

The overall programming model of OpenWhisk is illustrated in Figure 13: OpenWhisk.

The event is coming from some source which may be a message bus such as Kafka or other Pub/Sub framework, an IoT device, an API invoked, or something else. A trigger is associated with a particular event. The action is the user-defined software which is packaged in a container and associated with a rule for the trigger that should be invoking it. It is possible to chain and create composite actions for more sophisticated models. The action executes its job and finally produces some output.

OpenWhisk can be deployed in a variety of ways, including Kubernetes, through Ansible Playbooks, Docker-compose, and through Vagrant.

12.6.14 Knative

Knative is a project in CNCF (Cloud Native Computing Foundation) that enables serverless computing in Kubernetes, which could be deployed in on-premises infrastructures. It extends Kubernetes and leverages Istio, which is a service mesh technology, for this.

Figure 14: Knative overview shows the overall view of Knative. It provides three pillars of functions:

- Build. Provides the infrastructure and toolchain to generate the functions from the source code to generated containers that will be deployed in the Kubernetes infrastructure.
- Event. Provides an underlying event mechanism that enables delivery and management of events between producers and consumers.
- Serve. Provides a set of control primitives that defines how the serverless workloads are deployed in the infrastructure, which includes things like revision management, routing, and overall lifecycle management of the deployed functions.

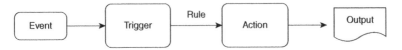

Figure 13 OpenWhisk.

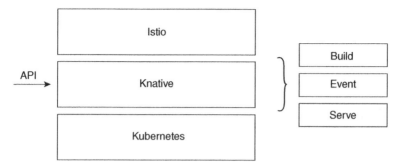

Figure 14   Knative overview.

12.6.15   IoT Devices

Thus far, we have focused the view more toward the infrastructure and software running in the edge compute device or server. In this section, we will shift our focus toward the leaf- or IoT devices. There is a large set of device classes in this space that are connected to the infrastructure over a wide variety of wireless or wired network technologies.

These devices are many times constrained in terms of compute, memory, and overall resources at the hardware level, many times with embedded, customized operating systems with small footprints, and many times based on a Linux derivative and sometimes traditional real-time operating system such as ThreadX.

Since these devices are constrained, it is often not possible, or desirable, to include them into an edge cluster due to the overhead and limited value of doing that. Instead, the data or control plane is integrated with the rest of the infrastructure over the network. We will discuss an especially important and commonly adopted technology for this – Device Twin.

12.6.16   Device Twin

A device twin is a digital and virtual representation of the state of a device and is a concept universally used in IoT space. The device twin is typically a representation of the device in the form of the JSON document and allows for characteristics and properties of the device to be stored such as

- Device-specific information such as hardware components, software, and operational versioning information.
- Metadata such as the location of the device.

The device twin information can be mutable or non-mutable, i.e. some information is read-only and cannot be modified, whereas other information can be modified. When some property is changed on the device twin representation, the device twin implementation will attempt to change the desired state of the twin to also be reflected on the physical device. In other words, the device twin always attempts to synchronize state to that of the physical device.

12.7   Conclusion

In this chapter, we introduced foundational concepts around cloud native technologies and edge computing and associated important concepts used today in industrial IoT deployments. This space is still not close to the level of maturity we have seen in enterprise, cloud, and CDN space and keeps evolving.

There are still many challenges faced in this space where technologies and solutions are emerging for.

- IT/OT convergence and associated aspects of ownership and management of deployments of devices through the lifecycle.
- Device management particularly in remote locations and areas with poor, intermittent connectivity or slow backhaul network links.
- Security challenges and threats faced in areas where devices are not always physically protected or locked up throughout the entire lifecycle of the device.
- Workload placement and optimization across heterogeneous network edges from local datacenters, far edge and on-premises to meet SLA (Service Level Agreements).

References

1 Cloud-native computing foundation (CNCF). https://www.cncf.io/

2 Fallon, A. (2021). What is cloud native computing foundation (CNCF). https://www.techtarget.com/searchitoperations/definition/Cloud-Native-Computing-Foundation-CNCF.

3 Cloud native computing foundation (CNCF) projects. https://github.com/cncf.

4 Raj, P., Vanga, S., and Chaudhary, A. (2022). Cloud-native computing: how to design, develop, and secure microservices and event-driven applications. https://www.wiley.com/en-sg/Cloud+native+Computing:+How+to+Design,+Develop,+and+Secure+Microservices+and+Event+Driven+Applications-p-9781119814764.

5 Frederic Desbiens (2023). What is an edge-native application? https://opensource.com/article/23/3/what-edge-native-application.

6 Edge-native applications, https://www.nearbycomputing.com/edge-native-applications/

7 Pachinger, F. (2022). Edge native applications are conquering the smart device edge. https://blogs.cisco.com/developer/smartdeviceedge01.

13

Explaining the Industrial IoT Standardization Efforts Toward Interoperability

13.1 Introduction

Interoperability is foundational to deliver any IoT system to the market at scale, and this is of utmost importance for the IoT O&G industry also. In this chapter, we are going to explore and describe multiple facets to interoperability requirements that has to be met in order to deliver to that end. Industrial standards help significantly reduce market fragmentation in general, to ensure that vendors delivering different parts of a system all stand behind a common framework with clear outlined conformance and certification tests for the various parts of the system to work together.

Standards in general have a precise scope in what problems they are trying to solve and seldom include objectives to deliver interoperability end-to-end, comprehensively covering all the aspects required for the system and its components to be fully interoperable.

13.2 Different Aspects of Interoperability

A distributed edge system is not fully interoperable unless all the parts of the system are themselves interoperable. We are going to explore and discuss different layers and parts of an edge IoT oil and gas (O&G) system that would require interoperability requirements whether delivered through an industry standard of

The Power of Artificial Intelligence for the Next-Generation Oil and Gas Industry: Envisaging AI-Inspired Intelligent Energy Systems and Environments, First Edition. Pethuru Raj Chelliah, Venkatraman Jayasankar, Mats Agerstam, B. Sundaravadivazhagan, and Robin Cyriac.
© 2024 The Institute of Electrical and Electronics Engineers, Inc.
Published 2024 by John Wiley & Sons, Inc.

general conformance and interoperability driven by the organization or third-party vendor or provider delivering the end system.

- **Network technology**: the network technology carrying the information between systems and providing the essential backbone and conduit of information must be interoperable. This requires interoperability from an inter- and intra-technology domain. The latter is usually guaranteed by the standard itself such as Wi-Fi, Bluetooth, 802.15.4, LPWAN (Low-Powered WAN) technology or 5G, and 4G, but the former requires diligence when it comes to how network technologies are ingested and bridged, for example.
- **Protocol:** the protocol that provides services with respect to retransmissions, packaging, serialization and deserialization, and security over the air must be interoperable. There are several industrial efforts in this space that we will investigate and discuss further. One very relevant standard here is OPAF (Open Process Automation Forum) and OPC Foundation. This also includes how protocols enabled with different underlying network technologies that carry and deliver interoperability as a whole to the system, i.e. what is the common denominator from the scope of the protocol – which could be IPv6, for instance.
- **Information model and data representation**: there is one thing having interoperability at the protocol layer, with interoperability ensuring that different network technologies can carry and deliver data. Data representation and the informational model are more concerned whether different parts of the system are represented digitally and what the data model looks like in order to monitor, control, and manage those virtual resources in the system.

13.3 ISA95

The ISA-95 or its official name ANSI/ISA-95 Enterprise-Control System Integration is an important component to the architectural definition of the standards and technologies presented in this section as they are referenced from a taxonomy perspective [1].

The ISA-95 standard provides a consistent terminology for the manufacturing process and all the actors involved from suppliers to manufacturers and is applicable to any IIoT (Industrial IoT) segment from discrete to continuous manufacturing. This standard defines the overall foundation and information models for operations and production and first and foremost defines a hierarchical or pyramid of levels of responsibility for the framework depicted in the diagram below. Figure 1 provides an overall view of the ISA-95 architectural definition.

Level 0 refers to the actual and physical processes in the manufacturing and involves data coming directly from equipment found on the factory floor. The

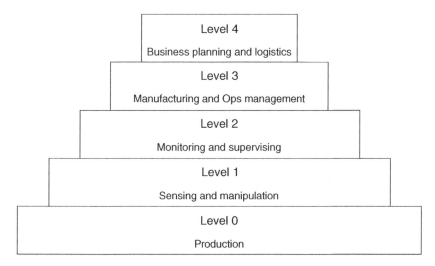

Figure 1 ISA-95 framework.

amount of data at this level can be enormous in size, depending on the complexity and size, machines, process, and size of the factory. The data emanated here are measured in very small time increments and have to be acted on in a very fast and deterministic manner many times.

Level 1 refers to the low-level sensing and actuation of factory equipment. Sensing here refers to things like temperature, humidity, and pressure sensing, to name a few. Actions sometimes must be taken on these sensing data, which is the manipulation aspect of level 1 responsibilities. This could be automatic regulation of pressure by manipulating a valve to release pressure when it hits a certain threshold, for example.

Level 2 is responsible for monitoring and supervision. Data from the manufacturing process are typically transmitted back over a SCADA (Supervisory Control and Data Acquisition) backhaul, which we will discuss in more detail in a subsequent section. The data received through SCADA can then be visualized through an HMI (Human–Machine Interface), where different parts of the manufacturing process can be monitored and actions can be taken based on the health and operational state of the factory floor (e.g. suspend a machinery part). Data visualized here can be in fairly raw form such as telemetry data from sensors, but could also involve visualization of transformed or analyzed data, inclusive of AI and machine learning, through multiple sources in the environment. The data acted on here are in the timeframe of minutes.

Level 3 is responsible for the overarching production and management, referred to as "Manufacturing Operations Management" or MoM for short. This level is responsible for making production optimizations through efficient and effective placement of workloads, detailed scheduling and dispatching of workloads, reliability assurance, etc. to ensure desired end results. The timeframe here can vary but could be in hours to minutes.

Level 4 is responsible for the overall business planning and logistics and manages the overall site production with integration to materials, supply chains, and ordering systems to ensure production goals are met through supply chain management systems, SCM (Supply Chain Management), and ERP (Enterprise Resource Management).

13.4   SCADA (Supervisory Control and Data Acquisition)

It is another central topic in the ISA95 and is part of the subsequent discussion of the OPC-UA standard. A SCADA system is responsible for collecting and delivering important data relevant to the manufacturing process from the production floor to backend infrastructure where the data can be monitored and acted upon, including visualizing in real-time.

The responsibilities and scope of a SCADA system are as follows:

- Control industrial manufacturing processes locally and remotely
- Acquire and monitor live, real-time, manufacturing data and telemetry
- Provide interaction and control through HMIs for operators (human and non-human)
- Provide bookkeeping and records of alerts, events, and logging and tracing of the system.

The system collects manufacturing data through PLC (Programmable Logic Controllers) or RTUs (Remote Terminal Units). A PLC is a microcontroller with a set of I/Os and allows for programmability, where the PLC can take certain actions on the output of the PLC based on the input received and provides the integration point between the physical world and manufacturing equipment to the digitalized one.

An RTU is a more sophisticated microcontroller that provides more control and flexibility than a PLC but comes at a higher price and typically integrates wireless communication to the physical end device in addition to being more rugged and therefore suitable for deployment in more harsh environmental conditions than a PLC.

13.5   The Choice of Network Technology

It is critical for the IoT deployment. The O&G industry has deployments in locations and environments, where it can be hard to operate with traditional technologies we often see in other segments and markets such as Wi-Fi, Bluetooth LE, or a time slotted technology in the unlicensed spectrum such as 802.15.4 or a derivative.

Wireless technologies are undoubtably increasing in adoption over its wired counterparts due to their flexibility in terms of installation and related cost efficiency compared to a wired installation. There are situations where a wireless technology may be challenging to deploy due to several factors.

- Operational range of the technology
- Reliability and robustness
- Interference sources

While technologies that operate in the unlicensed spectrum allows for privacy and cost-effectiveness, they do have to meet regulatory requirements in the country in which they are deployed. One important factor here is the transmit power, which ultimately determines the operational range of the technology. Generally, in wireless communication, the range of any technology is determined by a few different factors

- The transmission power of the transmitter often expressed in dB.
- The operational frequency the technology operates in.
- The receiving sensitivity of the receiver, which means how weak the signal can be for the receiver to disambiguate the signal from other electromagnetic noise in the environment.
- Antenna properties such as gain and height of the antenna.
- The environment in which the technology is deployed.

The operational range is often used when discussing the *path loss*, which is the loss of signal strength over some distance. There are models used in the industry based on empirical data that can help one calculate and assess the overall path loss and hence the operational range. These have traditionally been done in specific outdoor areas such as rural, suburban, and urban areas where attenuation factors have been considered when deployed in a more challenging environment compared to a less challenging one.

The reliability of the technology has to do with robustness characteristics that can be applied or configured with the technology to enhance its resilience to noise or environmental interference. Some of these are technology-specific, i.e. they are applied at the physical or link layer of the technology such as

- Operational frequency where higher frequencies have different attenuation patterns, in that they attenuate faster than lower frequencies in addition to also being more problematic when there is no LOS (Line of Sight) between the transmitter and receiver or penetrating obstacles.
- The bandwidth in which the technology operates on for a given channel. The larger the bandwidth, the more information can be transmitted per unit time.

- Guard interval, which is the interval between successive transmissions. A longer guard interval may help reduce over-the-air collisions and interference but reduces the effective throughput.
- Modulation techniques that define how bits are expressed and modulated when transmitted over the air. A higher modulation scheme provides more efficient use of the spectrum, but may introduce decoding errors at the receiver end.
- Forward error correction techniques which add redundancy to the data sent with the receiver being able to detect and correct bit errors that may have occurred during the decoding phase of the signal eliminate the need for the transmitter to resend the data.
- QoS support where traffic can be classified into different priorities can help ensure that the most urgent and important data are received over less important data. For example, critical control commands to an actuator may be far more important than some general telemetry data from a platform transmitted wirelessly.
- Transmission coordination which helps orchestrate transmission and reception of data throughout the deployment can help maximize the spectrum utilization and reduce retransmissions and errors that can otherwise occur.

Any wireless transmission is subject to interference in the environment, which can cause receiver errors and hence retransmissions. There are multiple types of sources for these such as the follows:

- Interference from the same or different technology operating in the same or overlapping frequency band. For instance, between Wi-Fi and Bluetooth Low Energy [2], both operating in the 2.4 GHz band with overlapping channels. End devices having built-in Wi-Fi and Bluetooth SoC (System on a Chip) have mitigations techniques built in to help overcome co-existence issues.
- Obstacles in the environment cause fading patterns in the wireless signal. These are typically categorized into slow fading (where there is a certain correlation distance in magnitude of meters between peaks and valleys), often following a Gaussian distribution with some standard deviation with respect to dB. Fast fading are more rapid changes to the signal, that are caused by refractions and reflections of the signal, causing it to arrive at the receiver at slight time differences, which often follows a Rayleigh distribution.

13.5.1 Wi-Fi

It is one of the most deployed wireless technologies in the unlicensed spectrum for scenarios that requires high throughput and relatively low latency. The Wi-Fi standard [3] has evolved tremendously over the last two decades. Wi-Fi operates in the 2.4 and 5.2 GHz spectrum, which means that it is susceptible to obstacles and range issues in environments, which is particularly the case for 5.2 GHz operation, which really works best where the AP (Access Point) and device are stationary, ideally with a full or partial LOS over a range of 20–30 m. The 2.4 GHz channels in Wi-Fi works better than 5.2 GHz channels if the signal must operate through walls of different materials, but even so drywalls can become a challenge over distance, and for the environment where concrete or more solid materials are used along with other materials, it can also become challenging.

Traditional range issues with Wi-Fi can partially be overcome with the use of multiple spatial streams and external directional antennas that can be used outdoors in scenarios where there is a longer distance between the AP and clients and can work in hundreds of meters, sometimes even toward kilometers. While Wi-Fi has power-saving modes built into the standard, it may not lend itself as a good technology for deployments that are battery-operated, unless the capacity in terms of kWh is adequate to eliminate support personnel replacement and serving of those devices.

13.5.2 LoRAWAN

It is becoming more popular and offers some strong advantages over Wi-Fi, in that the operational frequency is lower, in the 800–900 MHz range, which allows it to operate significantly better in challenging environments, e.g. enclosures with obstacles such as metal and concrete. It also offers a superior range over Wi-Fi in the magnitude of kilometers. Additionally, it can operate with very low power usage (from the end device perspective), allowing devices to run on coin cell or small batteries over an extended period. However, LoRAWAN [4] is not a high-throughput technology, and the bitrate is in the range of single- to double-digit kbps.

LoRA has a few different operational modes depending on how it will be used in the environment. Traditional use cases for LoRA involves smart metering and other sensing devices that are periodically transmitting some telemetry or sensing data to a backend powered by LoRA gateways, which are presumed to be connected to dedicated power sources, allowing them to listen to traffic continuously. The different modes of LoRA are referred to as class A, B, and C devices.

- **Class A**: the device can send data periodically to the backend LoRA gateway. The LoRA gateway can acknowledge the data and additionally piggyback a request to the device to perform some operation (e.g. maintenance, control function, or similar functions). This is the only opportunity the backend has when it comes to backend → device communication. In other words, the backend can only reach end devices immediately after receiving initiated transmission from the end node. The rest of the time the device is sleeping in a low-powered state, allowing it to run for a very long period without battery replacement.
- **Class B**: the device supports operation just like in class A, but in addition to this, it also has slotted time periods when it wakes up to sense if there is traffic intended for it. The duty cycle, the amount of time actively activating the RX (receiver) function, and the time it sleeps can be configured. The tradeoff here

is power consumption vs. timeliness in which the backend can reach out and communicate with the device.

- **Class C**: the device transmits data at its regular intervals as in a Class A device. The moment the device has ended a periodic transmission of sensor data, the receiver is activated, and the devices actively listen for incoming traffic from the LoRA backend. This provides the best performance with respect to latency between an issued command/request from the backend until the operation is completed by the end node but has the worst power consumption characteristics as it never really turns off the receiver functions and enters a low-powered state.

The LoRA network stack looks different from Wi-FI and other technologies, in that it does not provide common protocol stack layers such as IPv4 or IPv6 or 6LoWPAN, which can serve as a convergence layer between LoRA and the rest of the infrastructure. The LoRA gateway provides the ingestion and translation between general IoT data in the infrastructure and the application layer payloads that are transmitted from the LoRA gateway to the end devices. Since the payloads are very small, existing IoT protocols may be challenging to support from an application layer, regardless of requiring more compact representation of data for LoRA.

13.5.3 Time Slotted Technologies

There are a few standards embraced by the industrial IoT markets which offer a better utilization of the radio frequency spectrum. WirelessHART and ISA 100.11a are the most well-known ones, with IEEE 802.15.4 [5] 6Tisch being a contender trying to standardize aspects of the former two, which are considered more proprietary in nature. Other technologies often use a variant or mechanism to sense the channel they are trying to transmit on, then transmit, and if a collision occurs backing off for a random period before retrying. This is a grossly simplified description, but the key is that the decision in which a transmission or receiver operation is decided by the individual device. WirelessHART, ISA, and 6Tisch introduce the notion of a schedule to decide when devices should transmit or receive data. The scheduling function is managed centrally and communicated to the end devices. This allows for better usage of the spectrum as the likelihood for collisions with the network is effectively eliminated. In addition, this also offers power saving opportunities for the devices as they only need to be active when they must transmit or receive data. These technologies are low in power but can achieve greater range and resilience because they form a mesh network. Nodes in the network can be as follows:

- End nodes, where they only must be active when they are authorized to transmit or receive data.
- Router nodes, which can transmit and receive data, not just for themselves but also as a forwarding function up or down the mesh network to children or parents.

13.5.4 Cellular Technologies

There are a variety of cellular options operating in licensed spectrums, which may be required when it comes to operational performance meeting deployment requirements, particularly in challenging scenarios where coverage is challenging due to range and other factors using one of the unlicensed technologies discussed in previous sections. Beyond providing communication services between end devices and backend infrastructure, these technologies offer location services also, which can be provided by the device itself or provided by the mobile network, with accuracies down to a few meters. MIoT (Mobile Internet of Things) is a GSMA [6] term that refers to several 3GPP standardization efforts of IoT technologies in the licensed spectrum.

There are several options for cellular backend connectivity between managed devices, sensors, actuators, and other IoT equipment. Traditional cellular connectivity options supporting 2G, 3G, and 4G technology are still being used, but there are other options available that may better suit the requirements when it comes to power consumption and bandwidth.

Narrow-Band IoT (NB-IoT) was developed and standardized specifically for IoT communication and is a low-power wide-area (LPWA) technology, which has low data throughput requirements with infrequent communication patterns between the device and backend with typically small payloads. It is compatible with standard 4G networks and can offer tremendous range but with reduced functionality compared to a traditional 4G device, but retains the same security architecture. If correctly calibrated, NB-IoT-based devices can support the life cycle of up to 10 years for a wide range of different use cases applicable to the O&G industry. Devices operating NB-IoT can be remotely managed, including software and maintenance updates which is critical if vulnerabilities are discovered post-deployment. The size of such updates will definitely have an impact on the overall battery lifetime of the device, but with the right software architecture, the updates to the software should follow industry best practices, which include avoiding monolithic upgrades of software as these would require the entire software package to be updated, which adds to the power consumption due to the network traffic incurred and processing on the device to perform the update. Instead, a modular software design allows for targeted software components (libraries, frameworks, drivers, etc.) to be individually updated when needed.

Data communication between the device and backend occurs over the GPRS for most M2M (Machine-to-Machine) devices; alternatively SMS (Short Messaging Service) can also be used. SMS has several constraints over the GRPS such as a higher power consumption and limitations when it comes to the data payload maximum size (140 bytes).

There are support coverage extensions defined in 3GPP that allow for devices to operate in modes that extend their coverage beyond the normal calibration to include what is referred to as Robust and Extreme Coverage. These extended coverage operational modes enable coverage in areas that are otherwise impossible for operating on in the normal mode. There are also different power classes defined that provide better performance of the devices to the cost of power consumption.

LTE-M (Long-Term Evolution for Machines), also referred to as LTE Cat M, support voice and data communication with low power consumption, which may lend it suitable for battery-powered devices. Its standardization falls under 3GPP and was introduced in Release 13 with additional extensions in Release 14 and 15. It is tailored to support mission-critical scenarios, where QoS and real-time data communication are critical and seen more often in autonomous cars such as V2X and come with higher operational costs than that of a technology such as NB-IoT. For communication that requires over 100 bytes, it is often more efficient to send those over an IP-based protocol stack such as UDP (User Datagram Protocol).

LTE-M offers advantages over NB-IoT when it comes to firmware/software updates, voice communication, and mobile devices in addition to being better prepared to be enhanced as a technology with new usage models and scenarios over NB-IoT. Mobility here is referred to static deployments of an end device vs. a device which is not fixed in its installation and moving or deployed on a boat, shipping container, or other vessel that is mobile in nature. The attainable latency is also superior in LTE-M compared to NB-IoT, which can be critical for applications that operate under aggressive duty cycles and require very fast response times.

13.6 OPAF

The Open Process Automation Forum (OPAF) is an international forum of vendors, system integrators, academia, and semi-conductor manufacturers developing a standard for a generic and process control architecture. The OPAF is part of a larger consortium called The Open Group [7] whose objectives are to define interoperable standards for business aspects across many segments.

In this section, we will primarily discuss one of the standards emanating from the OPAF referred to as the O-PAS standard which is working toward defining and interoperable and vendor neutral architecture for process automation reusing existing standards and technologies and can hence be seen as a standard of standards. The OPAF has collaborated with the key foundational standards on which it is defined, such as

- OPC Foundation
- Industrial Internet Consortium (IIC)

- International Society of Automation (ISA)
- Distributed Management Task Force (DMTF)
- Control System Integrators Association (CSIA)
- Fieldcomm Group [8]
- NAMUR
- PLCopen
- ZVEI

Many of the control systems deployed to this point are largely proprietary, closed systems, which makes it difficult to integrate, extend, and interoperate them across vendors. This is one of the driving forces behind the OPAF initiative, which aims to deliver an open architecture which is vendor-neutral, paving the way for cross-vendor and ecosystem interoperability, in addition to achieving other objectives such as security and robustness criteria. Delivering these objectives helps in reducing the overall cost associated with systems as closed proprietary systems have proven to be expensive to be maintained over time, in addition to delivering additional value through enabling new scenarios and use cases, which may not be possible with those existing solutions. The open nature of the architecture opens up the playing field, allowing for the system to be better future-proofed and avoiding vendor or ecosystem lock-in, allowing system integrators and solution owners to have more control in solution composition.

The scope of the OPAF covers continuous, discrete, and hybrid manufacturing processes, where the oil and gas production is part of the continuous process block. The scope of what the OPAF will be focusing on can be classified into the following categories:

- Manufacturing Execution System (MES)
- DCS (Distributed Control System) Human–Machine Interface
- DCS I/O
- PLC (Programmable Logic Controller) Human–Machine Interface
- PLC I/O

The OPAF does not focus on SIS (Safety Instrumented Systems) and related I/O functions and is hence explicitly outside the scope of the standard. The OPAF takes inspiration from the ISA-95 (Purdue Model) to help define boundaries between responsibilities for the areas outlined above and specifies associated requirements with respect to interoperability, modularity, and portability.

The reference architecture in the OPAF envisions a fully distributed system where different components are connected to a real-time service bus to provide an advanced manufacturing and control solution offering automation, zero-down time maintenance and upgrades, extensibility, configurable redundancy, and resilience to meet different set of requirements with respect to availability.

The OPAF delivers the O-PAS standard, which is classified into seven distinct parts:

- Technical architecture overview that explains definitions and presents the set of interfaces defined throughout the system
- Security functionality and requirements that should be met for the components to be O-PAS-compliant.
- Profiles define the flexibility and composability of the system and define Winteroperability requirements from hardware and software interfaces, where each profile defines the set of components and related functionalities that should be supported.
- O-PAS Connectivity Framework is the core of the distributed nature systems, which defines the interoperable network fabric that connects the different parts of the system and is based on OPC-UA, which will be described in the subsequent section.
- System management defines all of the required functionality when it comes to monitoring and management of the system, from software to hardware ranging from the operating system, platform software, and applications.
- Information and Exchange models define common services and structures to pave the way for portability of applications.
- Physical platform defines the distributed control platform (DCP) and the underlying I/O system.

Figure 2 below is a simplified illustration of the general taxonomy and architecture of O-PAS.

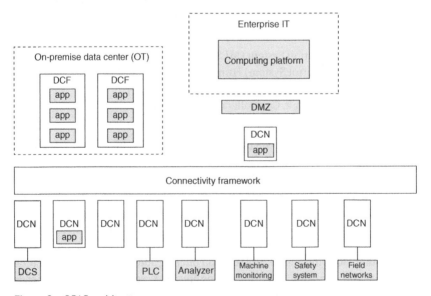

Figure 2 OPAS architecture.

The gray shaded boxes are outside the scope of what the OPAF and O-PAS are standardizing and defining. The O-PAS defines how those legacy or non O-PAS-compliant systems can be integrated and supported. The DCN (Distributed Control Node) is connected to the connectivity framework and paves the way for the interoperability and integration with other systems such as PLCs, analyzers, and wireless field networks over which sensors and actuators may be connected. These types of DCNs are called Gateway DCNs. The DCN can run applications, control management, and provide connectivity within and externally to the O-PAS deployment. An installation can contain a few DCN nodes to several hundred or even thousands, depending on the complexity of the continuous production.

The connectivity framework is the central portion providing the connectivity and over-the-air/wire connectivity and security to all of the components in the system. We will be discussing the connectivity framework in the next chapter, which is based on OPC-UA. The diagram describes the OT (Operational Technology)-level integration of an O-PAS system with on-premise data centers that may perform advanced real-time processing and monitoring of the production system as well as the enterprise IT side generally separated with a DMZ (Demilitarized Zone) providing isolation between business management platforms and solutions from the OT side.

OSM (O-PAS System Management) is not depicted in the diagram. It is responsible for lifecycle management of the entire O-PAS infrastructure from a manageability standpoint.

13.7 OPC-UA

The Open Platform Communication (OPC) foundation [9] defines the OPC-UA (Unified Architecture) standard, which is a successor to the older OPC standard which is significantly different. This chapter will focus on OPC-UA over the legacy OPC standard, which was focused on devices deployed with Microsoft Windows and hence is not platform-independent, as it significantly used Microsoft-specific technologies such as COM (Common Object Model). OPC-UA is not tied to a particular underlying platform architecture or dependent on the operating system or OS-related communication technology.

13.7.1 Objectives of OPC-UA

The standard has a broad set of objectives defining requirements to help pave the way for interoperability between systems in a platform agnostic manner, starting from the underlying protocols that it depends on for operation, security architecture, communication paradigms, data modeling, and extensibility both in security and information/data models.

OPC-UA is fundamentally a SOA (Service-Oriented Architecture), and the standard delivers functionality such as

- *Discovery* defines how devices and services are discovered in a consistent and interoperable manner in the environment. This can be local or remote devices (i.e. not on the same network segment where local discovery mechanisms can be used to discovery brokers).
- *Organized addresses* allowing for the data to be organized in a hierarchical manner to help represent arbitrarily complex systems that can be monitored and managed by clients.
- *Subscriptions and Events* allow for clients to subscribe to and monitor pertinent parts of a service and act on events triggered, which may require a consumer to take some action.
- *RPC (Remote Procedure Call)* enables data models to be encapsulated as objects, allowing clients to invoke methods implemented on a server as remote objects.

13.7.2 OPC-UA Architecture Overview

The OPC-UA architecture is structured in multiple levels. The very bottom part of the architecture defined the protocol mappings required for OPA-UC to effectively be able to run over different types of architectures, and this layer defined how the mapping to OPC-UA should be in order to extend and support new and emerging network technologies that the standard did not include in its scope initially, hence providing a level of abstraction from the carrying and foundational set of protocols delivering data from OPC-UA entities from the technology realizing the transport of data. Most commonly, the carrying protocol bindings are for IPv4 and IPv6 and would, by default, support a broad range of technologies using one of those as the data layer protocol. Figure 3 shows the architectural overview of OPC-UA.

The transport and communication paradigm is defined next. OPC-UA started with a client-server notion where the interaction was initiated by the client to a server it had previously discovered or had *a priori* knowledge of its endpoint. This provides a 1 : 1 communication pattern that could enable synchronous, asynchronous operations, subscription, and event indications.

Synchronous operations are ones where a client-initiated request is followed by the response from the server. The completion response from the server indicated that the requested operation has concluded at the time the response was issued by the server.

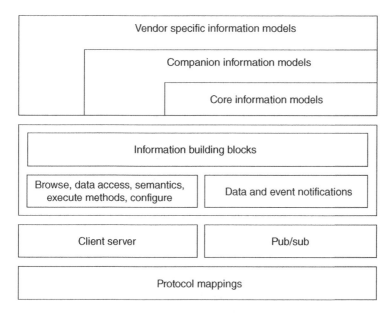

Figure 3 OPA-UA architectural organization.

Asynchronous operation is one where a client sends a request, which is acknowledged by the server, i.e. the request is inspected, validated, and authorized. The response from the server indicates that the operation is pending but has not yet been completed. A completion callback, or more naively client polling, would help infer when the submitted operation has been completed successfully or has failed.

The OPC-UA in later releases (release 13) introduced Pub/Sub as an alternative supported paradigm between entities in an OPC-UA system. Pub/Sub, or Publish Subscribe, is based on a communication paradigm that enables many-to-many communication (i.e. M : N), which is very useful and required for certain scenarios.

13.7.3 Publish/Subscribe Pattern

Pub/Sub is based on three essential parts to work: a publisher, a subscriber, and a topic. The topic is what publishers publish to and subscribers subscribe to. Without loss of generality, a topic can have multiple publishers and multiple subscribers, and the Pub/Sub system provides the delivery of data to subscribers as they are being published to topics of their interest. Figure 4 illustrates the entities in a Pub/Sub system.

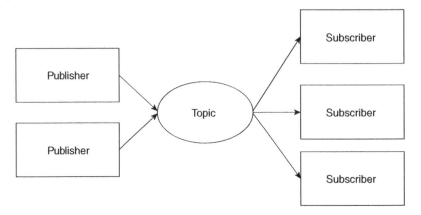

Figure 4 Pub/sub system.

Depending on the Pub-/Sub-specific standard or framework, the Pub/Sub infrastructure can offer a diverse set of features. A few common and desired features and characteristics one look for are as follows:

- Confidentiality, Authentication, and Access Control to ensure that data delivered between publishers and subscribers are secure, in that over-the-air communication is encrypted, and only authorized principles are allowed to publish and subscribe to topics. Fine-grained control here is typically also desired, i.e. you would have per topic access control where publishers or administrators can configure who is authorized to publish and subscribe to certain topics or related set of topics.
- Data Delivery Services to ensure guaranteed delivery and QoS with varying degrees of QoS based on topic and/or publisher and subscriber in terms of resilience, retries, and priority. Store-and-forward to ensure delivery can be made to subscribers who subscribe to a topic after critical data have been pushed, or have intermittent connectivity, and data retention, data stickiness, and similar functionality, which further helps fine-tune the policy and configuration of different parts of the Pub/Sub system.

13.7.4 OPC-UA Pub/Sub

The OPC-UA Pub/Sub supports MQTT, which is an OASIS standard [10] and AMQP. There exists many different implementations of MQTT, both commercially available and open-source with varying degrees of feature functionality, performance, and scaling characteristics, particularly in orchestration deployments over Kubernetes or K3s, for instance.

OPC-UA Pub/Sub defines two different mechanisms for Pub/Sub based on the infrastructure. Local networks not only allow for UDP broadcast but also unicast traffic in a format optimized for size interpretable by OPC-UA clients. The second one is a message queue broker, where MQTT or AMQP is used. The same data format, or JSON, can be used but does not offer the same efficacy when it comes to data representation over the wire/air.

The security model for OPA-UA using Pub/Sub looks different irrespective of the local vs. the brokered approach being taken. When using a broker based on MQTT or AMQP, standard TLS-based connections offering authentication and access control are used. For the local network case where traffic is broadcasted, or otherwise enabled without a dedicated broker, *a shared key* is used which publishers and subscribers can get access to through a SKS (Security Key Server). The SKS approach has a weaker security posture since shared keys are being used, which if compromised can become a huge problem even if the SKS provides authentication and authorization checks by a principal when a security key is requested.

More advanced features related to QoS and data retention are challenging to support in the local network case. For larger and more complex deployments, it is more desirable to use the brokered Pub/Sub support over either MQTT or AMQP.

13.7.5 MQTT vs. AMQP

The two defined integrated Pub/Sub protocols are MQTT and AMQP in OPC-UA, and there are some differences and tradeoffs between them, where the choice depends on desired features, performance, and scaling aspects of respective technology.

MQTT (MQ Telemetry Transport) [11] is an OASIS standard and typically operates over IP as the network protocol and supports reliable message delivery and bi-directional communication between end-points. It enables TLS for security under the transport layer, enabling confidentiality, integrity, and authentication capabilities between the clients and supporting enablement of authorization protocols such as oAuth. There is a wealth of implementations of MQTT from very lightweight implementations suitable for small, constrained devices to open-source, commercial, and heavier implementations offering additional robustness and scalability.

AMQP (Advanced Message Queue Protocol) [12] is also an OASIS standard and provides a higher level of control and flexibility to the application when it comes to the transport or routing of messages from publishers to subscribers. The model used in AMQP is based on Publishers, Exchange, Routes, Queues, and Consumers. The exchange in AMQP is responsible for taking incoming messages from publishers and route those to the intended queues they should go to. There are multiple types of exchanges, as well as attributes that control properties of the exchange (e.g. lifecycle management of an exchange and its name). There is a unicast, or direct exchange type, fanout exchange, topic exchange for Pub/Sub patterns, etc.

AMQP is a more complete messaging protocol, and although it emerged from the banking industry, it has been adopted and used in many other industry segments over time. The security architecture and comprehensiveness offered by AMQP is richer than what is provided through MQTTT. For example, AMQP offers more messaging scenarios and paradigms than just Pub/Sub and also a richer set of transactional primitives such as "Acknowledge a message" and "Reject a message" (although in the context of OPC-UA we are primarily concerned with Pub/Sub). With the right implementation of the standard, they both offer great scalability in native, standard deployments as well as clustered and orchestrated scenarios, but MQTT is superior when it comes to supporting very constrained embedded devices with very limited computational and limited resources. Extensibility is provided naturally in AMQP, whereas MQTT does not offer extensibility. Application developers would need to make choices of pub/sub technology based on requirements and anticipated roadmap that may have an impact when it comes to extensibility requirements.

13.7.6 Future Pub/Sub Integrations

For OPC-UA could include technologies such as DDS (Distributed Data Service) which has gained tremendous adoption in certain segments such as mission-critical and fault tolerance systems and AMR (Autonomous Mobile Robot), particularly those running an ROS (Robotics Operating System) where communication within logical software nodes on the AMR and between the AMR and external devices benefits from the DDS. Fundamentally, the OPC-UA architecture could define the proper architectural mapping and bindings for the DDS to be another choice for an OPC-UA system to work under a Pub/Sub methodology.

13.8 DDS

It is an OMG (Object Management Group) standard [13] that was co-authored back in 2001 by RTI (Real-Time Innovations), which is a software framework company. The first release by OMG was published in 2004. The DDS is highly distributed and does not have a centralized broker, such as AMQP and MQTT. This makes the DDS more resilient in some aspects as it does not have a single point of failure and therefore help in scaling over a brokered system. Additionally, the DDS provides a rich set of QoS (Quality of Service) functions in the network, which allows for data, topics, and actors in the system to operate under a QoS framework, where the system provides the ability to set priorities for communication flows with respect to timely delivery, delivery guarantees, and redundancy. The DDS provides excellent throughput and performances in terms of messages/second with very low latency and jitter compared to many other Pub/Sub technologies.

13.9 Integration with Telemetry and Big Data

OPC-UA clients (and servers) could also integrate with time-series databases and other database systems for purposes of exporting relevant information from their endpoints to offer comprehensive observability to the IoT infrastructure owners in both domains of IT and OT, respectively. The exported data could be used to provide valuable insights in terms of operational aspects of the manufacturing and production side as a source for analytics such as predictive maintenance.

Telegraf is a client-side technology in TICK stacks that would enable for critical telemetry data and metrics to be pushed to time-series databases for longer-term storage, analysis, and report generation that can provide insights to the organization as well as longer-term data retention. The data transmitted to an InfluxDB instance or other suitable backend storage and database technology can help provide anything from sensor readings, infrastructure state, and status with respect to performance outages, for example, to production and operational status.

Telegraf endpoints can be instrumented to include important information to the backend such as

- Metric name
- OPA endpoint details
- Security details (protocol, policy, and mode)
- Node information
- Identifier and type
- Tags – any additional required tags

13.10 IEC Standards used in the OPAF

There are several IEC (International Electrotechnical Commission) standards that have been included in the definition of the O-PAS 1.0 and 2.0 version of the standard. ISA/IEC 62443 is related to cybersecurity management to provide secure, robust, and fault-tolerant systems, with particular focus on Industrial Automation Control Systems or IACS. ISA/IEC 62443 contains technical reports

(TRs) targeted for different personas or audiences of the system such as hardware vendors, system integrators, to asset owners, which covers different domain areas with TRs in each one:

- General
  - Concepts and models
  - Master glossary of terms and abbreviations
  - System security conformance metrics
  - IACS security lifecycle and use-cases
- Policies and Procedures
  - Security program requirements for IACS asset owners
  - IACS protection levels
  - Patch management in the IACS environment
  - Requirements for IACS service providers
  - Implementation guidelines for IACS asset owners
- System
  - Security technologies for IACS
  - Security risk assessment and system design
  - System security requirements and security levels
- Component
  - Secure product development lifecycle requirements
  - Technical security requirements for IACS components

IEC 62541 is what is used for connectivity fabric and refers to the OPC-UA. A popular open source implementation for OPC-UA is open62541, which is developed in ANSI C and hence well-portable to multiple platforms and environments, is standards compliant and licensed under MPL v2.0.

13.11 RedFish

It is a management protocol standardized through DMTF [14] which can be used for in-band as well as out-of-band manageability. In-band manageability refers to the management of a system through a connection terminated in an agent executing in the operating system of the managed device, enabling administrators to remotely manage the device. Out-of-band manageability refers to a model where a connection is made to a discrete, or different processor and subsystem from the host processor such as a BMC (Board Management Controller), for instance. The latter provides benefits such as manageability of a device if the operating system on the host has stopped responding or gaining access and management prior to the operating system has started, e.g. accessing and managing BIOS settings via remote KVM (Keyboard Video Mouse). The former has benefits, in

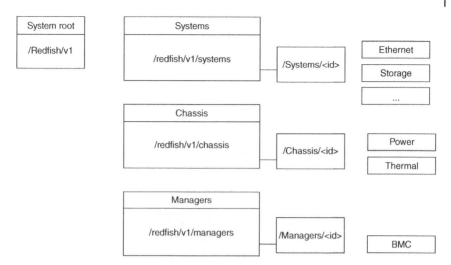

Figure 5 Redfish resource hierarchy.

that it is typically more responsive and faster, which suits it better when doing interactive sessions such as a remote desktop session.

RedFish is based on REST (Representation State Transfer) and is stateless communication paradigm popularized with HTTP and HTTPS. The payloads are represented in JSON, which makes them human-readable. RedFish allows for a backend infrastructure to reach out and manage, monitor, upgrade, and maintain compute systems deployed in the environment and is what OPAF and O-PAS has adopted as the management solution.

RedFish defines the interfaces to the services and the data model for the different parts of the managed system (such as storage and networking). The standard is extensible and allows for third-parties to extend with their own schemas, or extend schemas with new attributes and properties, which is tremendously useful as it allows for customization and management beyond what the standard originally outlined.

The RedFish standard defines different collections that can be accessed and managed:

- Collection of Systems, which provides details of a computer system and its subcomponents
- Collection of Chassis, which provides a logical view of the physical system such as power, thermal, and possibly location details
- Collection of Managers, which provides the view and access to the management entities on the device such as a BMC.

Figure 5 above illustrates the overall resource hierarchy and organization of restful resources exposed by a RedFish service. The standard REST operations GET, PUT,

POST, DELETE, PATCH, … are used in conjunction to view or modify different aspects of the system. For instance, a GET operation on <base URI>/redfish/v1/ systems would provide an array back of all the managed systems in the infrastructure with their corresponding IDs, they can later then be addressed and managed individually through their respective unique URI such as <base URI>/redfish/v1/ systems/<id> where <id> would represent a unique and valid system identifier.

13.12 The FieldComm Group

Itfocuses on technologies in the industrial automation space and consists of members from the industry, universities, and academia in general. The group has been around from 2015 and combined the assets from two pervious foundations; the Fieldbus Foundation and HART (Highway Addressable Remote transducer Protocol) Communication Foundation.

Section 13.5.3, we discussed time slotted technologies in the unlicensed spectrum where WirelessHART was mentioned in brevity. WirelessHART defines the over-the-air protocol and management of a schedulable time slotted mesh-based network that enables efficient communication between nodes in the WirelessHART mesh. The WirelessHART protocol delivers on some key design objectives, in that it is easy to manage and deploy and have self-organizing and self-healing characteristics. For instance, if a node goes offline in the mesh network, the management layer automatically attempts to heal by restructuring the parts of the mesh to ensure all nodes are available. The HART protocol is the older, wired definition of the protocol which is based on analog and digital signaling over 4–20 mA wires. The WirelessHART protocol uses a time-synchronized frequency hopping protocol that defines the time slots with a schedule to define the TX/RX (Transmit and Receive) operations for the nodes, whereas the HART protocol uses a token and the node who has the token is authorized for TX.

Foundation Fieldbus is the other protocol that came together when the FieldComm Group was formed. Fieldbus is a digital, bidirectional communication protocol widely used in process automation and closed-loop control systems that require real-time or have time deterministic requirements, for instance, tight duty cycles where several instructions have to be executed over a strict time period. The Foundation Fieldbus delivers two different implementations for different purposes.

- H1 is a low-speed communication bus over 31.25 kbit/s, which is used predominantly to connect field devices over twisted-pair wiring and is by far the most adopted.
- HSE (High-speed Ethernet) operates over higher speeds 100 Mbit/s and connects different host systems, gateways, and field devices using standard Ethernet.

What makes Foundation Fieldbus different from traditional wired instruments with point-to-point connectivity is that you would typically carry data from one type of a sensor or control to a single actuator. The Foundation Fieldbus enables for up to 32 devices to be controlled over single wiring. Additionally, each device can handle multiple types of sensors over that same wiring communicating multiple data points, reducing overall cost (wiring, instrumentation, etc.)

References

1 ISA Standards and Publications. https://www.isa.org/standards-and-publications.
2 Bluetooth. https://www.bluetooth.com.
3 WiFi Alliance. https://www.wi-fi.org/
4 Lora Alliance. https://resources.lora-alliance.org/
5 Time Slotted Channel Hopping. https://en.wikipedia.org/wiki/Time_Slotted_Channel_Hopping.
6 GSMA. https://www.gsma.com
7 The Open Group. https://www.opengroup.org/
8 Fieldcomm Group. https://www.fieldcommgroup.org
9 OPC Foundation. https://opcfoundation.org/
10 OASIS Open. https://www.oasis-open.org.
11 MQTT. https://mqtt.org/
12 AMQP. https://www.amqp.org/
13 OMG (Object Management Group). https://www.omg.org/
14 DMTF. https://www.dmtf.org/

14

Digital Twins for the Digitally Transformed O&G Industry

14.1 Digital Twins (DTs)

Is one of the key transformative technologies in digitalization of the edge and oil and gas (O&G) industry we will explore in more detail in this chapter, along with other already adopted and emerging technologies and tools. A digital twin is really a digital representation of a physical, real-world, object, process, system, or an entire supply chain. If designed and modeled correctly, it can help provide insightful details of how the real-world object that it is representing would behave and act under different circumstances.

The term digital twin was coined by NASA's John Vickers back in 2010 after a collaboration with Michael Grieves. Michael is the first one who applied the idea or concept behind digital twin, dating back to 2002.

A digital twin could be used to gain a tremendous level of understanding of a physical object (or a process) such as better understanding of characteristics such as specific weaknesses or strengths of the object. For example, a digital twin could be made of a wind turbine motor to understand how it performs over time with respect to wearing down of parts, corrosion, and when feeding the twin with input data from sensors such as wind, temperature, and humidity.

14.2 Digital Twins in Manufacturing

DTs can beneficially be used in manufacturing and production scenarios, inclusive of the O&G industry. Production design can benefit from digital twins during early stages of the development phase such as when exploring benefits and

The Power of Artificial Intelligence for the Next-Generation Oil and Gas Industry: Envisaging AI-Inspired Intelligent Energy Systems and Environments, First Edition. Pethuru Raj Chelliah, Venkatraman Jayasankar, Mats Agerstam, B. Sundaravadivazhagan, and Robin Cyriac.
© 2024 The Institute of Electrical and Electronics Engineers, Inc.
Published 2024 by John Wiley & Sons, Inc.

shortcomings of prototypes and could ultimately reduce the overall time and investment cost in terms of iterations to get a concept to product. This is achieved through digital twins and simulations that would help provide this insight.

14.3 Digital Twins in Process Efficiency

Data collected from sensors across the production line on a manufacturing floor could be in a digital twin that provides a virtual instantiation of the overall manufacturing process to help shed light on production inefficiencies and inconsistencies in production numbers over time to analyze and provide overall guidance of bottlenecks in the process that can be addressed to optimize the process.

14.4 Digital Twins and Quality Assurance

Product quality variance and yield may be realized through a digital twin to help improve overall consistency of product quality or inconsistencies. It may be required to instrument multiple discrete steps of the manufacturing and production pipeline to understand where improvements have to be taken in order to improve the average quality and reduce its variance.

14.5 Digital Twins and Supply Chain

Manufacturing and production is heavily dependent on consistency in the supply chain of materials and dependencies. During the pandemic, we witnessed supply chain disruptions across many segments and markets in the industry, which ultimately had a financial impact for the companies affected, including customers from consumer electronics, sanitization products, and food and produce. Consistency and predictability in the overall supply chain is of utmost importance to help ensure goods and services make it to customers in a timely manner. DTs instrumented to model entire supply chains can help improve agility, resilience, and predictability and identify weak spots in the supply chain.

14.6 Digital Twins and Predictive Maintenance

Unplanned downtime creates unforeseen production disruptions, which increases overall cost and has a significant impact on the overall bottom-line of any manufacturing and production line. The desire for a production line is to operate at maximum efficiency, but sometimes machines or parts break, which causes production to come to a stop, which would involve sending out technicians, understanding the root cause, secure replacement parts, and fix the issue. A DT could be used to monitor and understand the impact the environmental elements have on the part with respect to wear and tear. This insight could help predict when to plan for downtime based on the schedule, securing required replacement parts proactively and overall being able to minimize the impact of a part in a the production line having to be replaced or address any general maintenance required.

14.7 Industry 4.0

Is commonly used when speaking about DTs. Industry 4.0 is also commonly referred to as the Fourth Industrial Revolution or 4IR. The First Industrial Revolution began in the 1760s where humankind first used machines to create goods that were previously produced manually. The "Spinning Jenny" invented by James Hargreaves was a yarn-spinning machine. It started the limited use of steam engines and interchangeable machine parts and was predominantly hydro-powered.

The Second Industrial Revolution, also referred to as the Technological Revolution, took place in the late nineteenth century to early twentieth century, involving larger adoption of machine use and mass production and use of petroleum and electricity. The invention of the combustion engine and communication paradigms such as the telegraph were key to this period.

The Third Industrial Revolution is where we moved from mechanical and analog systems to digital electronics. IT (Information Technology) and computers were first introduced to automate previously implemented manual processes in manufacturing. This happened in the 1950s. The computer infrastructure automated many aspects but is still managed or operated by a human controller and introduced advanced telecommunication paradigms and use of PLCs (Programmable Logic Controllers).

The Fourth Industrial Revolution of Industry 4.0 is about intelligent computers to further automate and move toward a fully autonomous process when it comes to manufacturing. This involves smart machines and factories using information from the environment to help drive improvements in areas of efficiency, cost-effective opportunities, minimizing downtime, quality etc. The Industry 4.0 is built around technology advancements in multiple domains, including but not limited to

- Innovation in connectivity and communication from 5G, 6G, to unlicensed technology improvements such as Wi-Fi, BLE, and time slotted technologies.

Improvements in robustness, quality of service (QoS), throughput, latency, and cost pave the way for adoption of these to a larger degree than previously possible.

- Analytics through the emergence of Artificial Intelligence and Machine Learning made possible through innovations in models and AI frameworks, frameworks, and tools made available for different types of training reduced cost of computation.
- Innovation in human–computer interaction and interfaces ranged from traditional HMI (Human–Machine Interfaces) to virtual and augmented reality unlocking guidance and problem-solving remotely, which was otherwise not possible.
- Internet of Things (IoT) and Industrial IoT from standardization efforts to scalable and deployable solutions, which has enabled smart sensing and actuation technologies in brown- and green-field deployments, enabling even more virtualizing of the production environment such as virtual PLCs, DTs, and Predictive Maintenance as examples.

There are four types of DTs, and they differ primarily in their scope but not in the purpose, which is to provide a digital representation of an actual object or process.

- Component twin is a DT that represents a discrete and specific component of a system. This could be something as small such as a nut, bolt, and belt to understand how it behaves under stress or extreme temperatures.
- Asset twin is a compounded representation of multiple-component twins and represents a product. The example above with the wind turbine could be viewed as an asset twin.
- System twin is a representation that brings together multiple assets, or products, to form a more comprehensive fleet, which could yield insights into how the individual products work together as a larger system.
- Process twin follows on the same logical pattern of bringing systems together. A common example here would be a system representing the manufacturing line and the process twin representing the entire production, which could include supply chain, ordering, factory floor, and workers. Here, understanding of timings and inefficiencies may provide valuable insights.

14.8 Digital Twin Concept

Figure 1 below provides an illustration of the concept and relationship between a DT and the physical object that it is representing.

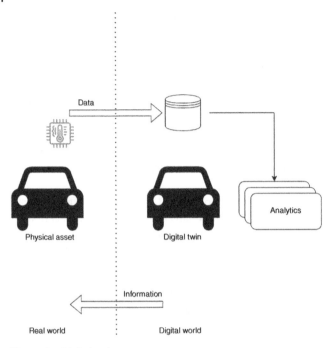

Figure 1 Digital twin concept.

The physical object, a car in this case, provides real-world simulation data from the real environment through sensors to the digital world where the data are stored and analyzed.

- Data are gathered from the physical asset through sensors, saved, and transferred over to the digital world.
- The data in the digital world can then be evaluated and analyzed and also visualized through monitor management systems. It can be run through simulations and parameters, and data can be adjusted to simulate different types of scenarios.
- Insights are provided back to the real world, which may be used to make certain data-driven decisions and interventions.
  - Optimizations
  - Diagnosis
  - Prognosis

14.9 Standards and Interoperability

There are multiple standard efforts in the DT domain. The ISO (International Standard ...) 23247 defines a "Digital Twin Framework for Manufacturing," which is regarded as a fairly new standard. The standard defines four distinct layers of the reference architecture.

- **Observable manufacturing domain**: this is outside the domain of the DT but describes what needs to be observed and modeled from the physical world, i.e. the manufacturing plant in this scenario.
- **Data collection and device control entity**: this layer ingests data from the environment and is the bottom-most layer of the DT itself. It provides the communication fabric and pre-processing and filtering of data, as needed from the environment and can control/actuate the OMEs (Observable Manufacturing Elements).
- **Core domain**: this is the domain that provides most of the essential functionality of the device twin in terms of operations and management. It supports multiple sub-applications such as simulation capabilities, analytics, and overall monitoring and control with support for extensibility through a Plug-and-Play capability.
- **User domain**: this layer provides the applications interacting with the digital domain through a human user and integrations to legacy applications (e.g. ERM – Enterprise Risk Management).

14.10 IDTA Standard

The Industrial Digital Twin Association (IDTA) is a German-based association that focuses on DT technology with over 90 member companies in the industrial space, with additional partners helping shape the success of Industry 4.0. The IDTA views DTs at the core of Industry 4.0. The IDTA views DTs as reusable and composable assets that can be reused and interoperated in the production environment like a USB peripheral. The IDTA has defined the notion of an Asset Administration Shell (AAS) which implements a DT and defines the foundation for interoperability through standardization. The AAS contains a set of sub-modules that defines various aspects of the DT such as states, properties, measurements, and capabilities that enable cross-vendor interoperability.

Currently, there are over 40 sub-models defined that are either fully ratified and published or in review. Below are a few examples.

- **Generic frame for technical data for industrial equipment in manufacturing:** This module provides interoperability with respect to describing an asset using a dictionary technology or key-value pairs, enabling cross-industry parties to understand the asset.
- **Time-series data**: This module standardizes on a format for time-series data emanating from the environment through real and virtual sensors that have to be normalized in order for interoperability and unambiguity.
- **OPC-UA server data sheet**: This module defines how an OPC-UA server can, and should, be used for interoperability within the context of a DT and AAS when it comes to the namespaces used, profiles, and facets.

- **Reliability**: This module is focused on system and functional safety and data reliability through data model definitions according to IEC and ISO standards such as IEC/CDD 62683-1 DB.
- **Wireless communication**: This sub-module defines the Wireless Communication System (WCS), which is a crucial aspect of Industry 4.0. The sub-module defines entire Transmitter and Receiver chains across the protocol stack from the physical layer such as modulation, bitrates, and RSSI (Received Signal Strength Indicator) for control systems to take appropriate actions based on the condition and health of the wireless channel.

The abovementioned list merely provides a sample of the more comprehensive list of sub-modules defined and supported. The complete list and current state of sub-modules can be found here [5].

14.11   Digital Twin Consortium

Is a consortium consisting of members from academia, government, and industry, and its objective is the overall definition and development of DTs. They are focusing on the establishment of a rich and broad ecosystem across a wide range of segments for DTs, actively looking to address technology gaps that exist and address proprietary aspects of existing technology through standardization to pave the way for vendor neutrality and interoperability. Examples of segments are as follows:

- Academia and research
- Agriculture, engineering, and construction
- Manufacturing
- Health and life sciences
- Natural resources
- Mobility and transportation
- Security

14.12   Digital Twin in O&G

What are the benefits of using DTs in the oil and gas industry? We have discussed the concepts of a DT and generally how it can help understand the characteristics of physical objects and complicated systems. In the O&G industry, any improvement when it comes to production can have a huge impact on the bottom line.

DTs can be used in a broad set of areas such as

- Monitor equipment status, state, and overall health.
- Minimizing production issues by detecting potential failures before they occur, hence eliminating unplanned downtime. This is commonly referred to as predictive maintenance.
- Diagnosing issues to support troubleshooting and analyze the root cause of various issues and problems.
- Identify process improvement opportunities to further increase efficiency, hence decreasing overall operational cost.

Even a small efficiency improvement can yield millions of dollars in savings.

14.13 DT Complexity and Trade-offs

While it may seem like the DT can indeed solve and provide very insightful learnings that can create opportunities from prolonging lifetime, improving quality, and generally making a product of process better – it is important to understand that the investment in the DT does not come for free.

One needs to understand the overall cost and benefits the DT may provide as the complexity of the DT is assessed. This aims to understand resources and dollars that go into the development of the DT overall. Additionally understanding the type of information and insights it will ultimately provide in terms of scope and breadth must be taken into consideration when pursuing the path of employing DT technologies in manufacturing or production of goods. Lastly, the choice of the technology platform is of utmost importance as there is variance in features, focus, and scope across commercially available DT platforms and required expertise needed internally to realize and integrate with the production environment.

14.14 Architectural Concepts

There is more than one architectural representation of DTs in existence in terms of overall scope and stack layers.

IBM's reference architecture of a DT consists of seven layers, from the bottom closest to the physical environment:

- **IoT stack**: ingests and provides input data to the DT through sensors and communication protocols.
- **Data**: data aggregation, asset state and structure, data lake, archiving, and storage services.

- **Systems of record**: integrates with IT/OT systems such as ERP (Enterprise Resource Planning), EAM (Enterprise Asset Management), and PLM (Product Lifecycle Management).
- **Simulation**: Multidimensional and predictive models and stochastic and deterministic simulations.
- **Analytics AI**: Analytics and models such as predictive maintenance and pattern recognition and optimizations. Feature extractions, diagnostics, and event pattern recognition.
- **Visualization**: Human-readable, 3D/2D models, AR/VR support, real-time dashboards, alerting, etc.
- **Process management**: Workflow, simulation tool-chains, and business process models

14.15 Simulations

A question that arises might be what is then the difference between a DT and a simulation. Let us spend some time looking at what a simulation is and the history of simulations. Simulation is defined as "imitation of a situation or process" or "the imitative representation of the functioning of one system or process by means of functioning of another" according to Merriam-Webster's dictionary.

The notion of simulations to simulate what may happen in the real world is not new and stems back to the 1940s where the Monte Carlo technique was first developed, which is a mathematical model to help predict possible outcomes under uncertain conditions based on a choice or action. The model was invented by John von Neumann and Sanislaw Ulam during World War 2.

A simulation is often constructed and built on a theoretical or mathematical model of a specific scenario or part of a product or process and often constructed and used in the design phase or post-production. The objective of a simulation is to generate the model defining the system or process and provide it with realistic and representative input data and observe the state and output of the model. If the simulation is an accurate enough representation of the system and the input data representative of the real-world input parameters, the model would be able to yield accurate internal state and output (Figure 2).

To simulate a system or a process, we need to start by defining a model that is an adequate representation of that system, object, or process. This can be accomplished by studying and performing measurements and observations in the real world and includes defining the set of input data that the model is dependent on. The real-world experiments can vary greatly in complexity based on what the simulation is set out to measure.

An example of a model that has been used in radio propagation simulations is the Okumura model. During the mid-1980s, Yoshihisa Okumura was trying to

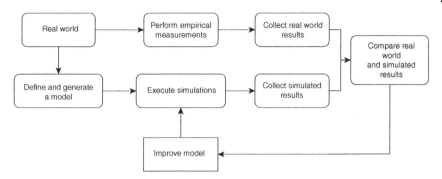

Figure 2 Simulation flow.

create a model for how radio signals propagate and are affected by attenuation as the distance from the transmitter increases. He traveled around in Tokyo and collected measurements throughout the city recording the strength of the received signals from different locations from the base station transmitter. He then was able to define a formula that could predict path-loss in an urban environment, which could then be instrumented as a mathematical model in a simulation environment.

- The frequency of transmission. Lower-frequency signals propagate longer than higher-frequency signals with the same transmit power. The constraints in terms of frequency range is 150–1500 MHz.
- The height of the transmitting and receiving antenna. Higher antennas enables longer transmission distances. The height of the antennas in this model ranges between 1 and 10 m.

The model also included a correction factor that could be added to the overall path-loss model, which enabled it to work not only in an urban environment like Tokyo but also in suburban and rural environments. This model is a close approximation for how radio propagation with respect to path-loss behaves in the real world. This model could then be used alone to generate simulations or be used in conjunction with a more complex system. For example, let us say we are developing a new mobile Internet infrastructure based on some radio technology that will operate across a few frequencies, and we would like to make sure that the system has an efficient handover algorithm. Handover is the process in mobile communication when a handheld device is moving geographically and must transfer the base station from one to another because of signal attenuation. We want the handover process to be as efficient as possible as this otherwise impacts the overall quality of the experience of the customers using the system. A customer using this system traveling on a train moving across the network performing multiple handovers would not want any interruption in the service but ideally a smooth and seamless handover experience. A simple simulation for this would need a few

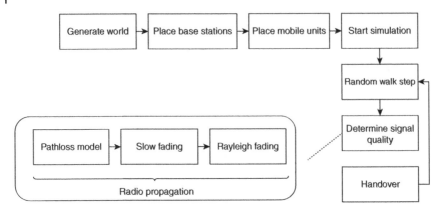

Figure 3 Handover simulation.

different models to be in place along with identifying the relevant external parameters that may have an impact.

- Pathloss model (as mentioned above) along with transmitters placed throughout the overall map/geography covered by the simulation.
- **Slow fading model**: An object moving in the environment will be subject to something called slow fading, which is due to the terrain such as mountains, trees, valleys, and other obstructions. The slow fading may have a certain correlation distance between peaks and valleys.
- **Fast fading model**: A radio signal will be received by the receiver multiple times with small variations in time as the signal has been affected by environmental elements. This causes the received signal to be subject to a statistical Rayleigh distribution.
- **Mobile unit**: A model that can instrument users adhering to some random walk model (from point A to point B) is required.
- **Handover model**: Performing algorithm required for a handover (i.e. what is the decision in which a handover should happen from one base station to another for the mobile unit).

With these, a rudimentary simulation could be implemented that simulated for many mobile users moving around. We can now measure the efficiency of the handover algorithm under these circumstances and work toward improving it during the early stages of the design phase to the point where we have confidence that we have a robust algorithm that would be able to operate meeting the overall objectives defined.

Figure 3 above provides a very high-level view of what such simulation system could look like.

- The world in which the simulation should take place is generated. This involves creating all the base stations and place them geographically on the map with a predefined set of geodesic coordinates. The world may be an urban, suburban, or a rural area. The mobile users could be of different types (e.g. travel by train, car, and bike) all having different travel velocities and different source and destination targets.

- The simulation is then started, and each mobile unit performs one cycle of the random walk which, based on the sample frequency (i.e. the desired granularity of each iteration of the simulation), will represent a unit of time. The higher the sample frequency, the smaller time interval between each successive point in the simulation. Higher values here may offer more details but would yield longer simulation times.

- Once the random walk has completed, each mobile unit in the simulation will determine its overall signal quality. In this very simple example, we are only basing a handover decision on the signal strength, which is not adequate in the real world as more parameters and quality metrics would need to go into that decision. Parameters that an actual handover algorithm may additionally use for evaluation might be the number of mobile units for each base station, the overall traffic load of base stations, and more relevant quality metrics such as BER (Bit Error Rates) and packet loss rates.

- The handover algorithm is then provided all the data for each mobile user and performs a decision as to whether a unit should be transferred to another base station. As every handover will result in some service interruption for the user, it is important to minimize the number of handovers while retaining adequate service for the user.

14.16   Digital Twins vs. Simulations

In contrast to DTs, simulations are not adaptable and are fixed or static in nature. The model which represents what the simulation will execute is generated and once outputs are generated can be compared with real-world data enabling it to be tuned, optimized, or adapted better to the real world. A DT can adapt its internal modeling of the physical object or process it is representing, allowing it to be dynamic in its nature, allowing it to converge toward a more accurate and better representation.

A simulation system traditionally does not operate on real-world input data but on a specific set of input data and ranges of those variables – hence it does not act on actual input data to the simulation framework the way a DT is. It is therefore more of a theoretical approach to how we believe the system or process will operate in the real world and for that reason limited to the specific design of its implementors under which measurements have been taken and the model created.

Finally, a simulation is typically used to test a specific subset of functionality of a larger system or process, hence not being able to provide a complete or thorough view or scope of the entire system or process. The DT can provide a lot more insightful, detailed, and accurate information.

Combining the use of simulations with DTs can offer many advantages in the development of a product or manufacturing process. In the early stages of research and development where an actual product does yet exist, simulation is all that is available. A simulation and model can be defined and used to help develop the first instantiation of a DT prototype that can be used with a simulation, using real-world data and iteration until the prototype behaves and produces expected results, enabling for the physical product or asset to be realized.

14.17   Digital Twin Products

There exists many solutions with different levels of sophistication and support in the DT domain.

- Microsoft Azure provides a digital twin offering and framework that enables customers to model their environment and objects using the Digital Twin Definition Language (DTTL), which is a data modeling language used based on nodes and edges. The DTTL was also defined by Microsoft. The Microsoft DT product offering enables live execution environments and integration of sensor data through IoT HuB. Output data and events from DTs can be fed to other Azure services such as Azure Data Explorer, Synapse Analytics, or Event hubs.
- IBM has a set of software assets and services for the DT domain such as the Digital Twin Exchange which is a marketplace for digital twins, many based on the Maximo Asset Health Insights, which is a single platform for customers to monitor and manage their assets. The software suite allows for insights through data collection to be analyzed to help provide preventive, predictive, and pre-scriptive actions that may have to be taken to help improve efficiency.
- Oracle's DT offering for the IoT segment has an implementation offering three types of digital twins they refer to as Virtual Twin, Predictive Twin, and Twin Projections. The virtual twin represents a physical object or asset representing and capturing all the required attributes to create a semantic model that can be operated using a declarative paradigm. The Predictive Twin builds on the virtual

twin, in that it captures an analytical or machine learning model that can be used for prediction purposes of the physical asset, such as predictive maintenance. Finally, the Twin Projections integrates the Predictive Twins insights to the business backend systems, providing more holistic and comprehensive insights and enabling actions to be taken on predictions proactively across, for instance, a manufacturing floor or site.

14.18   Digital Twins and Manufacturing in the Future

DTs is one of the rapid growth areas having broader and broader adoption across many segments of the industry and thanks to its versatility can be applied to almost anything and has become a general best practice, increasingly replacing historical data-driven approaches [1]. It is predicted that by 2025 almost 90% of all Industrial IoT platforms would have adopted digital twins in some or all parts of the industrial platform and reaching a standard within the industry by 2027 [2].

Other studies from 2022 from Altair are asserting that over 70% of business they have surveyed began investing in DT technology in the last year, and the key drivers are in areas of product time to market, accurate risk management, customer satisfaction, real-time control, and management and efficiency and cost reduction [3].

The overall market is predicted to be close to 16 billion in 2023 alone [4] and only keeps increasing year over year. The Fourth Industrial Revolution and the emergence of mature and solid digital twin technology will help companies further increase manufacturing and operational efficiency and help improve cost profit margins and is true for the O&G industry as well as many other market segments.

References

1 Bob Violino. Digital twins are set for rapid adoption in (2023). CNCBC. https://www.cnbc.com/2023/01/21/digital-twins-are-set-for-rapid-adoption-in-2023.html (accessed 09 September 2022).

2 Mark Crawford (2021). Digital twins for manufacturing. https://www.asme.org/topics-resources/content/7-digital-twin-applications-for-manufacturing (accessed 09 September 2022).

3 Why digital twins adoption rates are skyrocketing (2022). https://altair.com/newsroom/articles/why-digital-twin-adoption-rates-are-skyrocketing (accessed 09 September 2022).

4 David Immerman (2020). Why IoT is the backbone for digital twin. https://www.ptc.com/en/blogs/corporate/iot-digital-twin (accessed 09 September 2022).

5 AAS sub models templates. https://industrialdigitaltwin.org/en/content-hub/submodels (accessed 09 September 2022).

15

IoT Edge Security Methods for Secure and Safe Oil and Gas Environments

15.1 Introduction

Security is a foundational pillar in edge computing and Internet of Things (IoT), which can have more consequences in terms of impact and damage compared to other markets. Traditionally in security, we often look at three pillars often referred to as the CIA triad (Confidentiality, Integrity, and Availability). Let us examine those terms a bit more.

Confidentiality has to do with keeping information secure and ensuring that it is protected from malicious eyes. The objective is to keep information private through various techniques and prevent disclosure of the sensitive data or information. There is always an impact to what happens if the sensitive data are compromised or leaked and based on the severity of the impact, security architects decide how to best protect the asset.

Integrity is concerned with ensuring that the information or data is accurate from the moment it leaves its source to its destination. With accurate here, we mean that it has not been tampered or modified either intentionally or nonintentionally, in other words the *trustworthiness* of the data.

Availability refers to the general availability of a system or solution and its corresponding services to its consumers, i.e. all of the authorized principals of the system have access to the required services, related information, and assets.

In traditional IT Enterprise systems, the priority between Confidentiality, Integrity, and Availability was often in that specific order. Confidentiality is more important than Integrity, followed by Availability. Compromised data or assets could have a devastating impact on Intellectual Property (IP) or other company-related assets, whereas a Denial of Service (Dos) attack against personnel

The Power of Artificial Intelligence for the Next-Generation Oil and Gas Industry: Envisaging AI-Inspired Intelligent Energy Systems and Environments, First Edition. Pethuru Raj Chelliah, Venkatraman Jayasankar, Mats Agerstam, B. Sundaravadivazhagan, and Robin Cyriac.
© 2024 The Institute of Electrical and Electronics Engineers, Inc.
Published 2024 by John Wiley & Sons, Inc.

compute system equipment would be disruptive, but would at least not compromise company assets.

In the IoT domain, particularly when it comes to production or manufacturing of things, the relative priority of these pillars is many times slightly different, where *Availability* is generally more important than traditional IT due to the simple reason that a disruption in production has a direct impact to profit and cost for the organization. A production line that goes down due to an attack that would stop manufacturing may have rippling effects across multiple parts of the supply chain.

In this chapter, we will look at security from an IoT and Edge Computing perspective, providing an introduction to some foundational security definitions and principles and then moving on to exploring modern security architectures in this space to meet the aforementioned triad.

15.2   Protecting Data

When it comes to providing assurance and protecting data throughout its entire lifecycle, we often separate this into three different categories: protecting at rest, in use, and in transit. Data protection at rest refers to providing protection of data (e.g. through cryptographic services) when it is not actively in use, for example, when it is stored on non-volatile storage such as a flash disk or solid state drive.

Data protection in transit refers to providing protection of the data when it is transmitted between two endpoints over a network, providing assurance that only authorized principals can access the data so that anyone eavesdropping or snooping traffic between the communicating parties cannot access the data.

Data protection in-use refers to providing protection of the data when it is in use by software, ensuring that adversaries are unable to access the sensitive data, whether it being through uncontrolled access mechanisms of the software, misconfigured software libraries, frameworks, databases, or inadvertent access to process memory pages or cache.

15.3   Past Examples of Security Attacks

There exists many examples of exploits and attacks in the IoT space, and in this section we are going to look at a few of those to shed some light on the impact these attacks can have, not only in the economical perspective but also societal in general.

The Mirai botnet attack [1] emerged in 2016 and was used for numerous DDoS (Distributed Denial of Service) attacks across the globe. Mirai is a malware that exploited vulnerabilities in various IoT devices such as home routers, smart cameras, and thermostats to turn them into "bots" that could be used to orchestrate distributed attacks against selected targets. The compromised IoT devices had security vulnerabilities, where default usernames and passwords allowed for the malware to take control of the devices, injecting itself and continue spreading by searching for additional insecure devices to enabling the malware to continue spreading. The malware then performed numerous DDoS attacks across the Internet, most notably a famous cyber-security journalist but also Dyn, which is a major DNS (Domain Name Server) provider. The latter attack had an impact throughout North America and Europe, where major service outages occurred across popular websites and services such as Amazon, Comcast, CNN, Playstation Network, and Visa. The Mirai attack is an example of the *Availability* pillar mentioned above and the consequences of that attack.

The BlackEnergy attack [2] against Ukraine's power grid infrastructure took place in 2014 and was a combination of social engineering and skillful hacking to bring down a large portion of Ukraine's infrastructure which impacted over 250,000 people. The attack used the phishing technique where an email with an Excel attachment was sent out to many employees, and the Excel file contained a macro that was inadvertently executed by an employee that provided remote access to hackers to the IT infrastructure. The hackers then discovered that some of the critical industrial control systems on the OT (Operational Technology) side was connected to the same network they were able to gain access to the power grid control infrastructure and disrupt the power supply to over almost 250,000 people, affecting multiple distribution centers and even bringing down the support call center, making it impossible for customers to reach out and get status details and updates about the outage. This attack shows a severer impact of the *Availability* pillar, but this time much severer as it attacks foundational electrical infrastructure in a country.

The SolarWinds backdoor attack [3] in 2020 was one of the most sophisticated cyberattacks with a huge impact, some of which today are not entirely clear. This was a supply chain attack where attackers were able to compromise computer infrastructure to a vendor of over 30,000 customers and then leveraging the existing trust relationship that exists between the vendor and customers to deploy malicious software and compromise customer equipment. The attackers compromised a monitoring system called Orion that SolarWinds provides to its customers, which monitors the overall network and applications. Hackers were then able to produce what looked like a legitimate update, which included a small footprint Trojan horse. Impacted customers received notification about the software update from SolarWinds as it was deployed and installed on customer systems. The

Trojan horse activated a backdoor that enabled hackers to get remote access to the compromised systems. Through this backdoor, they were able to deploy software without detection, which enabled them to gain insights into vulnerabilities they could further exploit and went undetected for several months, gaining unauthorized access to emails and other sensitive information. It is estimated that over 18,000 customers were impacted, which includes an estimated 20% of government accounts.

Supply chain attacks are likely to see an uprise as it can deliver a high-impact attack, utilizing trust relationships between partners in the supply chain by identifying and exploiting one weak spot in the supply chain infrastructure. This cyberattack provides an example of exploiting *Confidently* and *Integrity* pillars.

The Colonial Pipeline Ransomware Attack [4] took place in May 2021 and was one of the first big impact ransomware attacks impacting the industry. The hackers were able to gain access to the internal network of Colonial through a VPN (Virtual Private Network) that employees are using for remote access purposes. The specific account was at one time tied to an employee at Colonial but was not actively in use; however, the password for the account had appeared in other leaks on the dark web, suggesting that the employee may have been using the same password across multiple accounts. The hackers encrypted the billing and customer data and asked for 75 Bitcoins, close to 5 million dollars to get the decryption key to Colonial to recover their data. The ransom was eventually paid, but the Federal Bureau of Investigation (FBI) has since recovered most of the Bitcoins.

These types of attacks are a real threat to corporations and are going to increase in rate and level of sophistication over time. The opportunities for hackers to seek financial gain through these types of attacks only increases as more and more of our critical infrastructure is becoming digitalized. This further understrikes the importance of a sound and solid security architecture to reduce the risk of potential attacks. Not only does paying the ransom money further motivate criminals to continue attacks, these types of attacks have a direct impact to the business continuity and the bottom line beyond, also impacting the everyday life of citizens. This cyberattack provides an example of exploiting *Confidentiality* and *Availability* pillars.

15.4   Security Foundation

Before we start to look at security from an Edge computing perspective and particular deployment scenarios applicable to the oil and gas industry, we need to introduce the basic elements that are required to harden and improve the edge security.

Symmetric ciphers are algorithms used to transform *plain text* to *cipher text* with the use of a *key*. The plain text can be transformed to the cipher text by applying the key to the algorithm, and conversely, applying the key again on the cipher text will produce the original plain text; therefore, these are referred to as symmetric ciphers. Only the intended parties have access to the key, and the cipher algorithm should have been designed in such a way that it is not feasible from a time perspective with the help of computed resources and cryptoanalysis to recover the plain text from the cipher text without having the key in its entirety.

Symmetric ciphers are used today everywhere when communicating between endpoints, along asymmetric ciphers which we will explain in the next section. A good algorithm would yield better protection against the attack the longer the key is; however, a longer key size also involves additional computing needed to perform encryption and decryption operations. Today, it is common to see 256 bit key sizes, and many platform vendors do provide fixed function units in hardware, which allows for the encryption and decryption steps to be offloaded to hardware, e.g. in the packet processing pipeline directly or when used with user space libraries, such as OpenSSL.

The National Institute of Standards and Technology (NIST) [5] continuously provides industry recommendations of encryption algorithms to use along with recommended key sizes. They also keep a list of deprecated cipher suites and when those should be taken out of production and deemed insecure.

- **AES**: Advanced Encryption Standard using 128-, 192-, or 256-bit keys
- **Blowfish**: A successor to the DES with key sizes ranging from 32 to 448 bits

A few challenges when it comes to symmetric ciphers are the following:

- How do you protect the key on a system?
- How often to you rotate or change the key?
- How do you agree on a key between two entities and subsequently share it?

Asymmetric ciphers are different from symmetric ones, in that they use a different set of encryption and decryption keys; there is one key for encryption and another key for decryption. Here, only one key is considered an asset, referred to as the *private key;* the other key is referred to as the *public key*. Asymmetric ciphers require a significant amount of computing compared to symmetric ciphers, which is why they are typically used to bootstrap communication between parties (i.e. authentication of the peers, cipher suite agreement, and negotiation on a symmetric key to use).

Asymmetric encryption algorithms use a mathematical relationship between the two keys and the resulting plain- or cipher text, which in theory would enable anyone from computing the private key knowing the public key; however, due to the length and complexity involved in solving the problem, it is infeasible and impractical for any computer system to do in the foreseeable future. One of the most famous and adopted algorithms is RSA (Rivest, Shamir, and Adleman) [6] named after its inventors, which is anchored in the fact that a computer can easily multiply numbers of arbitrary size, but it is tremendously difficult to factorize a number into its primes. We consider p and q as very large prime numbers, and the product q is defined as

$$n = p \times q$$

Suppose the public key is n, and if you are able to find out the two prime numbers n was multiplied by, you would get to the private key. For small numbers, this is straight-forward, for example 35, you would easily be able to factorize it to 7 and 5, but as the numbers become sufficiently large, this becomes infeasible for a computer to do. The magnitude of order for such key sizes are in the order of 1000 bits (e.g. 2048 or 4096 bits long), which means that you are dealing with numbers such as 2^{2048}, which would have 617 decimal digits.

Elliptic Curve Cryptography is another alternative to asymmetric ciphers based on elliptic curves (ECC) which are also anchored in computational complexity, in that it is easy to go in one direction, but calculating its inverse is not feasible. ECC ciphers have some advantages over RSA, in that you can get the same level of computational complexity with shorter key sizes, which is advantageous when it comes to the performance of the algorithm and making them more feasible for more constrained devices (e.g. Bluetooth Low Energy devices), although they exists and are adopted across the board today.

In summary, comparing the RSA and ECC, we found that

- ECC is more secure than RSA, so to reach equivalent security with RSA, you would typically need keys that are about 10 times the size. The 2048-bit RSA key corresponds to 224–255 ECC bit key length. The 3072 RSA key corresponds to 256–383 bits.
- RSA is a very well-established and adopted technology that has met the test of time, but vulnerabilities have been found and is likely to be phased out over the course of time.
- ECC works on elliptic curves and is projected to increase in adoption over time as the RSA is phased out.

Asymmetric ciphers are fundamental security technologies in PKI (Public Key Infrastructure), which include certificates which the public key is embedded into along with other properties and finally *digitally signed* by a trusted entity. The certificates are generated by the same trusted authority and commonly referred to as Certificate Authorities.

15.5 Cryptographic Hash Function

Hashing algorithms are key technologies used in security technologies from *Message Authentication Codes* (MAC), *Integrity Protection, Attestation, Proof of Work,* and *Digital Signatures* to name a few. A hashing algorithm operates over a message to generate a *"unique"* fixed-size checksum or *digest* over the data essentially representing a unique fingerprint to the message the hashing function has been executed over. We use the unique in quotation because there is a very small likelihood for collisions, which means that two different datasets would generate the same *digest*.

A hashing function is a one-way function; in other words, it is not possible to get the message from the *digest*. Different algorithms generate different fixed-size *digests,* which range from 128 bits (not deemed secure today) to 384 bits and larger.

Figure 1: An illustration of a hash function generating a digest for an input message.

There are a few important characteristics we are looking for from a robust and useful cryptographic hash function.

- Efficient and fast in terms of computation.
- Likelihood of collision should be minimal.
- A small change to a message input should result in a radical change in the digest itself; this is called the *avalanche effect.*
- The hash function should be 1-way.

How would this be useful? Let us take an example of someone sending a message m from Alice to Bob. Alice could along with the message also send the corresponding digest, $digest_A$, as a *checksum* for Bob. When Bob receives the message, he could compute the $digest_B$ and compare them. If they are not the equal, it means that the integrity of the message has intentionally or unintentionally, e.g. during transmission, been modified.

There are multiple different options for hashing algorithms today, and some have been found to have vulnerabilities and are hence deprecated or no longer

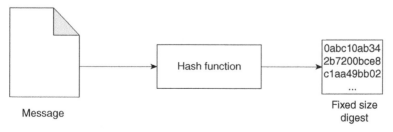

Message Hash function 0abc10ab34 2b7200bce8 c1aa49bb02 ... Fixed size digest

Figure 1 Hash function.

recommended to be used. A few of the most common ones used today are as follows:

- Secure Hash Algorithm (SHA). The SHA is a family of cryptographic hash functions that are one of the most popular used today in the industry. SHA-256 and SHA-384 are commonly used today.
- Whirlpool based on the AES and yields a 512-bit digest

15.6 Keyed Hash Message Authentication Code

The HMAC uses an underlying hashing algorithm to provide a special form of the **MAC** (Message Authentication Code) that helps verifying the integrity and authenticity of the data. A hashing algorithm would generate a value from some message, i.e.

$$\text{Digest} = \text{Hash}(\text{message})$$

The HMAC takes an additional shared secret known only between the trusted parties, which is then used in verification of the digest, i.e.

$$\text{Hmac} - \text{Digest} = \text{HMAC}(\text{secret}, \text{message})$$

The HMAC provides properties stronger than those of the hashing algorithm example used above as it would eliminate an attacker that knows the message to produce a legit message providing a corresponding digest. In the HMAC, the shared secret, known only to Alice and Bob, would be used as the second parameter and verified by Bob. It should be noted that this implies that the *shared secret* used between Alice and Bob has been communicated and shared prior to sending the message.

As with standard hashing functions, a small change in the key should have a huge impact to the resulting HMAC code. The HMAC algorithms are today commonly used with the family of SHA algorithms such as SHA-256, SHA-512, SHA-384, SHA3-224, and SHA3-384.

15.7 Public Key Infrastructure (PKI)

While public/private key pairs can be used in their raw form, it is more common to see them in place of a PKI system, where there are certificate authorities providing a certificate to communicating peers and their endorsement of the claimed identity of the other party. The challenge with public/private key pairs otherwise

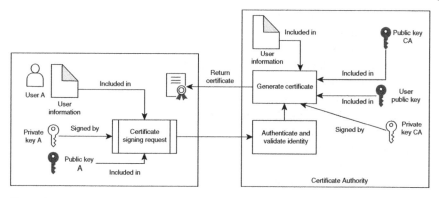

Figure 2 Certificate signing.

is to know who the communicating peer really is. Is it the intended party of someone masquerading to be that person?

Figure 2: Overview of a PKI system generating and returning a certificate from a Certificate Signing Request (CSR).

A user needing a certificate would first have to form a *certificate signing request* (CSR). There is a multitude of different CSR algorithms supported by the IT infrastructure today. The certificate authority needs some key information in order to issue a certificate to the user (or machine) that should be included in the certificate

- User information
- The user's public key
- Information to validate the user's identity
- Signed data that can be validated with the user's public key

The certificate authority (CA) will authenticate and validate the information contained in the CSR and generate a certificate, including the essential information provided by the user. It will then sign the certificate and return it back to the user.

It should be noted that there could be additional certificate authorities (i.e. a chain of authorities), each signing the request to form a chain of trust back to the root certificate authority. During authentication, the entire chain is validated, including validating key information in the certificate such as pertinent data and validity of the certificate (e.g. expiration date).

PKI can be realized mutually or one-way (server authentication). Mutual authentication with certificates is where both the client and server are using certificates to authenticate themselves before trust is established between them. One-way authentication is where only one end is allowing the other end to authenticate with a certificate, and the other is using alternative credentials such as *token* or username/password encoded in some form.

The latter is what we commonly see on traditional web services, where the server is authenticating itself to the client using the certificate (i.e. we know that we landed on www.amazon.com and not a phishing site because the certificate contains the DNS name for Amazon and has been signed by a CA the browser trusts, such as DigiCert). However, the client/user is authenticating itself using a different set of credentials derived from a username and password pair created when we set up the account. The former is what is more common in machine-to-machine communication, where both entities are using certificates to mutually authenticate themselves over a TLS (Transport Layer Security).

Using certificates signed by a mutually trusted party solves this problem; however, it comes with some challenges every business needs to address.

- Managing a large PKI infrastructure can be problematic and costly at scale
- How long are the certificates valid for and what does the certificate renewal process look like for devices and principals?
- What happens if the Certificate Authority is compromised and how CRLs (Certificate Revocation Lists) are managed and propagated effectively in environments with limited or intermittent connectivity?
- The certificate itself is considered public knowledge as it only contains the name, various certificate attributes, and endorsements along with the public key – how is the private key corresponding to the certificate stored?

15.8 Digital Signatures

These provides stronger properties than HMACs and are commonly used with certificates for signing emails, software packages, etc. A digital signature works with asymmetric keys and certificates instead of with symmetric keys such as an HMAC, and this difference provides an additional important property they provide over HMACs.

A digital signature would typically be applied in the following manner:

- A digest is calculated on the message (e.g. an email message).
- The message and digest are signed by the sending party by encrypting the entire content with its private key. The message and the corresponding public certificate (containing the public key of the sender) are transmitted to the receiver.
- The receiver validates the certificate from the sender and decrypts the content with the public key embedded in the certificate.
- The receiver calculates the digest of the plain text of the message and compares it with the digest that the sender has provided.
- If the sender's and receiver's digest matches, the message has not been compromised.

The additional security property introduced here is *non-repudiation,* which means that there is assurance from the receiver that the message really originated from the sender. However, there is nothing built-in here that would prevent an attacker from *replaying* a previously sent message. Those techniques can be built-in and extended by using nonces, timestamps, and other mechanisms.

15.9 Threat Analysis and Understanding Adversaries

When architecting and defining a security solution in general, it is important to conduct a thorough analysis of the system from a security standpoint, commonly referred to as a threat analysis. During this process, we analyze the security posture of the system to better understand the vulnerabilities of the system. In a threat analysis, we typically want to understand the following aspects.

- What are the security objectives of the system? In other words, what are we trying to provide assurance around when it comes to the security architecture?
- What are the non-goals? What are the aspects and scenarios that are entirely outside the scope of the threat analysis that we are not trying to address or solve?
- Identify what are the adversaries we are trying to protect against. There are multiple categories of adversaries that would have different access rights and be less or more difficult to mitigate attacks from, such as a user with non-admin access to the system, a user with root access to the system, and a remote network user.
- Identify what the assets are in the system. What are we trying to protect? This could be data, credentials, privacy-related information, etc. What type of protection is required of the asset? It could be protection against modification, disclosure, etc.
- Understand how those assets should be protected and what the impact is if one or more of the assets are compromised. What are the adversaries that could compromise the assets? What are the overall security threats in which those assets could be compromised
- Finally, which of the identified attacks are mitigated and which ones are not mitigated.

Doing the threat analysis, the system gains valuable information in terms of the overall posture of the system and will shed light on specific risks (i.e. unmitigated threats) that are deemed not acceptable and should be addressed, whereas some threats might be something the organization can live with because the *risk of those attacks being successfully carried out and the corresponding impact is sufficiently low.* There is no simple binary answer (yes vs. no) when it comes to the readiness and security of the system. It comes down to an informed risk assessment, where no security solution will be 100% proof from adversaries.

15.10 Trusted Computing Base

Is a key concept and is important to understand before we look at the different security aspects, challenges, and techniques for hardening an Edge IoT platform. The trusted computing base or TCB is the combination of all the hardware, firmware, and software components relevant to the security posture of the system. All the components in the TCB are providing the overall security of the system, so if one or more of these components are compromised or attacked, it may result in the entire system to fail or being compromised along with its assets. The TCB is responsible for enforcing system-wide security policies. It is very important to understand that while this is the set of components providing the trust, it does not necessarily mean secure. Trusted here refers to the fact that it is critical to security within the context of the overall system and does not mean that it is impenetrable by malicious software. In the subsequent sections, we will discuss more about how the TCB can be verified and attested by remote and local verifiers to ensure that the platform hardware and software is meeting compliance and associated integrity checks.

15.11 Edge Security and RoT (Root of Trust)

In security, we desire to have security anchored down into something immutable and trustworthy at the platform level and provides the foundation on which all security operations depend on; we refer to this as Root of Trust or RoT. IoT platforms provide multiple mechanisms in providing hardware RoT used for different purposes. These can come in multiple forms, such as one-time programmable fuses, ROM (Read Only Memory) for integrity, and verifying the authenticity of software components.

One of the key technologies used today is HSM (Hardware Security Module). There are a few different versions of HSMs, and perhaps one of the most common and adopted throughout the industry today is the Trusted Platform Module (TPM), and the TCG (Trusted Computing Group) [7] provides security specifications for multiple types of HSMs such as TPM and DICE (Device Identifier Composition Engine) [9].

The TPM provides a variety of security functions. In addition, a diverse set of security services can be built and defined to help address different business challenges such as disk encryption tied to hardware root of trust, platform attestation, a vault, etc. The TPM is meant to run as a secure element in a separate execution environment, different from the main host. Below is a summary of some of the essential functions the TPM provides:

- Provide a secure facility to store symmetric and asymmetric keys.
- Provide a facility to perform chained measurements through an operation referred to as "TPM_Extend," which paves the way for attestation. We will look at this in more detail in the attestation section.
- Provide cryptographic functions such as generating keys, sealing keys, and generating random numbers.

A TPM can be implemented and realized on a platform in multiple ways with different security postures.

- An implementation in software running on the host operating system. This is mainly done for testing and interoperability purposes and not used in any production deployment.
- A firmware-based TPM typically running on a co-processor adjacent to the CPU.
- A dedicated SoC (System-on-a-Chip) on the platform.

TPM as a dedicated SoC is generally a more costly solution but provides stronger isolation characteristics. Currently, TPM 2.0 is the major version used and deployed today.

15.12   DICE – Device Identifier Composition Engine

Is another security standard from TCG that came about after TPM and was targeted specifically for smaller, more constrained device types, while providing a good secure posture and ability to have security anchored in device RoT. The primary market for DICE was IoT for MCUs and small SoCs. It provides cost benefits over a traditional TPM, in addition to power and physical space benefits.

The DICE is primarily used to ensure that the boot integrity, the different software layers in the boot process, is authenticated and provides a trusted computing base (TCB) with a strong device identity, paving the way for attestation of the device's firmware and software. The DICE uses a UDS (Unique Device Secret) to uniquely protect each consecutive layer in the boot process cryptographically bound to the previous layer.

The DICE also provides a unique device identifier, referred to a CDI (Compounded Device Identifier). When the device comes out of reset, the device is computing a integrity check, or measurement, which is a digest of the first layer of the software on the device. The CDI is calculated by combining the UDS with the digest of the first layer software in an HMAC or hashing function. The CDI calculation could optionally include measurements of other states of the platform, such as hardware registers, and the device is required to keep disclosure and access to the CDI protected. If the CDI is compromised, a firmware update of the device would then yield a new CDI.

15.13   Boot Integrity

The key to providing a secure and trusted computing environment on the edge starts at the boot process and ensuring that all the software involved in the platform boot can be verified and is trusted. Secure Boot is an umbrella term that involves Verified Boot or Measured Boot.

Figure 3 Boot chain.

Figure 3: Illustration of chained trust in Secure Boot.

Verified Boot is anchored in some form of hardware RoT, whether it being Boot ROM, hardware fuses or Boot Guard. When the platform is powered on and released from the initial *Reset* state, the platform starts verifying the integrity of the very first lines of code executed, which is the Initial Boot Block (IBB). The verification of the IBB can be done against calculating the *digest* of the IBB and that of the hash value against the immutable hash value provided at the manufacturing time for the IBB. The IBB executes and before loading and executing BIOS, it first verifies its integrity.

The boot chain continues executing and passing control to the next one. The integrity of the next component in the boot chain is verified before it is executed. If the integrity of the next component cannot be verified, the boot process is halted and system boot process stops.

Measured Boot works similarly, in that it also verifies the overall integrity of the boot chain. The main difference is that during Measured Boot, the boot process is not halted immediately if the integrity of the next component in the chain does not meet its integrity value, and this is because the verification of the boot process is not done until the entire platform has booted up.

One of the main technologies commonly used in measured boot is a TPM, which here functions as the root of trust. The TPM contains a set of Platform Configuration Registers (PCRs). The primary purpose of these is to record measurements (i.e. hashes or digests) of the software being executed on a platform. A TPM typically consists of 16 or 24 different PCRs that are used to store different types of platform measurements. The values of the PCRs cannot be reset or modified by any software or firmware in the system. The way the values are stored in PCRs is through an operation called TPM_Extend.

The TPM_Extend operation calculates a rolling hash over the previous measurement stored in the PCR, i.e.

$$\text{digest}_i = \text{Hash}_{alg}\left(\text{digest}_{i-1} \| \text{data}\right)$$

A hash algorithm such as SHA-256 is used to calculate a new digest based on some data concatenated with the previously calculated digest value. This creates a cryptographically chained operation anchored down to the RoT of the TPM and the inability for any software component to store arbitrary values in PCRs, but rather must use the TPM_Extend function which internally performs the calculation outlined above. The first component storing data into a PCR would have $digest_{i-1} = 0$.

Once the platform has booted up, the different PCRs will have values in them as a result of different components performing their measurement operations through TPM_Extend, and the resulting set of these PCRs with the corresponding values can collectively be compared against the *expected* data, also referred to as golden values for the PCRs if the platform's integrity is met.

Disk Encryption is foundational to provide *security at rest* for any data deemed important to corporations. Disk encryption is accomplished through symmetric encryption, and only someone who has access to the symmetric key can gain access to the data. One of the issues with traditional disk encryption not tied to and hardware security module on the platform is that the security assurance is tied to the security of the filesystem of the operating system. An adversary gaining root or privileged access to the platform would be able to mount an encrypted volume and unlock its content.

The platform's TPM offers additional protection here, which not only allows for the symmetric key to be better protected by storing it on the TPM, but it also provides a mechanism to conditionally allow for it to be *unsealed* and provided to the host. This works by defining a set of PCRs and the value they should be satisfying and then sealing the disk encryption key to those. Before the disk can be mounted during the boot process, measurements of the targeted PCRs are taken, and when the request is provided to get the disk symmetric key, the TPM determines whether the PCRs are satisfying the authorized policy and then the key is provided back to the host to allow for the volume to be mounted and accessed by authorized principals on the host operating system. It should also be noted that the sealing is bound to a signed PCR policy, which can contain multiple valid PCR values allowing for software updates, which would otherwise modify the rigid PCR data, rendering the unsealing process to fail.

An authorized policy requires PKI infrastructure integration, as the policy must be signed and a public verification key and signature required to be provisioned during the onboarding process. It does provide a higher level of assurance over traditional disk encryption, or worse no encryption, as access to the data is only granted if the system is in a well-known good state. This prevents an attacker from gaining access by booting an alternative OS or login credentials.

15.14 Data Sanitization

Is critical to ensure there is a mechanism in place for remotely managed devices to wipe their data and ensure that sensitive information cannot be accessed. This would be a step implemented during the device decommissioning phase. When data are deleted, it is not really erased from the media it was stored on but could be recovered by a malicious entity if they get hold of the device.

The data sanitization process applies to any type of media such as providing storage services of non-volatile memory as USB flash drives, solid-state drives (SSD), and traditional hard drives. During the data sanitization process, the media, or part of the media, such as a volume, would be overwritten with random data multiple times to ensure the entire target is overwritten and not leaving any of potentially sensitive data behind.

Trusted Execution Environment provides a mechanism through which applications and associated data can run isolated from the rest of the system with assistance from underlying platform hardware. Data here could be anything, considered an asset or secret such as passwords, personal identifiable information (PII), and sensitive IP-related data or financial data. Intel SGX (Software Guard Extensions) is an example of this technology that we will use as an example, although other similar solutions exist. SGX provides multiple benefits from a security perspective, which we will discuss further in this section. First and foremost is that this allows for applications to keep its assets from other parts of the platform, even if the system has been compromised (BIOS, Operating System, etc.).

SGX provides what is called *enclaves* to applications. These are reserved non-addressable memory pages provided by the platform, which are encrypted by the platform. An application must be developed to take advantage of SGX and cannot automatically be instrumented with the technology. An application developer would create enclaves inside the application and determine what functions or code should run inside the application. When an application enters an enclave, the application gets access to the sensitive data or code inside the enclave in clear text, while the processor ensures that no other entity can gain access whether it is the operating system, a privileged user, the hypervisor, etc.

SGX can also be used to provision secrets to a trusted application inside of an enclave through remote attestation. This helps address an important and common problem, which is – how does one provide secrets to an application in the first place dynamically? The remote attestation server can help provision initial secrets or assets required by the application to run verifying the application's identity, ensuring that the application has not been tampered with (i.e. verifying its integrity) and that it is running on an Intel genuine SGX-enabled platform, and finally ensuring that the rest of the system is compliant.

15.15 Total Memory Encryption

Providing protection of system memory contents helps mitigate a variety of attacks but most notably hardware attacks where an attacker may try to gain access to DIMM (Dual In-line Memory Module) through custom hardware or attempting to persist the transient data the memory holds by spraying it with a liquid gas, for instance. AMD and Intel both offer memory encryption referred to as SME (Secure Memory Encryption) and TME (Total Memory Encryption) respectively. This means that data to and from the CPU are always encrypted with a transient key transparently to the rest of the system and users.

For multi-tenant usage models where guest VMs (Virtual Machines) are running on the system, there are additional technologies such as MKTME (Multi-Key TME) where multiple, different transient keys can be used to protect different parts of the memory. The hypervisor can manage the TME capability through using different keys to the different VMs transparently, hence providing stronger isolation characteristics and security posture between the virtual machines from an ephemeral memory protection standpoint.

15.16   Secure Device Onboarding

It is critical that edge devices can be securely onboarded in the infrastructure they are deployed in. When we talk about device onboarding, we are referring to the initial flow required for the device to be properly configured and have all the necessary security assets deployed for it to be functioning from Day 0 onward (Day 1 and Day 2 operations).

During this flow, it is necessary for the backend infrastructure to be able to authenticate the device itself. Is it a trusted device? What credentials are used to authenticate the device to the infrastructure and how were they configured to the device in the first place and where did that happen in the supply chain? There are multiple places where initial device credentials could be injected such as

- Device manufacturer.
- Intermediate supply chain issuer.
- IT/OT personnel.
- Installation technical.

There are multiple solutions to Secure Device Onboarding (SDO), and in this section we will briefly look at two options available today. The FIDO [8] (Fast Identity Online) Alliance is a non-profit organization composed of multiple leading companies in the industry driving standards that defines a set of technology agnostic security specifications. The FIDO has defined a standard for SDO referred to as FDO or FIDO Device Onboard, which is also hosted in LF Edge under Secure Device Onboard. The FDO specification aims to provide benefits around onboarding in multiple ways such as

- Zero Touch and Fully Automated Onboarding
- Fast and Secure

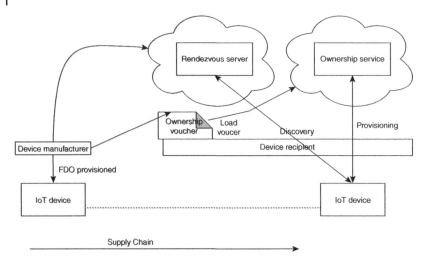

Figure 4 FDO overview.

- Open Standard
- Supporting multiple hardware and security technologies

  Figure 4: FDO Overview provides an overview of the FDO workflow.

- The device is initially provisioned with an FDO client software along with establishing a RoT-based key inside of the device, which is later used to authenticate the device. The RoT can take multiple forms and can reside in the TPM or other HSM. This step can happen at the manufacturing of the device or at a later stage in the supply chain. An ownership voucher is also created, which is provided along with the device, which will ultimately be provided to the rightful owner of the device who purchased it; it is an electronic credential that could be provided as a text file, QR code, etc.
- The IoT device is ultimately shipped and has gone through the supply chain along with its voucher.
- The owner of the device receives the platform and registers the ownership voucher with a rendezvous server and provided to the ownership service at the target infrastructure backend the device will ultimately use for its onboarding flow.
- The technician installs the device and connects it to the network. When the device is powered on, it will reach out to the rendezvous server (which would have been configured as part of the initial step). The rendezvous server acts as a DNS and provides a lookup service, providing the endpoint the device should connect to for the onboarding flow, the ownership service.

- The device connects to the ownership services and uses its RoT key to identify itself to the backend infrastructure, as well as the backend infrastructure authenticating itself to the device, proving that the device is connecting to a rightful owner of the device. The mutual authentication allows for a secure tunnel to be established between the ownership service and the target IoT device.
- Any additional credentials, configuration material, and other bootstrapping workflows required for the device to become fully operational can now happen inside the protected tunnel and is where the IoT owner would integrate its own unique workflows, such as enabling hardware rooted volume encryption or setting up remote attestation.

Microsoft Azure has an onboarding flow for IoT devices connecting to the Azure IoT hub [10] and the rest of the Azure services that can be enabled with HSM-supported devices in general and TPM 2.0 in particular.

Figure 5: Azure Onboarding illustrates this overall flow.

- The first step is for the device to be provisioned with the IoT Hub Client software which can communicate with the IoT Hub. It is also necessary for IT to take ownership of the TPM and extract a registration ID which is based on a digest of the TPM's Endorsement Key. Registration ID = SHA-256(Ek_{Pb}).
- The IT administrator registers the device in the IoT Hub portal as one of the devices enabling automatic provisioning and onboarding of.

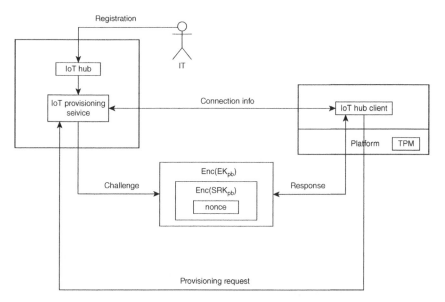

Figure 5 Azure onboarding.

- When the device is first booted up in the deployment, it will reach out to the Azure provisioning service and issue a provisioning request. It uses the Endorsement and Storage Root key of the TPM, $EK_{Pb,}$ and SRK_{Pb}.
- The Provisioning service verifies that the device is a device that has been registered and sends a challenge to the device by generating a nonce, which is encrypted using the public Storage Root Key and Endorsement Key of the device.
- The device receives the challenge, decrypts the wrapped nonce, and stores it in the TPM using its private keys for the Endorsement Key and Storage Key, respectively.
- The device finally reconnects and signs an SAS token using the nonce and then receives connection information from the IoT Provisioning Service.

In the last step, the device concludes the provisioning step, which then allows for the IT administrator to enable other custom workflows that should be triggered by registering for events emanating from the IoT Hub and Provisioning Service, respectively.

The onboarding flow comes with multiple aspects from a security perspective that should carefully be considered, and the complexity and supply chain dependence and interaction can increase depending on the adopted solution and technology.

- Avoid using static or shared credentials statically provisioned to devices they all use the first time they connect as this makes it challenging for the backend to authenticate a legitimate device vs. one that is using compromised credentials.
- Adopt a solution that uses some HSM for hardware RoT.
- Scalable solution that minimizes supply chain complexity and interaction if possible.
- Zero touch or close to zero touch solution to minimize manual steps required by technicians or IT administrators.
- Most onboarding solutions require a connection to a backend service; with a wired backhaul, this is straightforward. If the device is using a wireless technology such as Wi-Fi or private 5G, LoraWAN, that needs to be configured and provisioned before the onboarding flow can start – to what end does the provisioning technology selected help or is a custom or additional technology needed for this step?

15.17 Attestation

In the previous chapter, we introduced the concept of verified and measured boot, particularly we discussed how the TPM can be used to collect *platform measurements* that can later be verified to determine the integrity of the platform overall. This leads us to another important pillar often combined and integrated with verified

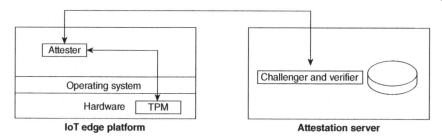

Figure 6 Remote attestation.

boot we refer to as attestation or remote attestation, as the entity performing the validation of an edge platform's integrity is typically a remote server.

Figure 6: Remote Attestation illustrates the interaction in a remote attestation workflow. All the devices the attestation server is responsible for managing have already been onboarded.

- The attestation server reaches out to the device whose integrity it needs to attest. The attestation request contains some type of challenge in the form of a *nonce (random number)*. The use of a fresh nonce helps mitigate against *replay attacks* to ensure that a compromised client device cannot replay old values. The request also contains the set of PCRs it requests the IoT edge device to collect and report back.
- The IoT edge device has a running component we refer to as attester on the platform. The attester requests for the TPM to collect and sign the set of PCRs coming in the request by using what is called an *Endorsement Key* whose public portion of the attestation server has as part of the onboarding flow of the node. The packaged response sent by the node back to the attestation server is referred to as a TPM quote. This quote contains an *Event Log* which includes all the software measured and resulting hashes.
- The attestation server verifies the TPM quote and the validity of the signature to ensure it is coming from the targeted device along with the set of PCR values. The attestation server has a set of golden values that it can compare against to ensure the device complies and meets the integrity requirements expected by the attestation server.

It should be noted here that there are additional software components one may wish to measure beyond what a standard Measured Boot process collects, as this process typically only measures boot chain up until the bootloader, kernel, and the kernel command line, and possibly the initram disk. This is possible with additional measurements done on user space components. An agent that measures additional user space components would typically target critical software running on the system, such as container runtime infrastructure, network manager, and device manageability agent. To anchor the user space measurements into a trusted and verified chain, it would typically start extending data into its own PCR (e.g. PCR 12) by using the resulting digest of the measured boot process as the seeding value.

One of the challenges for scale with these measurements is the static PCR an attestation server is comparing the golden values against. An edge infrastructure may have diversity in the type of platforms, version, and software running in the infrastructure, which would yield increased complexity for an attestation server to support. To add to that complexity, as nodes go through software updates, firmware updates, or OS updates, the corresponding measurements will change and the attestation server would need to support diversity of golden values across the fleet of infrastructure. It should be mentioned that the remote attestation is not a one-time request, but rather ongoing to ensure that the device maintains integrity and compliance throughout its life cycle. The scaling issue with supporting a diverse set of platforms with different software BOMs (Bill of Materials) is that measurement targets will yield a huge set of legit or golden PCR values. The authorized PCR policy discussed in the disk encryption section can be applied to the attestation workflows as well to overcome the scaling challenges that comes with the PCR brittleness.

Points to consider when enabling remote attestation for edge infrastructure:

- Establishing device trust and providing public keys to the attestation server as part of the onboarding and provisioning flow.
- Integration and interaction with the overall device management service and platform software updates from BIOS, Operating System, user libraries, and components.
- Management of golden values, heterogeneity of infrastructure types, software versions, and targeted measurements.
- Device remote attestation frequency and mitigation if compliance is not met. Policy decision vs. policy enforcement point in the system. What are the steps the IT/OT should take specifically to isolate and address potential security threats arising from attestation.

15.18 Defense in Depth

Is a well-known and adopted concept in computational security and is just as applicable to the IoT edge as it is for any other market. The principle is very simple, in that security with respect to data and information is protected at every layer in the software and network stack. If one layer is compromised by an adversary, another layer provides the necessary protection. This multi-layered approach provides security redundancy will strengthening the security posture against attackers and reducing the likelihood of a security breach. This is sometimes compared with medieval castles with multiple defense systems layered in front of the castle against attackers from moats to drawbridges.

Some examples of elements and relevant layers from a defense in depth perspective are as follows:

- Isolation of non-critical assets from public access unless necessary through air-gap or network isolation in addition to a network segmentation and isolation of devices that are independent and do not need to be reachable.
- Limit network port access to what is required for the business operations, possibly including source- or domain-level checks.
- Network firewall with policies in place that automatically blacklists remote IP failing to authenticate.
- Intrusion detection software that can detect anomalies when it comes to traffic patterns based on day and time between local and remote systems.
- Robust and thought through access control policies for authorized principals following the least privilege principle, which implies not providing more access rights to a user than what he/she requires to perform his/her task.
- Antivirus software and associated update/upgrade policies that must be met.
- Software BOM and overall system configuration meet compliance requirements defined by IT/OT.
- Device manageability service connected to the fleet with software update policies, ensuring that vulnerabilities in software can be addressed in a timely manner (e.g. CVEs).
- Password policies requiring strong passwords, updating at specific cadence, and requiring a layered security with MFA (Multi-Factor Authentication) based on email, OTP (One Time Password) key fob generators, SMS/Text, biometrics or another additional piece required for authentication.
- Enforcement of strong cryptography only where there is no possibility of negotiation and fallback to a weaker cipher.
- Not allowing for protocol where security in transit cannot be guaranteed with respect to confidentiality and integrity authenticating each user and device.
- Policies and process in place for revocation of certificates, keys, assets, remote device, and data sanitization.
- Policies and process in place to continuously monitor, verify, and ensure fleet integrity.
- Sound backup, recovery, and storage redundancy implemented to mitigate against malicious attacks or accidental loss of data.

15.19   Zero Trust Architecture (ZTA)

Is taking a disciplined approach to modern software architectures from a security perspective, incorporating many of the technologies and paradigms described in this chapter. The objectives here are to trust nothing and verify access to any resource in the network. This means that at every end point, every user, no matter where the call originates from, is always authenticated, and authorization checks are verified for the specific request or operation the principal is attempting to execute. It embraces the least privileged access described in the previous section and additionally attempts to minimize the impact of any type of breach of the system through this approach (i.e. limiting the blast radius of a partial system compromise).

15.20   Security Hardened Edge Compute Architectures

Would apply many of the technologies and principles described in this chapter. It is particularly important to understand the overall threat model and adversaries in the system and apply a security architecture that mitigates against the identified threats and adversaries. The key principals of security we described previously are as follows:

- Security at rest
- Security in use
- Security in transit

While the security technology anchored in platform hardware has been available for a long time, it can still be challenging to find end-to-end solutions that are correctly implementing them for IoT Edge deployments such as

- **Security hardened container runtimes with HW RoT**, such as verification of container images against trusted sources, private keys to remote container registries stored and access through an HSM, and dynamic and runtime integrity checks through attestation of the container runtime. Security policies in place to ensure limited access and privileges to the workloads. Integrity and attestation of the deployed containers and possibility for containers to run in a trusted execution environment.
- Security hardened orchestration control plane to ensure secrets, private keys between orchestration core services to agents running on worker nodes are stored in a vault backed by an HSM to ensure security at rest, secure mechanisms to transfer assets from orchestration agents to workloads/container images upon instantiation, and security in transit between workloads.

- Comprehensive device lifecycle management with clear security objectives and posture at every step from commissioning, provisioning, onboarding, monitoring and management, and updates to decommissioning of the device.

References

1 Jha, P., White, J., and Norman, D. Mirai Botnet attack. https://en.wikipedia.org/wiki/Mirai_(malware).

2 Miller, C. (2021). https://www.industrialcybersecuritypulse.com/threats-vulnerabilities/throwback-attack-blackenergy-attacks-the-ukrainian-power-grid/

3 Wikipedia contributors (2023). RSA (cryptosystem) https://en.wikipedia.org/wiki/RSA_(cryptosystem).

4 TCG. Trusted Computing Group website for TPM 2.0 Library. https://en.wikipedia.org/wiki/Colonial_Pipeline_ransomware_attack.

5 NIST. https://csrc.nist.gov/glossary.

6 Microsoft Azure IoT Hub Device Provisioning Service (DPS) Documentation.

7 TCG (2019). TCG Trusted Attestation Protocol (TAP) Information Model for TPM Families 1.2 and 2.0 and DICE Family 1.0. https://trustedcomputinggroup.org/resource/tpm-library-specification/

8 FIDO device onboard specification. https://fidoalliance.org/specs/FDO/FIDO-Device-Onboard-RD-v1.0-20201202.html.

9 TCG trusted attestation protocol (TAP) information model for TPM families 1.2 and 2.0 and DICE family 1.0. https://trustedcomputinggroup.org/wp-content/uploads/TNC_TAP_Information_Model_v1.00_r0.29A_publicreview.pdf." https://trustedcomputinggroup.org/wp-content/uploads/TNC_TAP_Information_Model_v1.00_r0.29A_publicreview.pdf.

10 Azure IoT hub device provisioning service (DPS) documentation. https://learn.microsoft.com/en-us/azure/iot-dps/

16

Securing the Energy Industry with AI-Powered Cybersecurity Solutions

16.1 Introduction

In today's era, everything is done via the internet and applications. There is no physical presence requested for most of the redundant processes. These redundant processes can be anything, varying from booking a gas or to applying visa. The range of redundant processes is actually unimaginable. This is possible only because of the collaboration between multiple fields, and the IT industry is at the helm of it. So, as a food for thought, if the futuristic life of a human is solely dependent on automation and technology, what would be possible threats and vulnerabilities that would be faced, specifically in the implementation and accessing of the non-renewable energy industry. An inventive vision for the future would be connecting the entire flow of the oil and gas industry to be automated from the extraction stage to the end-user consumption.

This chapter comprises various industries and fields under one roof. To be precise and on point, the overview of each industry and field is discussed, and then the collaboration of each is defined. The industries that will be corroborated for a better and secure future will be as follows:

- Energy Industry – Oil and Gas (Non-Renewable)
- Automated Factories
- Cybersecurity
- Artificial Intelligence-powered cybersecurity

The flow of these industries in collaboration with each other would be revolutionary [1]. This would pave the way for other such collaborations, such as space travel. An assumed flow of these collaborations [2] would be as described in

The Power of Artificial Intelligence for the Next-Generation Oil and Gas Industry: Envisaging AI-Inspired Intelligent Energy Systems and Environments, First Edition. Pethuru Raj Chelliah, Venkatraman Jayasankar, Mats Agerstam, B. Sundaravadivazhagan, and Robin Cyriac. © 2024 The Institute of Electrical and Electronics Engineers, Inc. Published 2024 by John Wiley & Sons, Inc.

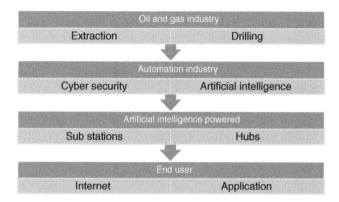

Figure 1 Flow of the various industries in collaboration.

Figure 1, where the renewable energy extraction is monitored and mined by automation of machines and is delivered to the end-user with ease via substations and hubs after due diligence of the required energy.

16.2   Energy Industry

The energy industry is humongous. Almost everything that needs power depends on the energy industry to sustain itself. It is an essential and critical market. The energy industry is categorized into two major categories.

- Non-renewable energy
- Renewable energy

16.2.1   Non-renewable Energy

This type of energy includes oil and gas. Non-renewable energy means that it is a natural resource, and the duration at which a non-renewable energy is produced via natural happenstance would not match the pace of consumption. Hence, non-renewable energy gets depleted with time. There are wars being waged for the ownership of mines and refineries all over the world. The non-renewable energy sector involves zillions of dollars for the processes and profits. The processes that are involved in processing of non-renewable energy sources like oil and gas are as described in Figure 2. These steps may vary from country to country because some countries directly import, whereas the other countries might mine and extract. The processes are country-specific, i.e. in accordance with the specific country's law.

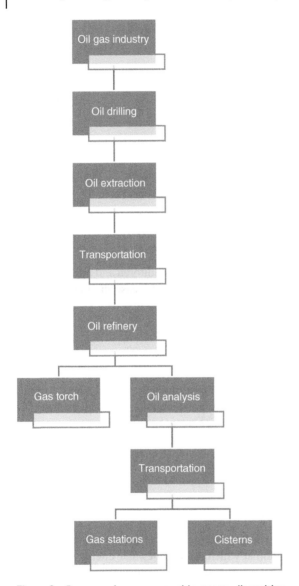

Figure 2 Processes in non-renewable energy disposition.

16.2.2 Renewable Energy

This type of energy includes solar, hydro, and wind energy. This can be easily produced with the current technology and machineries. Though renewable energy has its challenges, the current global scenario requires switching over to

Figure 3 Types of renewable energy.

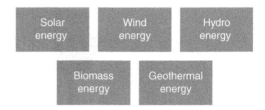

renewable energy due to various challenges such as global warming. Achieving sustainability in renewable energy is no easy feat. There are so many procedures and precautions that have to be carried out. The energy can be produced, but there are changes in the storage of renewable energy, which is quintessential in the field of renewable energy. The renewable energy processes also involve a series of steps from generation of power to storage and to consumption. The main sectors in renewable energy are as described in Figure 3.

Since this chapter deals predominantly in securing a workplace that produces non-renewable energy like oil and gas, it would be wrong not to perceive the importance of renewable energy.

16.3   Present and Future of Energy Industry Supply Chain

The non-renewable energy industry's supply chain is the place that requires automation to achieve the futuristic vision of digitization. The supply chain is how the resource is produced or found and processed and distributed for consumption. The supply chain as described in Figure 4 is the main target of digitizing the energy industry. Right now, there are so many challenges in the supply chain due to unforeseeable events such as natural calamities or war between countries. This disrupts the value of money and leads to a volatile economy.

If the supply chain of the oil and gas industry is being digitized [3], then there would be regularization, which would change the game in stabilizing the supply chain of the non-renewable resource. The present age of the supply chain of these energies is volatile due to multiple factors. If the due process is carried out and a system is implemented in place that is governed by automated refineries and distribution hubs, then there are less chances of disruption in the world due to this industry. Furthermore, non-renewable energy is categorized in the following types, as described in Figure 5.

This would be useful in categorizing the digitizing stages of the industry. The types involved in the supply chain are as follows:

- Upstream
- Midstream

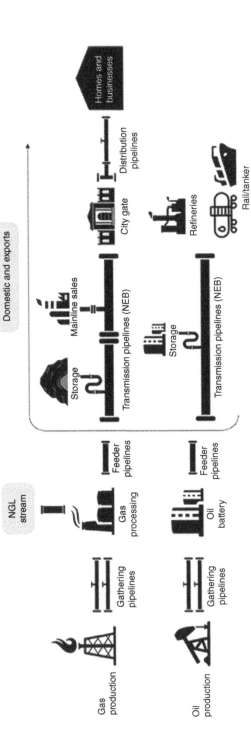

Figure 4 Supply chain of oil and gas.

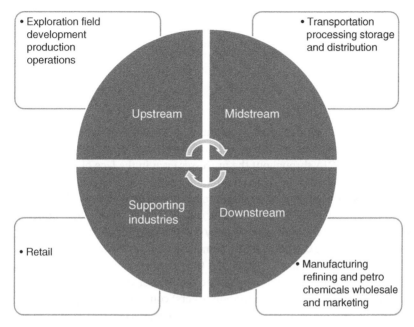

Figure 5 Supply chain categories.

- Downstream
- Supporting industries

If the supply chain categorization has been done conscientiously, then the digitization of the supply chain of the energy industry would attain ascendancy all over the world uninterrupted by any power.

16.4 Cybersecurity

Cybersecurity is the need of the hour. As technology blossoms, so does the threats. Comparatively securing a physical area is far easier than securing an intangible area. Hence, cybersecurity is a whole new ball game, where the bad guy's mind applies complex solutions or hacks to break into the system, while the good guys have to apply their minds twice as much to prevent these hacks and secure the digital space in their responsibility. Cybersecurity's foundation has three main factors:

- Confidentiality
- Integrity
- Availability

Figure 6 CIA triad.

These factors are often referred to as the CIA Triad [4] of cybersecurity, as described in Figure 6. This is the required model for securing any cyberspace.

Maintaining the CIA throughout the processes even in the face of any cyberattack is the ultimate goal in this scenario. To aide in this quest, like-minded cybersecurity experts have developed so many non-profit resources that would help the CISO (Chief Information Security Officer), CIO (Chief Information Officer), cybersecurity enthusiasts, and cybersecurity community in achieving this feat.

Cybersecurity can be categorized into many segments, as described in Figure 7. Some might seem similar to each other, but the operations that are carried out in each and every division differ. To streamline the field of cybersecurity, the good guys (either ethical hackers or penetration testers) come up with a specific set of instructions or procedures to follow. Supporting on causes like this, even government bodies have involved and drawn up a set of rules and regulations to adhere to the standards of security. These are often referred to as frameworks [5]. So, if any company has to be secured, it has to be that specific framework-compliant. This could weed out some cumbersome job of the cybersecurity officers to come up with new rules all the time. One of the examples for cybersecurity is the NIST framework, which is followed by the majority of the cybersecurity community and is a good standard that is recognized worldwide for cybersecurity. Under the categorization of cybersecurity, some of the segments are as below:

- Security Engineering
- Security Operations
- Governance
- Risk Assessment
- Threat Intelligence
- Framework and Standards

These divisions give an overview of the many facets involved in cybersecurity. These divisions touch the basics of the cybersecurity field. The role of the penetration tester even furthers these divisions. With these various divisions of cybersecurity, any field can be secured on the cyberspace with constant monitoring and updating of security measures from time to time. In the field of cybersecurity, it is always better to be safe than sorry because more than monetary loss that would be incurred following a cyberattack, the loss of trust of the consumers on the brand

Figure 7 Cybersecurity divisions.

is one of the most horrendous things that could happen for a company. This is often followed by the reduction of the company's stock price if the company is a public IPO. Numerous well-reputed companies have sailed this boat due to cyberattacks. One can even say that any kind of cyberattack or data breach can push a company's share into the bear market.

16.5 Digitizing of the Energy Industry

The digitizing of the energy industry is imperative. Only through digitizing can the cost of the energy industry be reduced, thus maximizing the profits and reducing the operational costs. For this to be achieved, the supply chain of the energy industry should be digitized. As seen in the beginning this chapter, there are four divisions involved in the supply chain management.

- Upstream
- Midstream
- Downstream
- Supporting Industries

Any industry is highly dependent on the data analytics. The logistics that the systems can offer can identify when and where something goes wrong and how to minimize cost overheads. This is what the energy industry requires now. Based on the consumption of oil and gas, demographic data can be used to regulate the transportation of the energy, which can in turn manage the supply and demand perfectly.

Digitization of the energy industry has so many perks. Starting from the upstream digitization, the processes involved in it are as furnished in Table 1. The types of digitization are minimal and are not limited to only these types of implementation.

The upstream digitization of the energy industry has so many perks. The types of digitization that can be implemented in upstream is not restricted to only software, rather it uses the whole nine yards of the IT industry with IOT, real-time data analytics, Artificial Intelligence, etc. If this digitization is implemented, everything will be in order. Thus, the implementation cost of the upstream energy industry can be regulated. Everything starts with the upstream; hence, the implementation should be implemented from the upstream first. This employs the top to bottom approach because if only one sector is digitized and the rest is undigitized, then it would be a moot point. The whole purpose of digitizing the energy industry is to regulate and digitize the process throughout all the processes.

Next is the digitization of the midstream of the energy industry. The midstream digitization takes care of the transportation of the oil and gas to the storage and processing of the oil and gas, as described in Table 2.

Table 1 Description of digitization in upstream.

S. No	Type of digitization	Description
1	Sensors	Sensors on the rig detect abnormal temperature
2	Integrated operations center	The IOC Engineer receives an alert and performs diagnosis via the interactive 3D model
3 (a)	Surveillance drones	Drones investigate the off-shore rig and share photos/live videos in real-time
3 (b)	Real-time request oil field services (OFS)	IOC identifies required services and issues a service request to OFS vendors, where the best bid is accepted in real-time
4	Real-time analytics	Predictive Data analytics determine maintenance needs based on surveillance data and integrated supply chain order parts
5	Smart devices	Engineers receive alerts and incident details on their smart watches / mobile devices and prepare for service
6	3D printers	Parts and tools required to fix the issue are printed in real-time using 3D Printers
7	Delivery drones	On-shore drones deliver parts from the warehouse to the off-shore rig
8	Tablet/smart glasses	Engineers utilize virtual models on tablets and augmented reality data on smart glasses to perform maintenance

Furthermore, the implementation in midstream digitization can lead to the following properties as well:

- Minimizing human intervention
- Digital workforce
- Network optimization
- Predictive maintenance

Next up is the digitization of the downstream energy industry. The downstream industry deals with refining and then distribution of the resources. Depending on the country in which the digitization is being implemented, the processes and rules might change. However, the basic requirements for any downstream digitization requirements are as furnished in Table 3.

The downstream digitization also further uses digital transformation in the following areas:

1) Business priorities
2) Foundational capabilities
3) Agile approach

Table 2 Description of midstream digitization.

S. No	Type	Description
1	Digital asset management	The SCADA System detects a leakage on the pipeline. Sensors are used to monitor the pipelines and machines in real-time
2	Integrated control room	The ICR technician receives an alert and performs diagnostics. The issue is isolated and maintenance schedules and reviewed to plan an optimal repair window
3	Biometric monitoring/GIS	Wearable devices monitor the field worker's location, safety, and job status
4	Drones/PIGs – pipeline inspection gauges	Drones and/or PIGs investigate the leak and share the real-time data and video
5	Analytics/simulations	AI user's surveillance and flow data to perform work simulations and impact analysis to determine optimal work prioritization and workforce allocation
6	Smart devices	Engineers receive alerts and incident details on their smart watches/mobile devices and prepare for service
7	Digital workforce management	An engineer initially scheduled for planned maintenance is re-assigned to higher-priority leak repair job
8	Dynamic inventory management	Dynamic inventory management supports logistical decisions to optimize the sourcing of repair parts based on the availability and leads-time in case of unplanned events
9	Smart trucks	Parts are delivered by the GPS-/sensor- enabled trucks to support real-time tracking and coordination of activities
10	Tablets/smart glasses	Engineers utilize guided workflows on tablets and augmented reality data on smart glasses to perform maintenance while collaborating with remote specialists in real-time

If the digitization of the supply chain of oil and gas is successfully implemented, then there is true evolution in the energy industry. This leads to the next part of the digitization, that is the securing of the digitized space. The minute the resources are exposed to the internet, the exposure is unimaginable. Thus, securing the cyberspace of the energy industry is imperative. Moreover, the security testing, which is the penetration testing, should be done ahead in all possible scenarios, and only then should it be released for the users.

Table 3 Types of downstream digitization.

S. No	Type of implementation
1	Data analytics
2	Mobility
3	Internet of Things platforms
4	Cloud computing
5	3D models/Digital Twin
6	Artificial intelligence
7	Drones
8	Cobots
9	Virtual/augmented reality technologies
10	Blockchain
11	3D printing

In the process of achieving cybersecurity under any area, the security engineer's pot of gold is the MITRE ATT&CK framework.

16.6 MITRE ATT&CK Framework

The MITRE ATT&CK framework is a system where every hack that has ever happened in this whole world is mapped to techniques, tactics, and to groups that have taken responsibility for the specific hack. This is a goldmine of information because before the MITRE ATT&CK framework [6], there was nothing of this sort. In a way, this is like the dictionary of cyberattacks. There are various factors that are taken into account for each breach. The security engineers can even determine the psychological profile of the hacker based on the techniques that are being selected for the hacking attempt on the organization.

This comes in handy for the CISOs and CIOs in guarding the organization against known vulnerabilities and threats. At least, when a hacker attempts to any of the hacks listed here, it can be defended if the mitigation steps that are also given in the MITRE ATT&CK framework are followed to the dot and implemented before any attack happens. This will save lots of money, time, and face in the future for the company.

The MITRE ATT&CK framework is offered for three platforms:

- Enterprise
- Mobile
- ICSs – Industrial Control Systems

Even though enterprise and mobile are intriguing concepts to be discussed about the MITRE ATT&CK framework, this chapter is in the interest of spreading awareness of the ICS MITRE ATT&CK framework that is relevant to this oil and gas industry digitization, which is as described in Figure 8. The headings that are marked in blue are the tactics, which are the verticals in the ICS cyberattacks. The subtitles in each vertical are the techniques that are used to achieve the tactic.

The tactics in the ICS are as follows:

- Initial Access
- Execution
- Persistence
- Evasion
- Discovery
- Lateral Movement
- Collection
- Command and Control
- Inhibit Response Function
- Impair Process Control
- Impact

One of the good things about the ATT&CK framework is the mapping of everything [7]. A hacker cannot simply directly jump to the impact tactics. The hacker has to go through everything stage by stage, at least from initial access to discovery and lateral movement. The hacker's destination after the lateral movement stage depends on the intention of the hacker. Till then, there is a clear route of how a hacker would be able to gain access to the ICS and techniques that would be used by any hacker with malicious intent.

Each of the techniques has its own page, where the technique samples are described in detail. Along with the technique, the MITRE [8] framework also maps the steps for mitigation and detection. Mitigation is displayed for how to neutralize that particular technique if it was executed by any hacker or to reduce the impact of the attack. Detection is displayed for how to detect the presence or the execution of that technique in the ICS system. Each and every tactic and technique is assigned a unique ID.

16.7 CVE

CVE is the acronym for common vulnerabilities and exposures. This is also a part of MITRE, where all the publicly disclosed vulnerabilities are listed as a dictionary. Every vulnerability is given an ID for the CVE database [9]. This platform is constantly evolving and updating the list of the vulnerabilities and exposures so far on the internet.

Initial access	Execution	Persistence	Evasion	Discovery	Lateral movement	Collection	Command and control	Inhibit response function	Impair process control	Impact
Data historian compromise	Change program state	Hooking	Exploitation for evasion	Control device identification	Default credentials	Automated collection	Commonly used port	Activate firmware update mode	Brute force I/O	Damage to property
Drive-by compromise	Command-line interface	Module firmware	Indicator removal on host	I/O module discovery	Exploitation of remote services	Data from information repositories	Connection proxy	Alarm suppression	Change program state	Denial of control
Engineering workstation compromise	Execution through API	Program download	Masquerading	Network connection enumeration	External remote services	Detect operating mode	Standard application layer protocol	Block command message	Masquerading	Denial of view
Exploit public-facing application	Graphical user interface	Project file infection	Rogue master device	Network service scanning	Program organization units	Detect program state		Block reporting message	Modify control logic	Loss of availability
External remote services	Man in the middle	System firmware	Rootkit	Network sniffing	Remote file copy	I/O image		Block serial COM	Modify parameter	Loss of control
Internet accessible device	Program organization units	Valid accounts	Spoof reporting message	Remote system discovery	Valid accounts	Location identification		Data destruction	Module firmware	Loss of productivity and revenue
Replication through removable media	Project file infection		Utilize/change operating mode	Serial connection enumeration		Monitor process state		Denial of service	Program download	Loss of safety
Spearphishing attachment	Scripting					Point and tag identification		Device restart/shutdown	Rogue master device	Loss of view
Supply chain compromise	User execution					Program upload		Manipulate I/O image	Service stop	Manipulation of control
Wireless compromise						Role identification		Modify alarm settings	Spoof reporting message	Manipulation of view
						Screen capture		Modify control logic	Unauthorized command message	Theft of operational information
								Program download		
								Rootkit		
								System firmware		
								Utilize/change operating mode		

Figure 8 ICS matrix of MITRE ATT&CK.

16.8 CWE

CWE is the acronym for common weakness enumeration. This is also a part of MITRE. This platform is for identifying weaknesses in software and hardware, which are developed and contributed by the community [10]. This serves multiple purposes for identifying, preventing, and mitigating the known weaknesses while developing a software or hardware.

16.9 CAPEC

CAPEC is the acronym for common attack pattern enumerations and classifications. This is also a part of MITRE. CAPEC [11] can be used to identify attack patterns and based on that, required countermeasures can be deployed to safeguard the organization. This is available for all for better understanding of the attack patterns and to strengthen the defenses against such attacks.

16.10 CPE

CPE is the acronym for common platform enumeration, which was also a part of MITRE but now is being transferred to the NIST – National Institute for Standards and Technology. This platform deals with identifying products, networks and platforms, etc. This is no longer in active status. This platform has been moved to archive status by the NIST.

16.11 Cybersecurity Framework

Everything is better with frameworks. To deliver quality, a predefined set of standards would always help. That is the same in the cybersecurity field as well. As far as this cybersecurity framework is concerned, securing of the digital assets and services is the priority. In order to do that, a series of steps have to be followed to deliver good on the promise of security. The stages involved in a cybersecurity framework is as described in Figure 9.

There are three components that are required by any cybersecurity framework [12] to be implemented. It does not matter which framework one is following as long as the framework satisfies these core three components. There are various organizations that have developed cybersecurity frameworks. The basis on which all those have laid on is these three components. The objectives of the

Figure 9 Cybersecurity framework stages.

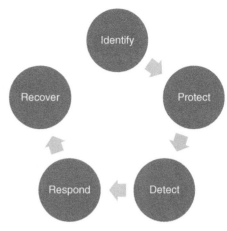

Figure 10 Objectives of the cybersecurity framework.

cybersecurity framework is a repetitive process, which is very simple. It follows the steps described in Figure 10.

16.12 NIST Framework

NIST is the acronym for the National Institute for Standards and Technologies. The NIST was developed by the United States of America for the reliable functioning of critical infrastructure, the President issued Executive Order (EO) 13636,

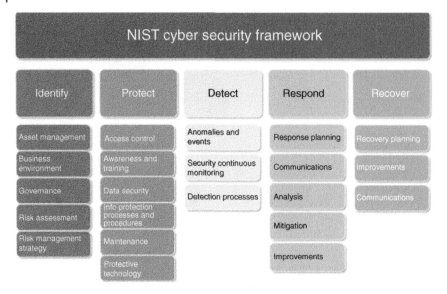

Figure 11 NIST cybersecurity framework.

Improving Critical Infrastructure Cybersecurity [13], in February 2013. This was created through the collaborative effort of both the government and industry. This framework as described in Figure 11. presents a voluntary guideline based on the existing standards and practices to manage and reduce cybersecurity risk. This was also designed to foster risk and cybersecurity management among internal and external organizational stakeholders.

16.13   Zero-Day Vulnerability

The worst case of cyberattacks is the zero-day vulnerability. Despite all these organizations doing their best to make the world a better place, there are few happenstances which are unavoidable. Zero-day vulnerability refers to the scenario when a hacker discovers a vulnerability in a system that is not in any kind of vulnerability database such as CVE, CWE, or ATT&CK framework. Zero-day vulnerability means the hacker has discovered an absolutely new route to hack into the system. Zero-day vulnerability leads to zero-day attack, which in turn leads to zero-day exploit as described in Figure 12. This is the worst case of cyberbreach because the exploit is unknown to the security engineers. This would cause a

Figure 12 Zero-day vulnerability cycle.

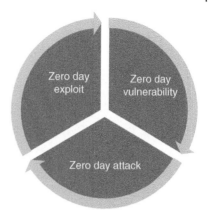

mayhem among the involved official parties because it would take longer than usual to mitigate this kind of threat.

After such a zero-day vulnerability is exploited, the security engineers rush to find out the root cause for the vulnerability and start developing a patch and update the system to neutralize the vulnerability. Once this happens, this event is recorded across platforms such as CVE, CWE, and ATT&CK. The constant updating of new vulnerabilities and mitigations is to ensure that the others do not suffer the same fate as the zero-day exploited organization.

16.14 Machine Learning

Machine learning is the concept where one teaches machines on predictive outcomes with classifiable parametric values [14]. Machine learning is one of the most sorted fields in the present time. There can be a lot achieved through machine learning. There are three main categories in machine learning. They are as follows:

- Supervised learning
- Unsupervised learning
- Reinforcement learning

There are different types of methods that can be implemented to teach a machine. The extensive methods under these three main categories are described in Figure 13.

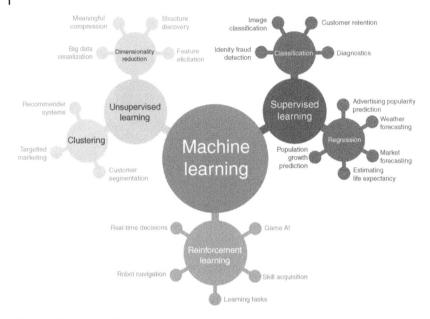

Figure 13 Machine learning methods.

Through any of these methods, one could make a machine learn to predict outcomes based on particular sets of input data. These input data and the process of outcome determine which type of learning is being implemented in the system.

16.15 Artificial Intelligence

Artificial intelligence is also a form of machine learning. Futuristically, artificial intelligence is where machines start to think and evolve based on past data and experiences. There are some misconceptions on artificial intelligence simply because of the power it holds over everything. Despite everything, if AI is used equitably, then rest assured everything would be fine. There are so many segments inside artificial intelligence, and the applications for the AI are perpetual.

Machine learning is a subset of artificial intelligence. Despite common knowledge, both have redeeming qualities that make both the concepts stand apart from each other. The differences between AI and ML are described in Table 4.

Table 4 Differences between AI and ML.

#	AI	ML
1	Overarching field	Subset of AI
2	The goal is to simulate human intelligence	The goal is to learn from data
3	Solve complex problems	Predict results based on new data
4	Leads to intelligence or wisdom	Leads to knowledge
5	Tries to find the optimal solution	Tries to find the only solution whether it is optimal or not

16.16   Fusing AI into Cybersecurity

Both cybersecurity and AI are innovative, inventive, and breathtaking fields in operation [15]. Collaborating AI into cybersecurity is arduous but fecund. There are six sectors in cybersecurity, in which AI can do wonders. The six sectors are as follows:

- Vulnerabilities
- Risks
- Threats
- Phishing
- Networks
- Malwares

These are the commonly exploited sectors through which the hackers gain foothold on to the systems. Now, deploying AI to constantly monitor these sectors would reduce the risk of exploitation almost to non-existent. The possible fusion of AI into cybersecurity in these sectors is described in Figure 14. Basically, if the AI takes care of these sectors in addition to human intervention now and then as a routine, the chances of a cyberattack being successful is null. In addition to this, the AI should be connected to the CVE, CWE, CAPEC, and the ATT&CK so that the AI database is automatically updated from these databases about the latest vulnerabilities, exposures, and exploits. This will prove effective in the long run when whatever the attack is being carried out the AI is ready with the mitigation process. Zero-day vulnerability is a thorn to the AI in cybersecurity, but as time progresses, even zero-day attacks can be prevented by AI.

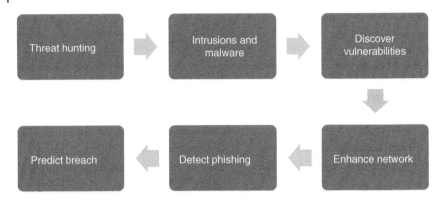

Figure 14 Fusing AI into cybersecurity.

16.16.1 Threat Hunting Enhancement

When implementing AI into securing the oil and gas industry, the important stage is to teach the AI what threats are and how to automate threat identification [16]. The AI should be able to hunt threats as and when they appear. AI's best aide in this would be the MITRE and the NIST framework. These frameworks are updated on each and every vulnerability, and the mitigation options are also given. If the AI can hunt the threats and identify the vulnerabilities, then create a mitigation process to neutralize the threat. Then, the AI fusing into cybersecurity truly is a wonder and an easily manageable task.

Threat hunting is now done via tools such as threat dragon. Here, the process and the modules are mapped and the users are defined. Accordingly, the tools help the pentesters to identify vulnerabilities and risks from external and internal entities. Mapping the entire process flow on to the tool helps the pentesters identify drawbacks or any break of access control in the flow.

Fusing of AI should be developed in such a way that the threat modeling is automated and based on previous experiences and data, and the AI can predict possible attacks for which mitigations can be carried out immediately.

16.16.2 Detect Intrusions and Malwares

The initial access of any threat in the oil and gas industry is the form of intrusions. The first stage of hacking is reconnaissance, where the malicious hackers would gather information as much as possible about the attack surface. This might range from ping scans to social engineering. To be frank depending on the psychology and skill of the hacker [17], it can be as simple as a Nmap scan to an expert of social profile assessment to guess passwords and stuff. There are two categories for detecting intrusions and mitigate. But it still is not enough, unless it is automated. Every time a patch is released, many other ways of exploiting a system is always found by the persistent malicious hackers.

To detect and prevent intrusions, the hosts use two functionalities. namely.

- IDS – Intrusion Detection System
- IPS – Intrusion Prevention System

Even with these systems in place, exploits happen regularly. The technology [18] is at the right place, but it is cannot withstand the attacks of the hackers. Hence, the IDS and IPS integration of AI should be directly linked to the CVE database. Every time an exploit is submitted to the vendors, a CVE ID is assigned to it. If the IDS and IPS are at the top of these details, then it might be difficult for any hacker to use a known CVE in an AI-powered security space because, even before launching an attack, the AI would figure out the intention of the process and shut it down.

Some may misunderstand that installing a firewall would safekeep the network or hosts. That is actually a myth. Though firewalls and antivirus could provide a degree of protection from the most commonly used tools and techniques, it cannot protect the system from every attack. For this, the antivirus should be constantly updated to the latest patches. Even then, there are users that turn off real-time protection so that the users can install some pirated software. So, firewalls and antiviruses are a necessity but not the end of the line.

Malwares are automated software that can infect hosts automatically, given time and space on the network. The world suffered a lot from the WannaCry ransomware. Though systems are equipped in identifying software that are not signed by developers or organizations, the users can override any precautionary measures that would be put forth by the systems because ultimately humans are responsible for any actions taken from the charted course by the AI. So, this should be considered one of the scenarios where even if a human tries to override the security measure, the AI should be smart enough to not allow it.

16.16.3   Discover Code Vulnerabilities

Bugs, vulnerabilities, threats, and exploits kind of sound similar but are different. But, the root cause of all these evils is the development of the software. If the software was supposed to do what it has to do without any aberrations, then these evils can be avoided. But, in accordance with to err is human, developers make mistakes all the time. The team of developers, testers, and managers is involved in the software development so that mistakes like this can be avoided. Discovering code vulnerabilities is like going to the place where it all starts and weed out any bugs or vulnerabilities. In order to do that, the inspection team should be top-notch and creative. Code vulnerabilities are found by inspecting the code written directly in developing the software. By inspecting the ground zero reality, losses and breaches can be avoided with minimal cost. Many organizations fail to do this. Only recent time hacks have shed light on the importance of cybersecurity.

The code used in development is the root cause of any vulnerabilities caused, which could be exploited by the hackers. This depends not only on the native codes but also the third-party codes that are being integrated with the software. Loose code sinks ships. Hence, code review is imperative in weeding out vulnerabilities.

16.16.4 Enhance Network Security

Network security is not an easy feat to be achieved. When an industry as big as the energy industry should be digitized, the extensive steps that should be taken have no end. It is a consistent process where the perspective should be every day is a new day to be attacked and how could the CISOs save the day with prevention rather than cure. Network security [19] largely depends on ACL – access control list. If all the stipulated hosts are within the limit of operability, then there can be no issues. When a host misbehaves and crosses borders between networks, bad things happen.

Network security covers the following fields:

- Data Protection
- Email Security
- Server Protection
- Mobile Security
- Network Protection
- Cloud Services Application

Proper implementation of network security can deter hackers from lateral movement [20] in the network. Lateral movement is a serious issue where the hackers try to penetrate the network as deeply as possible to create maximum impact. This would demolish the entire network of the industry. Such a situation should be deterred at any point. Proper security controls in the network infrastructure would deter hackers from breaching the organization.

16.16.5 Detect Phishing

While protecting everything from code to network, many organizations oversee or do not give importance to the most vulnerable asset in the organization, the humans, because no matter the unbreachable security controls implemented, if the hackers are able to gain the credentials of any user through which the impact can be maximized through lateral movement. No AI or any automated tool can identify this kind of a breach. Though there are controls for this such as IP restrictions, etc., it can be easily spoofed. The phishing attacks can be neutralized only by creating awareness among the human resource. The awareness should instill the idea of security always comes first in an organization more than one's hobbies, etc. This is how phishing always works. Due diligence will be taken by the hackers to find out the likes and dislikes of the employees. A malicious link arrives in the inbox of the employee's email. Out of passion, the employee just opens the link to see what the fuss is about. That is the start of the butterfly effect [21]. No amount of prevention and mitigation would work if the employees are not educated in security. Hence, the people in the organization should be able to detect phishing not only in email but also in human interaction, such as a phone call or delegated via an assistant. It is better safe than to be sorry. When any instruction is made delegatory which is out of the normal, it should be inspected closely and confirmed again with the authorized persons directly.

16.16.6   Predict Breach Risks

The job of an AI is to predict the future. If somewhere on the network a mysterious IP address appears, the AI should be on high alert and follow up on any misbehavior on the network, if any. This is how the intelligence of AI can secure the cyberspace of the energy industry. If the AI is able to analyze the data and predict the outcome [22], it can prevent a lot of cyberattacks for years to come. There are only a few AI tools that are available in the cybersecurity field.

16.17   Threat Modeling in AI

Threat modeling is an interesting sector in cybersecurity. Threat modeling [23] is the process of mapping out the steps to proceed with if a known threat is perceived in the vicinity of the organization. It might be a physical threat or a virtual through the wires. Either way, threat modeling always helps the case in securing the systems. The steps involved in threat modeling are as follows:

- Define Objectives
- Define Technical Scope
- Application Decomposition
- Threat Analysis
- Vulnerability and Weakness Analysis
- Attack Modeling
- Risk and Impact Analysis

There are various operations that have to be carried out in each step. Now automating the threat modeling by teaching an AI to learn how to recognize vulnerabilities, exposures, and threats is the key to a successful digitization of the energy industry. Through digitizing the energy industry, the exposure of the area to the general public warrants malicious intents toward the application of the energy industry. Every hacker that does have knowledge about this industry digitization would take a crack at the system. This is where the systems and humans, both in accordance, defend the system against these malicious hackers. The steps involved in threat modelling are as follows in Tables 5 and 6.

Table 5 Steps in threat modeling.

S. No	Steps	Description
1	Define objectives	Identify business objectives
		Identify security and compliance Requirements
		Business impact analysis
2	Define technical scope	Capture the boundaries of the technical environment
		Capture infrastructure|application|software dependencies
3	Application decomposition	Identify use cases|define app/entry points and trust levels
		Identify actors|assets|services|roles|data sources
		Data flow diagramming (DFDs)|trust boundaries
4	Threat analysis	Probabilistic attack scenario analysis
		Regression analysis on security events
		Threat Intelligence correlation and analytics
5	Vulnerabilities and weakness analysis	Queries of existing vulnerability reports and issues tracking
		Threat to existing vulnerability mapping using threat trees
		Design flaw analysis using use and abuse cases
		Scorings (CVSS/CWSS)|enumeration (CWE/CVE)
6	Attack modeling	Attack surface analysis
		Attack tree development|attack library management
		Attack to vulnerability and exploit analysis using attack trees
7	Risk and impact analysis	Quality and quantify business impact
		Countermeasure identification and residual risk analysis
		ID risk mitigation strategies

Table 6 AI-powered threat modeling.

S. No	Phases	Description
1	Automation	Eliminates repetition in threat modeling Ongoing threat modeling Scaled to encompass the entire enterprise
2	Integration	Integration with tools throughout the SDLC Supports the agile DevOps
3	Collaboration	Key stakeholders' collaboration: App Developers, Systems Architects, Security Team and Senior Executives.

The main challenge here is how to incorporate AI into the threat modeling. After the identification of the steps in threat modeling, the program should be developed in such a way that the steps are automatically fulfilled by the AI. The three stages in building AI-based threat modeling are as follows:

In each of these phases, there are subprocedures [24] that are involved in the successful integration of AI in threat modeling. Despite the challenges, if one is able to successfully integrate the AI to develop threat modeling for a system, then securing the energy industry is a piece of cake.

Threats are undiscovered vulnerabilities. These vulnerabilities should be routinely weeded out by penetration testers of the organization. If the vulnerability is not found and patched up, any hacker could kind of come upon it and exploit it. In reality, any unaddressed bugs after the development of the software is what leads to these kinds of vulnerabilities. So, the security should be taken seriously in the organization. Periodically, the organization should hire penetration testers from outside to test the organization's test. There are three types of tests that the penetration testers can go ahead with while testing out the vulnerabilities.

- Red Team: Offensive Testing
- Blue Team: Defensive Testing
- Purple Team: Both offensive and defensive testing

RED TEAM [25] mimics the behaviors of the hackers. The main aim of the red team is to crack the network and intrude the hosts to steal data.

BLUE TEAM [26] are the policymakers to secure the cyberspace in the energy industry. Blue teamers are responsible for the preventive measures put in place.

PURPLE TEAM [27] matches the desperate times, where the red team and blue team work hand in hand to maximize the returns and strengthening of the security of the organization in the energy industry.

If the AI is being trained in these kinds of testing as well, it would be very easy to protect and prevent attacks on an organization. For this to be a reality, the models in AI should be trained in various scenarios of cyberattacks, including all the tactics and techniques repeatedly. Each of the scenario can vary from each other to strengthen the training of the model. It is always good practice to train on 80% of data and test with 20% of data. Most cyberattack scenarios will be based on assumption. In the assumption that if such a vulnerability exists, then the kinds of attacks that can be launched by the hackers to exploit the organization can be observed. If the AI model is able to predict, it then a series of automated commands can be launched as a part of mitigation against this kind of attack.

16.18   Incident Response

Incident response [28] is in the event of an attack on how to proceed further with a proper mitigation process in place and documentation and reporting. There are basically three processes involved in the incident response sector.

- Before Attack
- During Attack
- After Attack Figure 15 Incident Response Wheel

Before Attack is the time to prepare for the scenario as to what has to be done where if any kind of attack happens. The first step of the incident response is to plan what to do next. What stage of the attack is it in now, etc. Next identify possible stakeholders

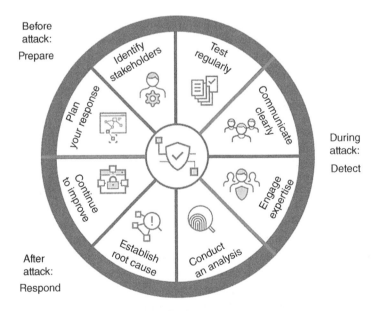

Figure 15   Incident response wheel.

as to people who are affected by this attack, etc. Once these are prepared, then based on the finding, tests are considered regularly under the assumption that the system is under attack. If the attack is launched, then it is the responsibility of the management to clearly convey that this is not a drill and a real threat to the organization. Once that is over, hire a cybersecurity expert to understand and mitigate the attack. It is already too late if there are no in-house security experts, and it is time to hire one and investigate what the problem is and how to solve it. Once the cybersecurity experts are on scene, there is an entire field called cyber forensics where the experts try to establish what happened and how it happened. Once the root cause is discovered, it is added into the vulnerability list, and then patches are developed to update the vulnerability. This cycle is done on a repeated basis to achieve minimum response time. These steps are clearly described in Figure 15, which clearly segments the phases and the methods involved in the incident response.

16.19   Fire Sale Scenario

With today's technology, anything is possible including turning fiction into reality. One such scenario is the FIRE sale that takes place in the film Die Hard 4.0. The world is now in the digitizing space of the resources. Once it is done, then the technology controls everything. The people who control the technology control everything by association. In this said fiction, there is a concept called FIRE Sale that is being performed by a group of hackers to take down the government and create chaos, thus collapsing a whole country.

The idea of FIRE sale means something like

F stands for financial,

I stands for infrastructure,

R stands for resources, and

E stands for environment

This means everything from which the government runs or relies must go down. A fire sale is a term describing a three-stage coordinated attack on a country's transportation, telecommunications, financial, and utilities infrastructure systems. The attacks are designed to promote chaos and foster a leaderless environment. The term "fire sale" is used because "everything must go."

Another example of an unbelievable scenario is where the Stuxnet worm did irreparable damage to the Iran nuclear program. The Stuxnet worm was an ultrasophisticated computer worm credited with taking down centrifuges critical to Iran's suspected nuclear weapons program. Hence, anything is possible that can be inspired by fiction. Recently, a group of hackers supported Ukraine in the war between Ukraine and Russia. These hackers demonstrated their opposition by taking down government websites of Russia. Cyber Warfare is not one of a fiction, but a reality.

16.20   Conclusion

The energy industry needs desperate digitization to reduce cost overheads. Due to COVID-19 and wars between countries, the energy sector took a downfall, and therefore there were so many volatile disruptions like the butterfly effect. Digitizing the energy industry with top-notch security is the key to move forward. The future of technology is blooming in various sectors. With the introduction of augmented reality and AI, nothing is outside the grasps of humanity.

AI-powered cybersecurity could protect the energy industry from malicious attacks. Once the system is in place, it can decide automatically on whether or not a vulnerability is a threat and what are the associated steps that have to be taken to mitigate the exploit. The Non-renewable energy is precious and also the reason for the depletion of natural resources on the Earth. By bringing natural resources and technology together, it would be like the best of both worlds where the resources are perfectly maintained with zero wastage by regularizing the supply and demand using the digitization. There are so many applications that can be applied with this kind of system. Caution has to be taken when handing over the resources to the computers. Though the computers take care of the resources, ultimately humans should have to power to override the computers and take a decision that is imperative at that specific time period.

References

1 Gezdur, A. and Bhattacharjya, J. (2017). Digitization in the oil and gas industry: challenges and opportunities for supply chain partners. 97–103. https://doi.org/10.1007/978-3-319-65151-4_9.

2 Bello, P. (2021). The role of digitalization in decarbonizing the oil and gas industry. *Paper presented at the SPE Nigeria Annual International Conference and Exhibition*, 2 August 2021, Lagos, Nigeria. https://doi.org/10.2118/207125-MS.

3 https://www.pwc.com/us/en/industries/energy-utilities-resources/energy/digital-in-energy.html.

4 Popescul, D. (2011). The confidentiality – integrity – accessibility triad into the knowledge security. A reassessment from the point of view of the knowledge contribution to innovation. *Proceedings of The 16th International Business Information Management Association Conference (Innovation and Knowledge Management, A Global Competitive Advantage)* (29–30 June 2011). Kuala Lumpur, Malaysia, pp. 1338–1345.

5 Srinivas, J., Das, A.K., and Kumar, N. (2019). Government regulations in cybersecurity: framework, standards and recommendations. *Future Generation Computer Systems* 92: 178–188.

6 Xiong, W., Legrand, E., Åberg, O., and Lagerström, R. (2021). Cybersecurity threat modeling based on the MITRE Enterprise ATT&CK Matrix. *Software System Model* 21: 157–177.

7 ATT&CK Homepage. https://attack.mitre.org/

8 MITRE Homepage. https://www.mitre.org/

9 CVE Homepage. https://cve.mitre.org/

10 CWE Homepage. https://cwe.mitre.org/

11 CAPEC Homepage. https://capec.mitre.org/

12 Krumay, B., Bernroider, E.W.N., and Walser, R. (2018). Evaluation of cybersecurity management controls and metrics of critical infrastructures: a literature review considering the NIST cybersecurity framework. In: *Secure IT Systems. NordSec 2018*, Lecture Notes in Computer Science, vol. 11252 (ed. N. Gruschka). Cham: Springer https://doi.org/10.1007/978-3-030-03638-6_23.

13 https://www.nist.gov/

14 Grève, Z.D., Bottieau, J., Vangulick, D. et al. (2020). Machine learning techniques for improving self-consumption in renewable energy communities. *Energies* 13 (18): 4892. https://doi.org/10.3390/en13184892.

15 Salloum, S.A., Alshurideh, M., Elnagar, A., and Shaalan, K. (2020). Machine learning and deep learning techniques for cybersecurity: a review. In: Hassanien, A.E., Azar, A., Gaber, T., Oliva, D., Tolba, F. (eds) Proceedings of the International Conference on Artificial Intelligence and Computer Vision (AICV2020). Advances in Intelligent Systems and Computing, vol 1153. Springer, Cham. https://doi.org/10.1007/978-3-030-44289-7_5

16 Kolokotronis, N. and Shiaeles, S. (2021). *Cyber-Security Threats, Actors, and Dynamic Mitigation*, 1e. CRC Press https://doi.org/10.1201/9781003006145.

17 Munaiah, N., Rahman, A., Pelletier, J., et al. (2019). Characterizing attacker behavior in a cybersecurity penetration testing competition. *2019 ACM/IEEE International Symposium on Empirical Software Engineering and Measurement (ESEM)*, Porto de Galinhas, Brazil, pp. 1–6. https://doi.org/10.1109/ESEM.2019.8870147.

18 Najera-Gutierrez, G. and Ansari, J.A. (2018). *Web Penetration Testing with Kali Linux: Explore the Methods and Tools of Ethical Hacking with Kali Linux*, 3e. Packt Publishing Ltd.

19 McNab, C. (2007). *Network Security Assessment*, 2e. USA: O'Reilly.

20 Mancini, S., Iacono, L., Hartle, F. et al. (2021). Introducing the common attack process framework for incident mapping. *International Journal of Cyber Research and Education (IJCRE)* 3: 20–27.

21 Diogenes, Y. and Ozkaya, E. (2018). *Cybersecurity-Attack and Defense Strategies: Infrastructure security with Red Team and Blue Team tactics*. Packt Publishing Ltd.

22 Yeo, J. (2013). *Using Penetration Testing to Enhance Your Company's Security.* Trustwave.

23 https://www.synopsys.com/blogs/software-security/threat-modeling-vocabulary/

24 https://blog.convisoappsec.com/en/code-review-and-pentest-what-they-are-and-when-to-use-them/

25 https://pentestmag.com/red-teaming-10000-feet/

26 https://gomindsight.com/insights/blog/red-team-vs-blue-team/

27 https://www.compact.nl/en/articles/purple-team-drive-defense-with-offense/

28 https://www.exabeam.com/incident-response/incident-response-plan/

17

Explainable Artificial Intelligence (XAI) for the Trust and Transparency of the Oil and Gas Systems

17.1 Introduction

Artificial intelligence (AI) is turning out to be an indispensable paradigm for businesses and individuals. AI is automating and accelerating a specific set of everyday problems such as classification, regression, clustering, detection, recognition, and translation. AI can classify whether an incoming email is spam or real. recognize a person's face in an image, understand a speech and convert it into text, create an appropriate caption for a scene, etc. The scope of AI is fast expanding. Industry verticals are keenly exploring and experimenting different things. Business processes are being automated and optimized through the smart leverage of all kinds of noteworthy advancements happening in the AI space. Increasingly, AI takes the centerstage in business operations across the globe. There is a dazzling array of integrated platforms, frameworks, toolsets, libraries, and case studies, and hence the adoption of AI algorithms and models has been picking up dramatically in the recent past. However, there are a few critical challenges to be surmounted before the widespread usage of AI models in mission-critical domains such as healthcare, security, retailing, supply chain, and infrastructure management. That is, business executives and IT experts insist for trustworthy and transparent decision-making by AI models.

This chapter is to explain the brewing challenges in the AI field and how they can be surmounted through competent technology solutions. Especially, how the fast-emerging explainable AI (XAI) is a set of methods and software libraries that allow human users to comprehend and trust the results created by AI

The Power of Artificial Intelligence for the Next-Generation Oil and Gas Industry: Envisaging AI-Inspired Intelligent Energy Systems and Environments, First Edition. Pethuru Raj Chelliah, Venkatraman Jayasankar, Mats Agerstam, B. Sundaravadivazhagan, and Robin Cyriac.
© 2024 The Institute of Electrical and Electronics Engineers, Inc.
Published 2024 by John Wiley & Sons, Inc.

models. Explainable AI is to describe an AI model and how it arrived at a particular decision. XAI is to explain the AI model's implications and potential biases. It helps in understanding model accuracy and fairness. Explainable AI turns out to be a crucial cog for mission-critical enterprises to embark on the AI paradigm with all the clarity and confidence. With the maturity of the XAI concept, the AI adoption happens in a responsible manner across industry verticals.

17.2 The Growing Power of Artificial Intelligence

Having understood the strategic significance of AI technologies and tools, business houses and AI organizations are keen on embracing this emerging digital technology to envisage sophisticated business workloads and IT services. The AI-powered empowerment helps enterprising businesses explore fresh avenues to increase both bottom- and top-line revenues. Businesses can easily retain their customers' loyalty by providing premium and path-breaking capabilities. Business houses can add additional clients through the delectable contributions of AI, which is famous for pinpointing hidden patterns in datasets. Creating actionable insights out of Big Data is the principal goal of the AI paradigm. By disseminating the discovered knowledge to business and IT applications, the target of making software applications and services intelligent in their operations, outputs, and offerings is being elegantly fulfilled. Besides software systems, all kinds of input/output systems such as wearables, mobiles, portables, hearables, and implantable, fixed, nomadic, and wireless devices in our everyday environment such as homes, hotels, and hospitals are readied to be smart. All kinds of devices, instruments, equipment, appliances, wares, utilities, machineries, etc. in our personal, social, and professional places are being astutely empowered to exhibit a cognitive and conscious behavior through the incorporation of the distinct AI power.

Thus, in the increasingly digital era, there is a massive amount of multistructured digital data. When AI models are supplied with digital data, the key aspect of knowledge generation happens. With the AI-enabled transformation, every tangible object in our midst becomes smart, every electronic device becomes smarter, and every human being, who is being digitally assisted and augmented by multiple devices, will become the smartest. To fulfil the dream of setting up and sustaining intelligent environments and enterprises, the contributions of AI methods are becoming hugely critical. Precisely speaking, AI is the core and central aspect for the dreamt intelligent societies. Leading market analysts, researchers, and watchers have forecasted for a huge market for the AI phenomenon. Technology experts, evangelists, and exponents are in unison in articulating and accentuating the wherewithal of AI in smoothening and strengthening the journey toward the vision of an industrialized and intelligent world.

17.2.1 The Growing Scope of AI for the Oil and Gas Industry

AI solutions are extensively used for solving complicated problems across mission-critical industries. The AI journey is mesmerizing. We have generative AI; industrial AI; and real-time AI; explainable, edge, and efficient AI; responsible, trustworthy, and transparent AI, etc. Google has introduced Compose AI in the Gmail application. Thus, there are a series of innovations in the AI. Enterprises keenly leverage the distinct capabilities of AI to envisage and realize enterprise–class systems to retain their consumers and to obtain new customers. In short, AI aids enterprising businesses across the globe to produce and offer pioneering and premium services to their esteemed end-users. A plethora of innovations, disruptions, and transformations is being enabled through AI. For AI to survive and thrive, the main contribution is none other than the huge accumulation of digital data. AI has laid down a stimulating foundation for making sense out of data heaps quickly. That is, transitioning digital data to information and to knowledge is being provided by AI algorithms and models. In the digital era, AI helps make more value out of data. AI supports optimization and automation of business processes. AI provides the visibility into business operations, and such an understanding helps decision-makers and experts to strategize and plan with all the clarity and alacrity. Important sectors such as the oil and gas industry are gaining more through AI. There are powerful AI solutions by Oil and Gas Solutions – beyond.ai for the oil and gas industry. In other chapters, we have detailed how AI contributes immensely to accelerating and automating a variety of upstream, midstream, and downstream activities. The role of AI in the oil and gas industry is growing rapidly with more understanding on the transformative power of AI technologies and tools. There are several product vendors and solution providers unearthing and providing competent solutions for empowering the oil and gas industry, which is incidentally facing numerous challenges from green activists.

17.2.2 Explainable AI for the Oil and Gas Industry

In high-risk and high-value industries such as oil and gas, healthcare, and finance, it is essential to know what the machine is learning from the data and articulate its findings. But AI models (machine and deep learning models) must explain why they arrive at a suggestion. The sections below will explain why the XAI is important. For AI to be a trusted advisor and human-like decision-maker, XAI turns out to be an important contributor.

17.3 The Challenges and Concerns of Artificial Intelligence

In the previous section, we have pronounced the growing greatness of the AI conundrum. There is a surge in embracing the AI mechanism across the world for a variety of reasons. A variety of business operations are being augmented and automated through the AI distinctions. With the flourishing of AI-centric processing units such as hardware accelerators, end-to-end integrated platforms for AI model engineering, evaluation, optimization, and deployment, facilitating frameworks, specialized engines, enabling toolkits and software libraries, and other enablers, the AI adoption and adaptation are picking up fast. Every process and data-intensive problems are being attempted to be solved through a host of innovations in the AI space. AI brings in the much-needed efficacy and affordability. The hugely complicated process of transitioning raw data collection into information and into knowledge to be consciously used by decision-makers, leaders, stakeholders, and executives is simplified through the inherent strength of AI algorithms and models. Briefly, making sense out of data mountains is being facilitated by AI techniques and tools. Producing intelligence at the right time at the right place for the right people is the core and central aspect of the AI paradigm. There are case studies and proof-of-concepts (PoCs) in plenty illustrating the utility and universality of the AI phenomenon.

Amid surging popularity, there are a few critical concerns being expressed by AI practitioners. With a series of remarkable advancements in the AI landscape, it becomes difficult to comprehend how a particular AI model has arrived at a decision. AI models, especially deep neural network (DNN) architectures and models, are hard to retrace how the inference is made. Therefore, AI systems are termed as a black box. It is opaque even for data scientists and engineers to understand the route taken by AI algorithms to land in an inference. There are several hidden layers in between the input and the output layer. It is incomprehensible how the trained, tested, and validated AI model reacts on fresh data. It becomes a tough affair and assignment for professionals to interpret and explain the result of black box AI models. There are the issues of biases in taking decisions. AI researchers are therefore keen on bringing forth enabling techniques to explain why and how an AI algorithm makes certain decisions on new data. Thus, in the recent past, with the increased usage of AI systems across several mission-critical domains, the new notion of explainable AI (XAI) is grasping the attention of many in the AI field.

Machine learning (ML) algorithms are being leveraged to detect whether a person is infected with the COVID virus. Epidemiologists are extensively using AI to monitor the spread of a disease in a particular locality. Drug discovery is being speeded up through the leverage of AI methods. For example, AI helped arrive at appropriate vaccines quickly for COVID virus attacks. It would have taken several years for genetics and molecular biology experts to come out with vaccines without the services of AI. Thus, AI is occupying the prominent and dominant spot in many fields. The AI adoption is expected to grow dramatically in the years to unfurl. Therefore, the need for XAI libraries and toolsets is to grow in the days to come.

17.4 About the Need for AI Explainability

With the interest on embracing and employing newer technologies to avail deeper and decisive automation in our personal and professional assignments is growing steadily, there is a need for evaluating the pros and cons of each technology. As discussed above, AI is the most promising technology for the future of the world. AI has all the power and virtue to thrive in the years to come. Yet, there is a need to keep an eye on its functioning and contributions across industry verticals. When AI is being used for advancing and automating healthcare operations, it is imperative to analyze and understand how AI arrives at certain decisions. Because even if there is a small deviation in AI's recommendations, there is a possibility for risks for human lives. Therefore, technocrats and business leaders insist on evolving viable methods to gain enhanced understanding of how AI algorithms and models work to decide something. AI inferences must be unambiguously explained to end-users and service providers. Deep neural networks in the AI domain are typically black-box systems. There is no transparency on how DNN architectures calculates and throws out inferences. The challenge at hand is to move toward gray-box systems, which are visible for human eyes and minds to perceive what is happening in the hidden layers of DNNs.

XAI ensures utmost transparency and can support iterative models that ensure correct outcomes. The long-pending goals of correcting of diagnoses of diseases and right medication are being fulfilled through XAI in an exemplary manner. When AI is used for the identification of high-risk patients for specific diseases, it is vital to understand which factors or features contribute for the identification decision. XAI comes handy in taking a few more relevant facets and facts into consideration to explain the AI decision. The scope of XAI expands further. With the unique contributions of the XAI paradigm, other verticals are exploring and experimenting with the strategically sound power of XAI. Financial services providers use XAI to review fraudulent transactions and claims to sharply reduce operational costs.

Drug manufacturers leverage XAI to fast-track the process of end-to-end drug discovery to control a pandemic. Thus, with enhanced trust on AI, timely and automated intervention is to pick up in the days to come. Human involvement is bound to decrease with the delectable advancements being accomplished in the XAI space.

As accentuated above, AI is not an infallible technology. Due to assorted reasons, the performance and prediction power of AI models may degrade. That is, AI models may result in biased and unfair decisions. XAI has the wherewithal to point out the discrimination; thereby any untoward incident can be stopped forthwith. Wrong data can lead to wrong decisions. The AI model features may be wrongly selected and weighted. Thus, possibilities for arriving and articulating incorrect inferences are there.

In short, explainability is a sincere attempt through technological advancements to infuse trust in AI predictions. The aspect of explainability is to provide a basis for understanding both the inputs and outputs of a model. With unsupervised machine learning, there is extraordinarily little insight into the input machines used to train their models and algorithms. Adding a human in the loop for training and validation provides an insight into inputs, but it can be biased and difficult to scale.

17.4.1 What Is AI Explainability?

Explainable AI (XAI) models occur when humans can understand the AI results. This stands in contrast to black-box models in machine-to-machine learning, in which there is no insight into what machines learn and how that information impacts results.

How is AI explainability measured? Measuring AI explainability requires a human subject matter expert to understand how the machines arrived at the outcome. The ability to re-enact and retrace the results and check for plausibility is also required, as is the need to perform an AI impact assessment.

How are AI explainability and interpretability different? Interpretability is related to how accurate the machine learning model aligns cause with effect. Explainability relates to the knowledge or what is known about input parameters so that model performance can be justified or explained.

17.5 AI Explainability: The Problem It Solves

The core advantage to advancing AI explainability is to reach the goal of trustworthy AI. This requires models to be explainable so that the AI program's actions and decisions can be understood and justified. It reveals the reasoning behind the prediction, so the information is more trusted. You can know what the model is learning and why it is learning it. You can see how the model is performing, which increases transparency, trustworthiness, fairness, and accountability of the AI system.

It is easier for stakeholders to trust and act upon the predictions made by explainable AI.

17.6 What is the AI Explainability Challenges?

As calls for more AI regulation increase, trust in AI models needs to be achieved. This includes AI governance to track models and their results. The European Commission has already proposed creation of the first-ever legal framework on AI for its member states, which addresses the risks of AI and positions Europe to play a leading role globally.

More trust can be placed in model predictions and results when explainable AI is achieved. If problems are identified, you can see where they exist and address them directly by changing a rule. This transparency, enabled by symbolic AI, helps minimize model bias because rules can be monitored and managed.

This stands in stark contrast with black-box ML models that use autonomous machine-to-machine learning. Explainability for these models is inherently difficult and costly to manage. This is because machine learning uses massive quantities of training data and complex models and no clear understanding of the rules used to train the models. This requires a great deal of work to understand what the model is doing.

17.6.1 Mitigate AI Risk with Explainability

Many companies are moving to a risk-based approach to AI to counter this challenge. This involves categorizing AI systems based on risk level and creating guidelines and regulations to mitigate risks based on the level of impact.

Hindering the widespread use of explainable AI is the difficulty of creating them. Explainable AI models can be challenging to train, tune, and deploy. The use of subject matter experts or data scientists to train models is time-consuming and does not scale. Another challenge with explainable AI is the inability for machine learning models to track how pieces of data relate to each other once fed into the algorithms. For AI to be used at scale, a combination of machine learning and human oversight is needed, and that is where symbolic AI can help. It provides the best of autonomous learning with human subject matter expertise to scale explainable AI.

17.7 The Importance of Explainable AI

Software engineers can easily retrace to understand and explain how a particular software application outputs a result. But that is not the case with AI engineers because of the extreme complexity of AI models. The functioning of AI models is not easy to understand. The human understanding of software packages and AI models is essential to gain the confidence of these automated systems. To give 100 percent control to AI systems, it should be easy and essential to know how the AI computation happens. Such a grasp on any booming technology allows us to measure, manage, and enhance it further to tackle varying demands. Precisely speaking, the overwhelming approach to fulfil the emerging requirements such as interpretability and explainability of black box AI models is to move over to white box AI models. Omitting the aspect of explainability in life-critical domains such as healthcare poses a threat, and hence it takes an incredibly careful and calculated move on embracing AI. Notwithstanding, AI is penetrative, pervasive, and persuasive.

AI is increasingly ingrained with our daily walks and works. Nevertheless, understanding how AI arrives at a decision is hard to fathom. Especially understanding the decision-making routes of complex AI models is a tough affair, and hence sophisticated AI models are being termed as black boxes. The requirement at hand is to transition complex AI programs into gray or glass boxes so that humans can understand how and why AI algorithms arrive at such a particular decision. The decision pathways are visible to humans. The AI systems can explain the decision-making process, and humans can easily interpret AI's decisions. The functioning of AI programs is transparent to users, and hence the trustworthiness of AI systems can increase sharply. All kinds of biases and deviations can be eliminated so that the decisions are fair and acceptable.

The aspect of explainability has acquired a special importance in the mission-critical domains such as healthcare. Any imprecision in AI models can result in an irreparable situation. For boosting the confidence of users on AI systems, the trust on AI must increase drastically. AI systems should not be influenced by age, gender, culture, race, abilities, etc. AI decisions are crucial for healthcare services providers and patients. All should be able to avail the distinct benefits of the ever-growing AI systems. There should not be any partiality and preferences. As patients are not typically proficient in AI technologies and tools, there is a need getting imposed on AI systems to explain their decisions clearly yet concisely. Thereby, all kinds of decisions and deeds across industry verticals hereafter will be information-backed. The intuition-based deals will go away soon.

Precisely speaking, explainability turns out to be the cornerstone to securing and strengthening users' trust in AI systems. By knowing an AI system's internal processes, end-users can easily comprehend its pros and cons clearly. A clear-cut understanding of any AI system helps remove any fear about it and to know its distinct features and functionalities. Such a deeper understanding can facilitate to envisage and incorporate fresh capabilities toward delighting patients and caregivers. There are powerful tools emerging and evolving for enabling a standard reporting of AI models. The report can include the type of model, the details on the dataset on which the AI model gets trained and tested, what is the performance level, and any impending risks.

As described in the other chapters, there are easy-to-understand and use frameworks and tools to help the system's users understand and interpret the results. The tools provide scores for each factor and feature, which influences the outcome. This helps reflect the patterns identified by the model in the data. That is, the significance of distinctive features in arriving at decisions can be quantified. In computer vision examples, heatmaps (referred to as saliency maps) come handy in expressing and exposing which parts of pixels of the image have contributed more for the result. This gives the cue that a certain factor has played a crucial role in shaping up the prediction. But there is no clarity on why that factor has induced the result. In other words, there is still no answer for the why-question. For example, chest X-ray images typically help doctors diagnose diseases. In those images, certain regions would have played a role in influencing the final prediction of the AI system. But the doctor does not know why those portions influence the result. That is, the humans ought to get the answer for the question why.

Explainable AI models bring forth viable solutions and structures so that data scientists and users can correctly understand and explain why they came to a particular decision. The model's analytical logic and internal operations are fully transparent and interpretable. That is, even non-experts can easily get to know the model's internal functioning.

Thus, explainable AI plays a significant role in mission-critical fields. As inscribed in the beginning of this chapter, there are severe issues emanating out of AI models' bad predictions. AI models recommend an action, but the action does not make any sense for people. Therefore, experts and industry leaders favor for strengthening the brewing idea of explainable AI. Explainable AI models self-explain the reason behind the recommendation. In short, the aspect of explainability is very much indispensable for people to have the confidence and clarity on the decisions. Trust is essential for any technology domain to flourish. Any mistake by AI systems can be life-threatening or affect the livelihood of people. Thereby, the field of XAI draws a greater attention as it can provide explanations in natural language and facilitate easy interpretation. Such an empowerment does a lot of good for decision-makers and end-users alike.

The AI adoption is increasing day by day. A myriad of industry verticals is immensely benefiting out of all the praiseworthy improvisations happening in the enigmatic AI space. Especially in the healthcare domain, AI is used to accelerate and automate a variety of manual tasks, which are typically error-prone. AI models with the explainability feature can contribute in many ways, including detecting anomaly or outlier by scanning medical images and other sensor data, pinpointing useful patterns in data heaps, in empowering caregivers with medication recommendation, and helping in articulating and accentuating risk-mitigation mechanisms. Experts have come out with certain scenarios wherein the XAI capability is being insisted. When fairness in AI decisions is indispensable, the importance of XAI is critical. If any decision results in an irreparable situation, then the aspect of XAI is crucial. AI learns and comes out with new hypotheses. For establishing the truthfulness of any hypothesis proposed by AI models, subject matter experts (SMEs) depend on the power of XAI.

17.8   The Importance of Model Interpretation

Any ML model has a predictive function, which untangles distinct relationships, patterns in input data, and exposes it for automated systems, decision-makers, and executives. The model brings forth a function between the independent (input) variables and the dependent (target or response) variable(s).

When a model finds something useful and usable in the incoming data, it helps people and systems take decisions correctly. The model interpretation is all about simplifying the process of understanding and explaining how and why these decisions are arrived at. The overall idea behind this is to ensure the transparency of the working condition of the model. Also, the common people must easily understand the model's recommendations, suggestions, inferencing, etc. Like software systems, models too have a few important non-functional requirements (NFRs). A model must have a better interpretability just like maintaining a reliable performance level. We all know the prominent performance indicators/metrics of ML models are precision, accuracy, recall, etc. However, performance metrics are typically fixed at the beginning. However, metrics do not tell everything about a model's predictions. The performance may degrade due to assorted reasons such as model concept drift and data drift. Hence, it is paramount to understand how and why a model takes certain decisions.

17.8.1   Model Transparency

An ML model is getting created by leveraging an ML algorithm on data features. That is, an ML model is a representation of mapping inputs to potential outputs. In other words, an ML model is a mathematical formula or equation to accurately map input independent variables (inputs) to dependent variables (responses). The bewildering idea behind model transparency is to gain a clear-cut understanding about how models are being derived and what are the things that directly or indirectly influence model decisions. The coefficients in linear equations/models, weights of a neural network, etc. play a vital role in shaping up model decisions.

XAI experts have pointed out a series of steps to be considered to simplify the process of model interpretation.

17.8.2 Start with Interpretable Algorithms

As indicated elsewhere, the smartest way to embark on model interpretation is to use easily interpretable algorithms like linear and logistic regression. Tree-based algorithms are also recommended. Some experts even recommend k-nearest neighbours (KNN) and Naive Bayes.

17.8.3 Standard Techniques for Model Interpretation

The interpretability is to guarantee the much-needed fairness, accountability, and transparency. Such an additional facility boosts the confidence of decision-makers and end-users to leverage AI models more for a variety of business and personal requirements. The proven techniques are given below.

1) There are exploratory data analysis (EDA) and data visualization techniques simplifying the complex task of model interpretation.
2) There are also dimensionality reduction techniques, which reduce the feature space for visualizing and understanding what factors are influencing a model to take certain decisions. The well-known reduction techniques include principal component analysis (PCA) and self-organizing maps (SOM). Further on, there is another method of **nonlinear dimensionality reduction** (also known as **manifold learning**). This aims to project high-dimensional data onto lower-dimensional latent manifolds.
3) There are well-known model performance evaluation metrics such as accuracy, precision, and recall. Further on, there are the ROC curve and the AUC algorithm for classification models.

17.8.4 ROC Curve

This is a receiver operating characteristic curve illustrating the performance of a classification model at all classification thresholds. This plots two parameters: True Positive Rate (TPR) (TPR is synonymously termed as recall) and False Positive Rate (FPR). Area Under the ROC Curve (AUC) is an efficient and sorting-based algorithm to compute the points in an ROC curve. The AUC provides an aggregate measure of performance across all classification thresholds.

Similarly for regression models, the standard performance evaluation metrics are mean squared error (MSE) and root mean squared error (**RMSE)** and mean absolute error (MAE) for regression models.

17.8.5 Focus on Feature Importance

Features are critical for making correct prediction. Therefore, the domains of feature engineering, optimization, and selection are getting a lot of attention these days as they must achieve the intended success. Understanding and articulating those features, which contribute for taking certain decisions or making recommendations, or performing inferences or articulating specific outcomes, etc. ultimately help in model interpretation. Skater (https://oracle.github.io/Skater/) is an open-source unified framework to enable model interpretation. Skater supports algorithms to demystify the learned structures of a black-box model both globally (inference based on a complete data set) and locally (inference about an individual prediction).

The following is a standard example of a feature importance plot from Skater on a census dataset.

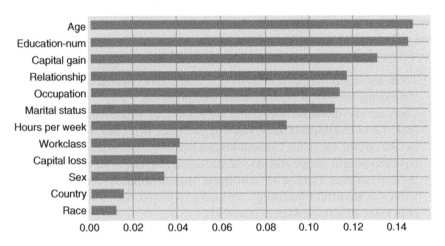

Here Age and Education-Num are the top two features, where Age is responsible for model predictions changing by an average of 14.5% on perturbing the Age feature. A feature's importance is calculated based on the increase of the model's prediction error after perturbing the feature. A feature is important if perturbing its values increases the model error. A feature is unimportant if perturbing its values keeps the model error unchanged.

17.8.6 Partial Dependence Plots (PDPs)

A partial dependence plots the marginal impact of a feature on model prediction by keeping other features in the model constant. The derivative of partial dependence can describe the impact of a feature. PDPs can also show whether the relationship between the target and a feature is linear or complex. PDP is a global method. That is, the PDP method considers all instances and makes a statement about the global relationship of a feature with the predicted outcome.

17.8.7 Global Surrogate Models

Previously, we have discussed how to simplify model interpretability by leveraging easily interpretable ML models, feature importance scoring methods, and PDPs. But there is a need to interpret complex models. One viable and venerable solution approach is to use global surrogate models. A global surrogate model is an interpretable model that is trained to approximate the predictions of a black box model, which can be any model derived out of any training algorithm. That is, global surrogate models are model agnostic.

The process is to create an interpretable surrogate model from the base and black-box model. A surrogate model is model-agnostic since it requires no information about the inner workings of the black-box model. Just the relation of input and predicted output is used. Tree-based models are found to be good for building surrogate models. The main steps for building surrogate models are explained in this book (https://christophm.github.io/interpretable-ml-book/global.html).

1) **Choose a dataset**: This could be the same dataset that was used for training the black-box model or it can be a new dataset from the same distribution. It is also OK to choose a subset of the data.
2) For the chosen dataset, get the predictions of the base and black-box model.
3) Choose an interpretable surrogate model (linear models or tree-based models).
4) Train the interpretable model on the dataset and its predictions.
5) Measure how well the surrogate model replicates the prediction of the black-box model.
6) Interpret/visualize the surrogate model.

Criteria for ML Model Interpretation Methods – Christoph Molnar has authored a few well-written books on interpretable machine learning (https://christophm.github.io/interpretable-ml-book/)

- **Intrinsic or post hoc interpretability**: This is leveraging intrinsically interpretable ML algorithms such as linear, parametric, and tree-based algorithms. Post hoc interpretability means first creating a black-box model, which is prepared using ensemble algorithms or deep neural network (DNN) architectures. Then, it is all about applying interpretability methods (feature importance scoring and other methods such as partial dependency plots (PDP), etc.
- **Model-specific or model-agnostic**: Model-specific interpretation tools are specific to intrinsic model interpretation methods. Model-agnostic tools are relevant to post hoc methods. These agnostic methods usually operate by analyzing feature input and output pairs. That means these methods do not access any model internals such as weights and constraints.

There are two types. Model explanation (that is, global explanation for the whole model) and explanation, for instance specific prediction (that means local explanation). Global explanation is to focus the whole model and to explain the operational logic. Local explanation is to explain the model's prediction for individual instances. Explanations come in the form of rules, feature importance, counterfactuals, plots, visualization, etc.

Global interpretability is to understand and explain model decisions based on interactions between the predictor features (input) and responses (output). Gaining a deeper understanding of feature interactions and importance facilitates the global interpretability. However, visualizing features with more than two dimensions and analyzing their conditional interactions are beset with many practical challenges. Therefore, the focus gets turned toward understanding subsets of features. Such modularization leads to global interpretability.

For local interpretability, it is sufficient to understand prediction decisions for a single datapoint, that is, the focus is on that datapoint, and the idea is to understand the prediction made by the model by looking at a local subregion in the feature space around that point.

Local data distributions and feature spaces may give accurate explanations. The LIME framework (discussed below) can be used for model-agnostic local interpretation. The best practice is to use both global and local interpretations to explain model decisions not only for one instance but also for a group of instances.

Primarily, deep learning (DL) models are inherently complex due to the participation of hundreds and even thousands of hidden layers in between the input and the output layers. Thus, deep learning architectures are being termed as black box systems. Further on, to finetune the prediction accuracy, ensemble models are being used. Ensemble models are being built by aggregating multiple AI models. Therefore, AI experts recommend leveraging simple AI algorithms such as decision trees and regression models, thereby the opaqueness of AI decisions can be avoided. The model performance is good when going for simple and well-understood algorithms. How XAI emerges as a solace and silver bullet for industries to benefit from the extreme power of AI is vividly illustrated in the image below.

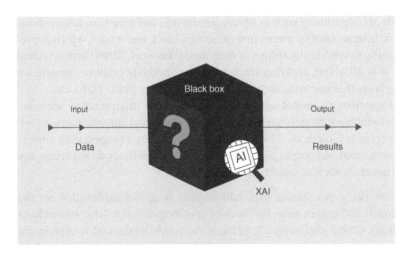

As machine and deep learning models are being represented through complex mathematical formulae, there is a need for competent approaches to reduce the model complexity. When the model complexity decreases, then understanding models' decisions and recommendations become easy to data scientists, business executives, and even end-users. As indicated above, calculating feature importance and tool-based model explainability are being seen as the way forward to nullify any bad implications of AI models.

17.9 Briefing Feature Importance Scoring Methods

In the literature, there are several proven and potential feature importance scoring methods such as mean decrease impurity (MDI), mean decrease accuracy (MDA), and single feature importance (SFI). These methods are good at producing a feature score, which clearly articulates the feature's impact on the predictive quality of ML models. However, it is important to know a specific feature's contribution in influencing the prediction. In the recent past, there came a few ML techniques that facilitate the detection of a feature's contribution. Such methods simplify and streamline the easy interpretability of the model's decision. The insights being shared by such ML techniques empower business leaders and data scientists to understand and reveal the criteria used while arriving at decisions. Financial and insurance industries are acquiring immense benefits through such measures. An in-depth explanation can be given to customers, who seek the basis on which certain decisions (approval or rejection) are being taken. Similarly, pharmaceutical and healthcare industries get the boost through the self-explanatory capability of recent-day and tool-assisted machine learning models.

Besides understanding the importance of every feature in decision-enablement, the aspect of feature selection contributes to easy interpretability of AI decisions. Throwing out insignificant features is important. Feature selection also tackles the curse of dimensionality. The problem of model overfitting is also minimized through appropriate feature selection. Thus, feature engineering involving selection, optimization, etc. turns out to be a constructive factor. Fortunately, there are a few competitive feature selection techniques. There are articles on different feature selection methods, and there are Python implementations for each of the methods. Experts also have articulated the pros and cons of each of the methods based on their practical implementation difficulties.

In short, the key motivation for XAI is to transition black box systems into glass box systems. Such an empowerment through the smart leverage of XAI libraries and packages helps data scientists check and fix biases. The much-needed fairness in arriving at decisions is getting fulfilled through XAI. There are two widely interested explainability algorithms: the LIME and SHAPE algorithm.

17.10 Local Interpretable Model-agnostic Explanations (LIME)

The key motivations for the LIME method are given below. This is based on the research paper published and made available at this page (https://arxiv.org/pdf/1602.04938.pdf). First, we accept that a linear ML model is more interpretable than a complicated ML model. However, the following linear model is not easily interpretable as it has several variables.

$$
\begin{aligned}
Y = {}& 1.50 + 3.3X_1 + 25.4X_2 + 312X_3 + 32X_4 \\
& + 436X_5 + 9.33X_6 + 2.3X_7 + 4.9X_8 + 0.3X_9 + 40.2X_{10} + 4.33X_{11} \\
& + 6.1X_{12} + 873X_{13} + 1.3X_{14} + 4.5X_{15} + 73.2X_{16} + 0.53X_{17} + 0.61X_{18} \\
& + 9.2X_{19} + 453X_{20} + 8.32X_{21} + 25.4X_{22} + 7.31X_{23} \\
& + 30.32X_{24} + 23.6X_{25} + 5.32X_{26} + 3.0X_{27} + 90.1X_{28} \\
& + 4.2X_{29} + 893X_{20} + 53.0X_{31} + 13.3X_{32} + 2.5X_{33} + 7.6X_{34} + 6.35X_{35} \\
& + 5.13X_{36} + 0.32X_{37} + 49.1X_{38} + 3.2X_{39} + 1.8X_{40}
\end{aligned}
$$

Second, for an individual prediction, only a few variables play a significant role in arriving at a prediction. Other variables are not contributing much for an individual prediction. The interpretation should make sense from an individual prediction's view. This is being termed as local fidelity. Globally prominent features may be irrelevant in the local context. Thus, even if a model has hundreds of variables, only a few variables directly contribute to a local (individual) prediction.

Having understood the growing requirements, the authors have proposed LIME, which makes it easy to interpret any model. LIME enables local fidelity (locally interpretable). The authors of LIME insist on building two types of trust for confidently using a model.

- **Trusting an individual prediction:** Users must trust an individual prediction to act upon that prediction with all the confidence and clarity.
- **Trusting a model:** Once the model is trained, tested, and validated, then it goes to a production environment. There are performance evaluation metrics such as accuracy, precision, and recall. If there is a need to visualize the performance of any multi-class classification problem, we use the AUC (**Area Under the Curve**) ROC (**Receiver Operating Characteristics**) curve. This has emerged as an important evaluation metric for checking any classification model's performance. ROC is a probability curve, and AUC represents the degree or measure of separability. AUC tells how much the model is capable of distinguishing between classes. With the higher AUC, the model is good at predicting 0 classes as 0 and 1 classes as 1.

LIME is a popular algorithm for model explanation. LIME explanations are primarily based on local surrogate models. As mentioned before, surrogate models are interpretable models that are typically learned on the predictions of the original black-box model.

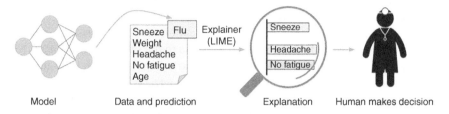

| Model | Data and prediction | | Explanation | Human makes decision |

Local interpretability can answer the question "Why is the model behaving in a specific way in the locality of a data point x?"

LIME tests what happens to the original black box model's predictions when some variations or perturbations are made on the dataset of the black-box model. Typically, LIME generates a new dataset consisting of perturbed samples and the associated black-box model's predictions. On this dataset, LIME then trains an interpretable model based on the proximity of the sampled instances to the instance of interest. A high-level workflow as articulated by the book author is given as follows.

- Choose the instance of interest for which you want to get an explanation of the predictions of your black box model.
- Perturb your dataset and get the black box predictions for these new points.
- Weight the new samples by their proximity to the instance of interest.
- Fit a weighted, interpretable (surrogate) model on the dataset with the variations.
- Explain prediction by interpreting the local model.

This supplies all the insights on how much each feature has contributed for the ML model for an individual prediction. This algorithm operates and delivers in a model-agnostic manner. That is, this algorithm can be applied for any black-box ML model. This also works for different neural network (NN) architectures. This algorithm works by leveraging the prominent technique of surrogate models. That is, instead of explaining the black-box ML model directly, it uses a local surrogate model to provide required explanations.

The surrogate model is being prepared in a phase manner. First, the LIME algorithm creates a new proxy dataset by making some permutations to the feature values of the available dataset. Second, each of these samples is assigned a particular weight value that is proportional to its similarity with respect to the instance, which must be explained. Finally, leverage a surrogate machine learning

(ML) model, which is an easily explainable model like decision tree and linear regression on the weighted proxy dataset.

The LIME algorithm creates a local surrogate model, which can be used to supply model explanations. This technique works even when there are complex black-box models. Further on, there is no need to train the local surrogate model on the same features that were used to train the black-box model.

The main limitation is the need for the creation of a local surrogate model that must explicitly define the similarity metric to the sample, which ought to be explained. Determining this similarity metric is not easy, and the metric values vary from case to case. If the similarity metric is not carefully chosen, then there is a possibility for instable explanations. That is, the explanation of identical samples may differ from each other a lot. There are several implementations of this algorithm using multiple programming languages. The Python-centric LIME package is available for practitioners. Interested readers can get sample LIME implementations at Introduction to Explainable AI(XAI) using LIME – GeeksforGeeks and Explain your model predictions with LIME|Kaggle.

17.11   SHAP Explainability Algorithm

Shapley Additive Explanation (SHAP) is a model explainability algorithm. This operates on the single prediction level not at the global machine learning model level. The algorithm derives its technique from the cooperative game theory. That is, it can determine the payout for each player within a cooperative coalition.

The payout for each player is being calculated based on the magnitude of the SHAP value that is associated with that player. The magnitude value is identified by the player's contribution. The Shapley value represents the feature's contribution on the final prediction. Thus, when the SHAP algorithm is used on a particular instance, there will be many Shapley values. Each value is associated with one of the instance's features.

The final prediction can be calculated by adding a baseline value, which is constant for each machine learning classifier/regressor. The magnitude and the sign of the Shapley value can determine the contribution of a feature in arriving at the final prediction. High absolute Shapley values indicate that the feature is important. The sign of the Shapley value conveys the direction. For a binary classifier, a negative sign means that the feature has pushed down the prediction to 0 label. Besides knowing the importance of individual features, it is important to know the importance of a collection of features. This holds good for image data.

Here is an illustrative example.

There is a trained, tested, and optimized machine learning (ML) model to predict apartment prices. For a certain flat in an apartment, the model predicts 300,000 USD, and we need to explain this prediction. The apartment flat has a size of 50 m^2 and is located on the 2nd floor in the apartment complex. There is a park nearby, and cats are fully forbidden. The average prediction for all apartments is 310,000 USD. The question is to understand how each feature value contributes to the prediction compared to the average prediction.

The effect of each feature is the weight of the feature times the feature value minus the average effect of all apartments. This is a linear model. For complex models, the solution comes from the cooperative game theory. Shapley Additive Explanation (SHAP), which is based on the concept of cooperative game theory, is the way forward. The SHAP algorithm is model-agnostic. However, its computational time is on the higher side. Especially if it must handle complex ML models with many features, the time complexity is high. The workaround for this limitation is to leverage a package that allows one to define the used machine learning model.

The Tree Explainer is an implementation of the SHAP package. This does a respectable job for tree-based algorithms such as random forest, decision tree, and Gradient Boost. The DeepExplainer package can explain deep learning (DL) models, which are complex. Here is an example. This package can provide the explanation of a DL image classification model that was trained to recognize a series of animals. The SHAP explainability algorithm can detect which pixels together caused the algorithm to decide a certain animal class. Further on, the SHAP algorithm can provide explanations for natural language processing (NLP) problems. This is through the package's Kernel Explainer (https://pub.towardsai.net/ explain-your-machine-learning-predictions-with-kernel-shap-kernel-explainer-fed56b9250b8). This is model-agnostic and can handle any type of ML algorithms. Thus, the feature importance techniques (LIME and SHAP) emerge as a new category of enabling tools for data scientists.

Here are some domains wherein the power of explainable AI is keenly sought.

- **Healthcare**: This is a life-critical domain, and hence any wrong decision can have a bad and sad impact. Explainable AI comes handy for surgeons, doctors, clinicians, and caregivers as it can explain the medication recommendations to patients by providing insights into the variables that took part in the decision-making process.
- **Finance**: This is another prospective domain for explainable AI. Important financial decisions are being taken through AI models these days. Herein, any mishap can result in catastrophic situations. Thus, the AI model must shed light on what prompted it to arrive at such a decision.
- **Recruitment**: Selecting appropriate resumes for further review is being automated through AI. Now the capability of explainable AI can tell us why a candidate's resume was selected or not selected. This nullifies any sort of bias and unfairness.
- **Fraud detection**: There is no doubt that fraudulent and suspicious transactions must be correctly captured and alerted in time. Similarly, fake product identification is another crucial element. On the other hand, any wrong identification can be a huge inconvenience for service providers and consumers. Therefore, for boosting the confidence on AI decisions, the aspect of explainable AI is a definite value-addition to nullify any bias and discrimination.

17.11.1 Advance the Trust with Symbolic AI

This is another recent approach for incorporating the much-needed trust in AI algorithms and models. With the widespread adoption of AI capabilities, regulatory actions, on the other side, are being confabulated. AI service providers are forced to take right and relevant measures to embed more trust into their AI initiatives. The simple and straightforward way to accomplish this is by using a rule-based approach (symbolic AI) to develop any AI model.

With symbolic AI, the relationship between concepts (logic) is predefined, and hence understanding how the model works and contributes becomes easy for anyone. Rather than creating explainable AI models, the rule-based approach is better. The control is with the AI engineer on the AI model. With the rule-based approach, any undesirable or biased results can be easily traced back to the root cause and insightfully fixed. Thus, to develop and run fully explainable AI models, without an iota of doubt, the rule-based approach is the way forward. As inscribed above, certain sectors insist for explainability. Any compromise on the aspect of explainability may be catastrophic. Without formal explanations, even legal complications or financial repercussions may emerge to haunt AI solutions and service providers. But for complicated and sophisticated business problems, symbolic AI turns out to be a difficult affair. Therefore, the automated feature engineering is being widely recommended. Thus, the hybrid AI approach is preferred. In a hybrid approach, explainability can be imparted in multiple ways as indicated below.

- We can use symbolic AI to make the feature engineering process explainable, and symbolic AI helps extract the most relevant features and use them to bring forth competent AI models.
- The other recommendation is to annotate a massive amount of unstructured data, and then a human well-versed in data engineering can validate them for their correctness. Finally, training an AI model can be initiated.
- Another possibility is to develop an AI model to automate the process of creating well-defined and comprehensive symbolic AI rules by taking annotations as the input.

In summary, with AI, it is possible for engineers to enable machine and deep learning (ML/DL) algorithms to extract actionable insights out of data heaps. AI algorithms have the wherewithal to find useful and usable patterns in datasets. The knowledge discovered gets disseminated to business and IT systems to do right and relevant tasks in time. Thus, insight-driven jobs can be accomplished in an automated manner. Also, multiple systems get integrated, and there are business processes for orchestrating multiple services and systems to do complicated tasks with all the clarity and alacrity. Thus, AI-powered automation in association with integrated systems and optimized processes is being seen as a meaningful change for the forthcoming era of knowledge.

17.12 Conclusion

Previously, we were using easily interpretable machine learning algorithms such as linear, logistic, and polynomial regression decision tree. But nowadays, we have bigger problems at hand. This forces us to leverage high-end machine and deep learning algorithms to create highly accurate problem-solving models. For example, random forest (RF) is an ensemble ML algorithm, which is a combination of multiple decision tree algorithms. The accuracy increases significantly with the RF algorithm, but everything comes with a price. It is quite difficult to interpret the internal functioning of these hugely complicated algorithms, and there is no clarity on how the algorithm arrived at such a suggestion, recommendation, detection, summarization, knowledge discovery, etc. When we humans take a decision, we can clearly explain on what basis we arrived at the conclusion. What prompted to take such a result can be meticulously articulated, but that is not the case of automated algorithms and systems. This insists on embedding the interpretability and self-explanatory capabilities into the algorithms to gain back the trust of people on AI-powered systems. Calculating the feature importance is one simple and straightforward approach to simplify the interpretability of AI decisions.

Bibliography

1 Pandey, D.S., Raza, H., and Bhattacharyya, S. (2023). Development of explainable AI-based predictive models for bubbling fluidised bed gasification process. https://pdf. sciencedirectassets.com/271496/1-s2.0-S0016236123X00143/1-s2.0-S0016236123015843/ main.pdf.

2 Mohaghegh, S.D. (2021). Explainable artificial intelligence thrives in petroleum data analytics. https://jpt.spe.org/explainable-artificial-intelligence-thrives-in-petroleum-data-analytics.

3 Srivastava, S. (2023). How AI Is revolutionizing the oil and gas industry – nine use cases and benefits. https://appinventiv.com/blog/artificial-intelligence-in-oil-and-gas-indsutry/

4 Aslam, N. (2022). Anomaly detection using explainable random forest for the prediction of undesirable events in oil wells. https://www.hindawi.com/journals/acisc/2022/1558381/

5 Kau, J. (2023). Explainable AI in manufacturing industry. https://www.xenonstack. com/blog/ai-manufacturing-industry.

18

Blockchain for Enhanced Efficiency, Trust, and Transparency in the Oil and Gas Domain

18.1 Introduction

Upstream operations typically represent the exploration, discovery, and extraction of raw minerals. The midstream operations include the safe transportation of the deposits identified. The final part is the set of downstream activities such as refining the minerals to make end-products such as petrol and diesel.

The oil and gas industry has always been a story of boom and bust. As the world is vociferously insisting on taking concrete steps for meeting the sustainability goals, there arise a few noteworthy challenges brewing for the oil and gas sector. Market researchers and analysts have forecasted that with the growing middle class population, there will be an increase in energy consumption with more vehicles on the road. New energy sources and types must be identified to meet the growing energy needs of the world. Energy conservation methods must be worked out and implemented with strictness. Energy wastage and pilferage must be stopped forthwith through the meticulous and measured use of pioneering technologies. The supply chain must be highly optimized to cut down operational expenses through a host of strategically sound digital innovations and disruptions. The growing supply chain complexity is being minimized through the improvisations at the digital ecosystem. Pipeline monitoring and management must be tightened up. Several promising and potential steps are being unearthed and given the utmost attention through the leverage of proven digital intelligence techniques and tools.

However, the happy news for the industry is the grandiose arrival and adoption of many path-breaking digital technologies. Profound and far-reaching innovations come handy in lowering oil prices, with an intensive focus on affordability,

The Power of Artificial Intelligence for the Next-Generation Oil and Gas Industry: Envisaging AI-Inspired Intelligent Energy Systems and Environments, First Edition. Pethuru Raj Chelliah, Venkatraman Jayasankar, Mats Agerstam, B. Sundaravadivazhagan, and Robin Cyriac.
© 2024 The Institute of Electrical and Electronics Engineers, Inc.
Published 2024 by John Wiley & Sons, Inc.

agility, and adaptivity. Extracting actionable insights in time out of exponentially growing data heaps is being seen as a first and foremost thing for the oil sector to flourish in the years to unfurl.

18.2 The Brewing Challenges of the Oil and Gas Industry

Experts have articulated the problematic challenges of the oil and gas industry. For the ensuing digital era, it is pertinent and paramount for the oil and gas industry to digitize crude oil transactions. Companies trade petrol, diesel, and gas and use homegrown proprietary and customized systems to track, organize, and record data. There are therefore disparate and disaggregated datastores. These are primarily centralized systems and hence hacking and stealing confidential, customer, and corporate data becomes an easy affair. Similarly, there are several crucial issues with current and conventional business workloads and IT services of the oil and gas industry.

Now with the wide and wise adoption and adaptation of the blockchain technology, there are good prospects and potential for reducing the transactional and operational costs of the industry. Financial irregularities can be eliminated, any mismanagement can be avoided, any problem resolution can get accelerated, any deviation can get nullified, data security guaranteed, and new possibilities and opportunities are bound to emerge and evolve. In this section, the principal limitations are being illustrated to enable the readers to understand how blockchain has come as a solace for the O&G industry.

The oil and gas industry vertical must find creative and cognitive ways for identified and hidden problems. Path-breaking climate change detection and mitigation technologies are needed to sustain the fragile environment.

18.2.1 Harmful Unconventional Resources

New oil and gas resources are being explored and unearthed. These unconventional resources include shale gas, oil sands, and coalbed methane. However, they are quite hard to access. Advanced technologies enable oil majors to use these resources. There are serious doubts and worries about the short-term and long-term implications of using these resources. Businesses and the human society may be affected with these unconventional resources. Though such resources may open new opportunities, the implications of using them may be harmful as per the statements of the experts in the oil and gas domain.

18.2.2 Rising Consumer Demand

As stated somewhere else in this chapter, with the continued increase of middle-class people across the globe, the number of vehicles on the road is steadily increasing. Therefore, the demand for oil and gas resources is increasing. This necessitates for oil companies and national governments to invest their time, talent, and treasure to produce more oils and to explore for additional oil fields. Further on, the midstream and downstream activities are becoming expensive. The pricing for these can vary drastically if there is a high demand during short supply. These are being stated as critical challenges for the oil and gas industry.

18.2.3 Social and Environmental Responsibility

With a renewed interest in sustainability, there is an additional pressure on the oil and gas industry. Our fragile environment is becoming spoiled through the increase in the number of vehicles that are being run on oil and gas. Herein, the oil and gas industry must embark on a few breakthrough technologies that can nullify the adverse effects of greenhouse emission of vehicles. The damage to the environment must be reduced.

18.2.4 Increased Demands for Renewable Fuels

This is definitely a paradigm shift. The world is veering toward the concept of renewable fuels, which have the power to ensure the stability and sustainability of our living environment. With renewables, it is possible to minimize the quantity and quality of air pollution. There are other critical advantages of renewable energy sources, which are classified as environment-friendly.

18.2.5 Increasing Costs for Maintenance

Establishing and sustaining oil farms, pipelines, and refineries are definitely expensive. By leveraging machine and deep learning (ML/DL) algorithms and models, it is possible to gain the capability of predictive maintenance. Thereby, oil companies are investing heavily on deploying advanced sensors and other IoT devices. Further on, by meticulously capturing all kinds of digital data getting emitted by various IoT devices, integrated data analytics platforms hosted in centralized cloud environments come handy in fulfilling the much-needed predictive maintenance; thereby, the life of business-critical assets, equipment, and machineries is getting improved.

Now with the concept of edge computing being nourished by product vendors, network providers, cloud service providers, and other contributors in the fast-growing ecosystem, real-time data capture, processing, and analytics is being facilitated to attain timely and actionable insights so that the cognitive and correct maintenance of various participants is ensured. Thus, a few proven and promising digital technologies and tools are being touted as the greater enabler of the oil and gas industry. Without the smart leverage of digital techniques, asset management and maintenance are a costly affair.

18.2.6 Decreasing Performance

The machineries, wares, utilities, appliances, instruments, and equipment being used in oil fields and farms are bound to have their performance level decreasing gradually. Thus, it is imperative to keep up the performance of each of the assets to maintain the desired output levels. The traditional techniques are found to be inadequate in guaranteeing the desired performance. This lacuna is being touted as one of the primary problems of the oil and gas sector. This yearns for fresh and futuristic approaches leveraging state-of-the-art technological systems.

18.2.7 Inefficient Supply Chain

Intrinsically the oil and gas industry is becoming complicated with the participation of many diverse partners, resellers, suppliers, service providers, etc. Many human professionals ought to be involved to ensure proper coordination among different and distributed players. This setup is risky and expensive. We need competent technological paradigm to surmount the problem of growing supply chain complexity. As described below, blockchain is being pronounced as the way forward to solve this complex problem.

18.2.8 Lack of Transparency

The oil and gas industry is riddled with the transparency issue for a prolonged period now. With the growing complexity, ensuring the much-insisted transparency becomes a tough assignment. With deeper visibility, it is possible to achieve the trust and transparency. Any kind of misgivings can be proactively pinpointed and nullified.

18.2.9 Lack of Proper Audits and Maintenance

Auditability is being seen as an important ingredient for the oil and gas industry to survive and thrive in the extremely tough time. The auditing process must be given an utmost importance to keep everything in proper order and place. For any oil major to attain the intended success in its long and arduous journey, the auditing process must be continuously updated. Though it is termed as a costly affair, considering the criticality of fuels for the betterment of the human society, different mechanisms and methods are being worked out.

18.2.10   Poor Waste Management

Without an iota of doubt, the chemical waste, which is large, is a dangerous thing for the humanity. Therefore, the aspect of waste management gets a special mention. If not properly done, then the chemical waste turns out be a real horror for people. It is hugely harmful for the living environment. Governments have formulated strict regulations for waste monitoring and management.

We have discussed the widely known problems of the oil and gas industry. There are a few technical, business, and society concerns about the oil and gas industry. Therefore, the industry is veering toward the idea of digital transformation. There are several trend-setting digital technologies facilitating the much-needed business transformation. Blockchain is one of the niche digital technologies having the wherewithal to bring in desired and delectable changes in how the oil and gas industry is functioning. This chapter is dedicated to explaining the practical details of the blockchain technology and how it comes handy in solving and surmounting some of the identified problems in an elegant manner.

Blockchain comes with a slew of ground-breaking features and functionalities to empower oil companies to face all sorts of brewing challenges in a confidential fashion. For example, the exemplary automation facilitated by the emerging concept of smart contracts is going to be game-changing for the oil and gas industry. A lot of manual and error-prone activities can be fully and confidentially automated through the cognitive leverage of smart contracts, which run on blockchain network infrastructure.

As explained below, blockchain provides a fully transparent ledger to be accessed by everyone. When the logs are open for everyone to view, then it is easy to pinpoint any deviation or discrepancy in time. In addition, ledgers are immutable. That means any one can view the data, but no one can manipulate the data there. Also, there is no central authority to monitor and manage the whole thing. Through the aspects of decentralization and distribution, the trust needed to do transactions is being established and enforced in a stringent manner.

18.3   About the Blockchain Technology

The blockchain domain is penetrating every worthwhile industry vertical. There is a greater understanding about the innate power of blockchain. The much-insisted digital transformation initiatives are being accelerated and automated through the participation of blockchain. Security, affordability, and simplification requirements are being fulfilled through the appropriate usage of the noteworthy advancements happening in the blockchain space. Oil and gas companies are under immense pressure to cut down expenses and to explore fresh avenues to earn more revenues. The complex processes of the oil and gas industry are being optimized through the smart leverage of blockchain. Pioneering and premium services are being visualized and delivered to clients and consumers to keep up the loyalty along with greater transparency.

Blockchain has shown a lot of promise for the oil and gas industry. Blockchain ledgers can keep up all kinds of transactions safely. Immutability is one important contribution of blockchain. That is, any kind of illegal and unauthorized manipulation of transaction data is not permissible. However, the data access, assessment, and availability are being facilitated by the internal mechanism of blockchain. Automated contract execution is being enabled by blockchain. There are several improvisations in the domain of blockchain's smart contracts. The trust being traditionally established and enforced by intermediaries is being guaranteed by blockchain. The direct trading and transaction between participating parties are being streamlined through the inherent influence of blockchain. The trust and transparency being facilitated by blockchain have laid down a stimulating foundation for sharply improving the efficiency and hence the productivity.

Blockchain in association with competent AI technologies and tools records, tracks, and executes contracts and intimates the concerned in case of any fraud instances. Equipment sensors duly confirm the fulfilment of contract agreements. This triggers a bevy of activities in the form of notification, billing, payment, etc. In short, oil and gas companies and their royal customers are to avail a growing array of distinct benefits in the form of high-speed transactions, heightened accuracy, and the much-insisted security. The ingrained sensing and perception features are to succulently empower oil companies to visualize and realize value-added services. The goal of sustainability is also being facilitated through the blending of the IoT devices, blockchain ledgers, and AI models.

18.3.1 The Power of Blockchain

Blockchain is an immutable digital ledger of transactions. That is, blockchain can record the transaction of anything of value. The transaction data are fully secured through proven and potential cryptographic methods. The transaction information is stored in blocks, and then a chain of sequentially interlinked blocks is being formed. Such an arrangement is beneficial in multiple ways. This blockchain database is formed and maintained in multiple computers. Therefore, this is being termed as decentralized ledger. That is, no single entity is in control of that blockchain ledger. Any misadventure with that blockchain database can be blocked. There is a need for attaining the consensus of all the participants to bring in any change in the transaction data and structure. Due to the participation of many, this is termed as the blockchain network, which can assure the trust needed by participants to do financial transactions. The much-needed transparency is guaranteed. All the contributions of intermediaries are getting automated through such a computer network. The operational costs and complexities are nullified through the blockchain network. Further on, due to the automation, the time needed to accomplish interactions and collaborations gets sharply reduced. This fast-evolving idea is being seen as a silver bullet across industry verticals.

Blockchain is being recognized as one of the prominent digital technologies to bring in the much-needed bona fide business transformation. Every enterprising business is embarking on digital transformation initiatives to be right and relevant to their partners, employees, and end-users. There are several other digital technologies such as artificial intelligence (AI). The recent entrant of blockchain is being seen as a positive signal for the struggling oil and gas industry.

As articulated in the previous chapters, the oil and gas environments are stuffed with multifaceted sensors, actuators, rigs, machineries, robots, drones, pumps, and pipes. There are specialized platforms near oil fields, and they are fit with several advanced equipment, engines, wares, devices, etc. All these digitized entities individually and collaboratively generate a massive amount of useful and usable data. Blockchain safely and securely stores the data. As we all know, there is a cool convergence between the IoT and blockchain. That is, IoT sensors and devices data is preserved in them through the aid of blockchain. Oil and gas assets store their data and are empowered with smart contracts. That means devices can activate certain things based on events, which fulfil certain contractual obligations. With the distinct assistance of blockchain, the oil and gas (O & G) industry is destined to be hugely beneficial and impactful.

Blockchain technology can bring in decisive and delectable transformations; thereby, supply chains and energy markets become smart. Other auxiliary aspects such as waste management becomes simplified and streamlined.

18.4 Blockchain-Powered Use Cases for the Oil and Gas Industry

The oil and gas industry is down with many intermediaries, a lot of manual operations, and the associated costs. For realizing the goal of the digitally transformed oil and gas industry, automation is the way forward. Luckily, there are many innovative, disruptive, and transformative technologies to fulfil the automation target. Using a distributed ledger is being presented as the most promising way forward to ward off lingering challenges and concerns of the industry.

Blockchain in the oil and gas industry facilitates near-real-time recording of transactions and enables a deeper visibility among all the participants. With such enhanced transparency, the risk factor is sharply less. Blockchain databases hosted in oil and gas devices and assets simplify data integration, which comes handy in eliminating the catastrophic risks such as double spending, fraud, and manipulation. As indicated above, the operational efficiency of oil majors is also fulfilled through blockchain. Blockchain assists in continuous monitoring, measuring, managing, and maintaining oil resources and assets.

18.4.1 Digitization and Digitalization for Operational Efficiency

All kinds of oil and gas industry assets (physical, mechanical, electrical, and electronics) are meticulously digitized through the application of powerful digitization and edge technologies (sensors, actuators, beacons, bar codes, stickers, RFID tags, smart dust, specks, microchips, etc.). As we all know, the O & G industry is becoming complex because of the participation of multiple stakeholders including business partners, suppliers, retailers, transporters, IT organizations, and skilled workers. Typically, each of the suppliers maintains their own ledger strictly in compliance with their precepts and policies. This may instigate the possibility for duplication.

Blockchain facilitates embedding the contract code in the transaction database for automating various tasks. The contract gets executed only when it is validated by all the participating parties. As we know, all ledger transactions get fulfilled through a consensus. Blockchain-enabled transactions are immutable through the intrinsic security measure. Blockchain methodically tracks each transaction.

Smart contracts running on blockchain infrastructure enhance the operational efficiency of oil and gas companies. Blockchain eliminates intermediaries so that required actions are carried out between the concerned parties directly. Each node in the blockchain network contains the same information. Thereby, any kind of mutation attempt can be nipped in the budding stage itself. As certain activities are getting automated, operating expenses for oil companies are reduced. And, the quantity and quality of transactional errors get significantly reduced. Thus, blockchain is seen as a silver bullet for the digital era.

18.5 Blockchain for Improved Trust

There is a dazzling array of safety-critical industrial equipment in the oil and gas industry. An oil and gas company must employ highly trained and competent people to operate the equipment safely. A company's blockchain network can securely record and store the employee or contractor certification, such as H2S training, first aid, and welding. The advantage of embracing blockchain is that the verification of certificates and standard operating procedures (SOPs) can be done by all members at any time. That is, records cannot be changed; thereby, the much-needed trust is established and ensured through the blockchain network.

18.5.1 Transparency in Energy Markets

As widely accepted, blockchain is the way forward to ensure transparency. Energy providers are keenly exploring and experimenting with blockchain. Blockchain-enabled platforms empower energy suppliers with the much-needed visibility, controllability, and auditability. Such a setup enables consumers to make correct purchase decisions. Environmentalists can explore the ways and means of achieving greener societies by lessening the carbon footprint. Blockchain contributes immensely to data collection, storage, analysis, and control in the energy industry.

18.6 Sensor-Enabled Invoicing

As indicated above, the fusion of blockchain into IoT edge devices is to open scores of hitherto unheard and unknown possibilities for providers as well as consumers. Specialized engines and equipment put up in oil plants and a host of pipeline sensors together enable to track output quantity and to prepare invoices to their customers accordingly in real-time. Thus, automated invoice production and processing are being realized through blockchain-enabled sensors and devices. Multifaceted sensors embedded in multiple locations ensure accurate and real-time billing for the work getting accomplished.

The O & G industry is using a growing array of multifaceted IoT devices, equipment, machineries, actuators, and sensors to capture a variety of decision-enabling and value-adding data from multiple sources. Typically, the data originate from these IoT entities and from the assets on which these IoT edge devices are being embedded externally and internally. By attaching edge devices, all kinds of assets, merchandises, and artifacts put up in important environments such as oil wells. platforms, and refineries are slated to become digitized. Every activity in oil fields gets minutely monitored and managed. Oil-related data get captured and castigated through centralized and cloud-based data analytics platforms to extract actionable insights. In the recent past, edge devices and servers emerge as the computational resources to perform proximate data processing to emit real-time insights. Thus, IoT sensors and actuators bring forth additional astute capabilities for the oil and gas industry.

18.6.1 Hydrocarbon Tracking and Easy Record Keeping

The blockchain technology is used to track regulated substances effectively at each stage of the supply chain process. Record keeping must be carefully done internally. If there is any error, the industry is bound to suffer exorbitantly. Considering the importance, the industry is veering toward blockchain, which could keep a record safe and secure. Blockchain ledgers cannot be manipulated, and hence they become an integral and beneficial part of the growing oil and gas industry. without manipulation; therefore, it is an integral part of the whole oil and gas industry.

18.7 Transportation Tracing

Trains and trucks are being used to transport petrol, diesel, and gas products across cities. Monitoring and managing this increasingly complex network are neither straightforward nor simple. Databases are isolated, and hence integrated data management is a difficult proposition. Also, errors can sneak into databases. The use of IoT sensors and actuators increased significantly to monitor transport vehicles. Capturing the IoT data streams and sending it to blockchain databases are being prescribed as the way forward to solve such a complex process. The recent phenomenon of smart contracts is being touted as the game-changer for the oil and gas industry.

Smart transportation is crucial for the intended success of the oil companies. Highly advanced technologies are needed for ensuring that. Tracing the transportation vessel is also important. Blockchain assists and assures data tracking in real-time. Data transfer can be accelerated through blockchain. Advanced cryptography is being used to share encrypted data at high speed.

18.7.1 Crypto-backed Oil and Gas

Blockchain empower the oil and gas organizations by offering cryptocurrencies and tokenization to transpose the value inside them without the services of financial institutions such as banks. The digitization idea is gripping the oil and gas industry. All the assets are systematically digitized. In addition, the oil and gas industry processes and transactions are being fully digitized to function generously. Blockchain has all the power to radically change the industry. With blockchain, the security of transaction data is fool-proof. The transparency requirement is also easily fulfilled. Cryptography-enabled transactions are unbreakable and impenetrable. Consumers and investors will gain a lot with this. The trust guaranteed by banks thus far is being replaced by blockchain. This will open hitherto unheard opportunities and possibilities for the sector.

18.7.2 Compliance Automation

For oil companies, there are many compliance requirements (environmental, taxation, etc.). The blockchain network stores all the transaction data across geographically distributed nodes, which can be accessed in real-time. Such a facility empowers regulatory authorities to access and assess transactions to check whether the compliance requirements are fully adhered. The much-published cryptocurrency is another interesting option for the oil and gas industry to flourish in the days to unfurl.

As nations across the world keep changing the regulations and tax amount, the idea of smart contracts can overcome this perennial challenge. That is, any change can be easily accommodated.

18.7.3 Environmental Compliance

This turns out to be an important requirement for the oil and gas industry today. Blockchain ensures everything in the supply chain is trackable and traceable within producers, distribution partners, transporters, refineries, and retailers.

18.8 Data Storage and Management

The O & G industry data are inherently complex, and hence collecting, stocking, and sharing data across different stakeholders are neither straightforward nor simple. Blockchain simplifies data storage, maintenance, and trustworthiness by providing a single repository of all information. This data repository is being meticulously monitored, managed, and secured by keeping a copy of it in multiple computers. Therefore, besides easy and quick data accessibility, high availability of data through quick recovery is guaranteed. As data get stored in geographically distributed places, users can connect the nearest server to get the data.

The number of IoT devices is growing rapidly, and the IoT device data size is also exponentially increasing. Due to the substantial number of IoT devices in a mission-critical environment, monitoring every IoT device and its operations is neither straightforward nor simple. Further on, ensuring the security of IoT devices and their data is a tough assignment with the current architectures and technologies. It is here that blockchain comes to the rescue. Blockchain-powered data storage guarantees an utmost security for data. Blockchain through an additional layer of abstraction leverages the distinct advancements in the field of cryptography for assuring heightened security.

18.9 Digital Oil and Gas: Strengthening and Simplifying Supply Chain

Digitization and digitalization technologies are continuously advancing. Every physical, mechanical, and electrical system gets digitized. With the arrival of powerful connectivity methods and modules, all kinds of digitized elements get connected with one another in the vicinity and with cloud-hosted software applications, operational and transactional datastores, and analytics platforms. When digitized entities interact with one another and with cloud applications, a massive amount of multi-structured data gets produced. There are integrated data analytics platforms. In the recent past, with the adoption of machine and deep learning (ML/DL) algorithms and models, the aspect of data analytics gets automated. Thus, digitization technologies enable the generation of a humongous amount of digital data, and the smart leverage of digitalization technologies helps extract actionable insights out of digital data.

The oil and gas industry has collected and stored big data. When data get subjected to a variety of deep and purpose-specific analytics, prognostic, predictive, and prescriptive insights get emitted. The intelligence uncovered is to improve the health and safety of various assets of the oil and gas industry. The operational expenses are reduced, while the productivity increases. In short, with the knowledge discovered and disseminated, a whole lot of sophisticated and complicated automation requirements is being accomplished across industry verticals, especially in the supply chain aspect of the oil and gas industry.

The supply chain is complex with the involvement of multiple partners. This needs to support invoicing, documentation, and compliance requirements. Undoubtedly, the heterogeneity and multiplicity-induced complexity is growing steadily. Blockchain through its unique and inherent competencies mitigates the rising complexity. Blockchain also optimizes supply chain processes.

As described above, the flourishing concept of smart contracts is a direct off-shoot of blockchain. The IoT systems are increasingly comprising IoT devices, sensors, smart contracts, and blockchain ledgers. By embedding smart contracts and blockchain ledgers in IoT devices such as oil and gas machineries, the supply chain aspect of the oil and gas world can guarantee utmost transparency, heightened trust, security, and immutability. Thus, the confluence of competent technologies and tools comes handy in eliminating a lot of limitations and in envisaging breakthrough products, solutions, and services.

Shell, one of the oil majors, has indicated that it is optimizing its current processes, creating new value propositions, and exploring new markets through the smart leverage of the blockchain technology. For more details, please visit this page (Blockchain Technology|Shell Global).

The futuristic and fabulous concept of digital oil and gas can play a pivotal role in shaping up the development of autonomous robots, which can do multiple activities with all the precision. Thereby, the safety and security of people working in rough, tough, and remote oil fields are fully ensured. Intelligent robots can handle all kinds of assets and equipment in oil production environments, transport pipelines, and refineries with all the care; thereby, any untoward damage can be avoided. Thus, digital innovations and intelligence are being given utmost thrust by oil and gas companies to sharply reduce cost throughout the value chain. Blockchain can help monitor the system all the time. So, issues are easily trackable through the blockchain platform. Blockchain can ensure that the data transfer happens without getting bogged down by manipulation.

18.10 Commodity Trading

Oil and gas commodity trading is becoming a complicated affair. The current systems being used for trading are outdated and inefficient with the participation of multiple isolated ledgers. These decentralized systems are tough to maintain and prone to hacking. Blockchain comes as a solace for this tricky situation. Through blockchain, commodity traders are blessed with high reliability and easy auditability, ready access to data, etc. Blockchain, the popular distributed ledger technology (DLT), can drastically reduce the amount of time spent on reconciling price and volume differences among traders by supplying the same data to all parties at the same time. Blockchain can also improve trade accuracy, increase scheduling and back-office efficiency, etc.

Distributed ledger technologies and smart contracts together enable oil companies to establish a direct trading with a consumer or a retail energy supplier through autonomous trading agents. The agent can pinpoint the best deal in the marketplace as per the consumer's demand. The agreement gets securely recorded in the blockchain database. The agreement gets automatically executed at the specified time of delivery. Payments happen automatically. All the transaction details are made available to all.

The O&G industry is overly complicated with the involvement of multiple parties. A tremendous amount of poly-structured data gets generated and stocked. A lot of lengthy documentations and complicated transactions have become the new normal in this industry. Automation is the way forward in such a complex scenario. The steadily maturing concept of smart contracts is being seen as a viable and venerable approach to significantly decrease paperwork, optimize the process, improve efficiency, and achieve cost reduction.

18.11 Land Record Management

One of the critical aspects for oil and gas companies is to meticulously manage land records. Now with the insightful adoption of blockchain, keeping land records becomes a simple affair. Any forgery and fudging action can be nipped in the budding stage itself. Blockchain ensures the tightest security for data. Land records with ownership, value, and transfer details have become immutable.

There is an enormous number of land records to be maintained. The investment also comes around millions of dollars. Therefore, there is an insistence for identifying workable ways and means of ensuring the highest security for records. Also, the movement of land records must be closely monitored and managed. It is cumbersome to safely manage all the paper documents. However, with the introduction of blockchain, the land record management becomes easier.

18.12   Financial Reconciliation

As widely known, blockchain is the irresistible technology for the finance domain. In the case of traditional auditing, the confirmation of transactions and accounts happens only at the end of the reporting period. However, with blockchain in place, auditors can continuously receive the immutable record of transactions. Further on, blockchain takes away the need to input accounting details into multiple databases. Blockchain helps in automatically reconciling among disparate ledgers. The real-time data availability being enabled by blockchain is being seen as a change in thinking in the financial industry.

Oil and gas firms across the globe do more than a million transactions every day. Buying and selling are the main activities. In these complicated processes, there are possibilities for human errors to creep in. There are several ancient and archaic procedures. Thereby, the efficiency is getting affected. Thus, the movement toward employing highly impactful technologies has been acquiring traction in the recent past. Blockchain is being seen as a promising paradigm to do away some of the critical challenges and concerns being associated with the oil and gas industry. Especially, the concept of smart contracts is getting a lot of attention these days. Smart contracts can self-execute the tasks when certain conditions or events get fulfilled. Payment gets automated through smart contracts when the required parties fulfil certain things. Thus, there are a lot of things to get automated with the help of sophisticated technologies.

Blockchain is all set to become a smart contract facilitator, trusted gatekeeper, and transparency purveyor. Blockchain has the wherewithal to improve the efficiency of the oil and gas companies by reducing operational costs and delays. Duplication gets eliminated while increasing transparency.

As the operational, analytical, and transactional data get dispersed across multiple systems and the manual reconciliation is mandated to bring in any change in data, there is no need for any intermediary to facilitate trustworthy interactions between different and distributed parties. Blockchain sharply increases efficiency in the aspect of supply chain management, which is extremely complicated in the oil and gas sector. Transactions can be completed easily, securely, and quickly between suppliers, financial institutions, and buyers without any intermediary. Data integrity is fully guaranteed through blockchain.

Experts point out that blockchain can help improve oilfield asset visibility. With improved visibility, the time to be spent resolving oilfield asset issues is reduced. By synchronizing with other digital technologies such as the IoT, advanced and automated data analytics through AI algorithms and models, digital twins, event-driven architecture (EDA), 5G, and edge computing, many more advancements can be envisaged and accomplished in the years to come. Digital technologies can keep costs in check and maintain profitability.

18.13 Oil Wells and Equipment Maintenance

Any leakage of potentially hazardous substances is being seen as a catastrophic risk for human life. This insists the importance of acquiring the capability to do predictive and preventive maintenance of equipment. Blockchain comes handy in methodically fulfilling such requirements by providing status data in real-time. Blockchain ledgers deployed in IoT devices proactively trigger alerts maintenance professionals with all the right and relevant details so that the maintenance activities can be started immediately with all the confidence and clarity. With proper maintenance, equipment lifetime increases significantly, and any damage to property and people gets nullified.

18.14 Waste Management and Recycling

It is critical to have a well-oiled waste management system in place. This is important for the oil and gas industry. Wastes, if not handled properly, can be a potential hazard and harmful for our fragile environment. Experts point out that through a series of smart contracts, it is possible for oil companies to do efficient waste management and resource optimization.

The gravity of the problem increases because oil companies purchase a large quantity of raw materials and not to reuse these products through the recycling process, and hence the amount of waste materials goes out of hand. With blockchain, it is possible to track these products and the waste that is getting produced in oil company premises. Such technology-inspired tracking enables decision-makers and executives to tackle this genuine problem.

18.14.1 Management of Wastewater

The oil and gas production companies produce millions of gallons of water. This wastewater often contains dangerous metals and other contaminants. Therefore, it is imperative to carefully handle wastewater to maintain the environment equilibrium.

Blockchain comes handy in creating records on how the waste is produced, transported, and disposed. Such records help in keeping a check on wastewater management.

18.15 Tracking Carbon Footprint

Oil plants and refineries emit a large quantity of carbon during their everyday operations. This pollutes and damages our living environment severely. Therefore, it is mandated by governments and environmentalists that oil companies must expedite and explore sustainable and smart techniques to reduce the carbon footprint. Blockchain through its unique decentralization method can track carbon emission quite efficiently and offer actionable insights so that the concerned officials can make use of it proactively to reduce the carbon footprint.

Reducing costs and having cleaner working environments are vital for oil companies. Blockchain offers an escape route. The blockchain technology can accurately estimate the carbon footprint of an oil field.

18.16 Improved Pipeline Inspection

Establishing and inspecting pipelines are a complicated task as it involves a lot of coordination among multiple players and local companies. Regional companies must meticulously inspect and share the correct information with the concerned in time to keep the business running without any issue. Any forging or fudging of data is very catastrophic. The coordination is error-prone, and hence the O&G industry is jumping into the blockchain bandwagon. The inspection service providers can upload the data into the blockchain platform. Thereby, the real-time knowledge visualization gets fulfilled. The company executives can view the inspection data in real-time.

Typically, a lot of paper works is needed for making and managing contract documents. Also, paper works are mandated for transportation, fulfilment, delivery, etc. in this industry. Therefore, the arrival of smart contracts is being seen as a good signal. Smart contracts can reduce paperwork. The productivity increases and the costs decrease. The coordination among local companies for doing pipeline inspections is made smooth. The inspection information must be submitted to the concerned to get the go ahead for any corrective action. As the inspection documents can be forged, the usage of smart contracts has come as a solace for the industry.

18.16.1 Error Reduction

Humans make mistakes plentifully. Even a small error in the oil and gas industry can be catastrophic. Thus, human resources must be supported through a host of advanced technologies to eliminate any risk. Blockchain has all that is needed to reduce human-made errors. Blockchain systems automate a few activities. With deep and decisive automation, the number of errors is bound to decrease drastically.

The O&G industry uses oil ticketing certification standards for effectively maintaining data. Human errors can sneak into these paper tickets knowingly or unknowingly. These mistakes demand for a series of reconciliation tasks; thereby, a lot of money and time gets wasted. As blockchain intrinsically supports transparency, any mishap or mistake can be pinpointed easily; thereby, a lot of unnecessary works gets eliminated.

Blockchain readily assists in creating and sustaining immutable and auditable transactions between multiple participants. And hence several concrete benefits are being accrued by oil majors across the globe.

18.17   Other Miscellaneous Advantages of Blockchain

As reported before, blockchain has the power to automate and accelerate several upstream, midstream, and downstream activities. Oil industry operations are being speeded up, the much-needed optimization requirements are being fulfilled, and the sustenance of our living and working environments are being guaranteed through the smart leverage of blockchain and its allied technologies and tools. Industries and governments are exploring and expediting the usage of blockchain for a variety of futuristic use cases. Compliance tracking across supply chains, pipeline inspections, tracking transport vehicles, etc. is being streamlined through blockchain, which is being presented as the trend-setting technology for data management. Any misgivings can be proactively and pre-emptively identified and nullified. As the oil and gas sector is brimming with big data, blockchain is being pronounced as the resilient, robust, and versatile technology.

18.18   Blockchain Challenges

Blockchain is quite new in the IT industry. The decentralization and distribution characteristics make blockchain ubiquitous. Cryptocurrency is the first and foremost application of the blockchain technology, and there are some widely indicated concerns in the case of cryptocurrencies across the world. Smart contracts are an emerging paradigm within the blockchain space. But smart contracts are in the very primitive stage. It must grow faster to be utilized readily and rewardingly. Scalability is the biggest challenge of the blockchain paradigm. For recording and processing millions of transactions, there is a need to leverage massive IT infrastructure modules (server machines, storage appliances, network components, and security solutions). Because of many participants in decision-enablement, the number of transactions per minute is comparatively less in the case of the blockchain world. There are several innovations and implementations to achieve higher throughput. Also, as the world is continuously batting for greener technologies for environment sustainability, blockchain experts, evangelists, and exponents are working overtime to unearth and articulate technologically competent and relevant solutions. There is a need for standard organizations to bring in a kind of universality and interoperability among blockchain systems, solutions, and services. Thus, there are a few critical challenges to be surmounted to make blockchain penetrative, pervasive, and persuasive too.

In summary, blockchain can bring forth numerous value additions to the oil and gas industry. Forming and sustaining a private blockchain network for an oil company to facilitate transaction life cycle management across all its stakeholders is gaining momentum these days with the faster maturity and stability of blockchain technologies and tools.

Blockchain enables buyers, sellers, and their banks to share transaction details simultaneously as they look at the same ledger. From the time an order is confirmed until its fulfilment, blockchain contributes by improving the process.

18.19 Conclusion

Blockchain is an impactful technology for the oil and gas sector. Blockchain guarantees the trust, transparency, security, immutability, efficiency, and scalability for new-age oil and gas companies. By applying artificial intelligence (AI) algorithms and models on blockchain data, next-generation competencies can be built and released by oil and gas majors to retain their clients and to attain additional customers. With the strategically sound combination of AI and blockchain, the oil and gas industry is to benefit immeasurably. Further on, the brewing challenges and concerns of the oil and gas sector are to be eliminated through the fruitful convergence of blockchain and AI. Cognitive systems and networks are being envisaged and implemented through the shrewdness of innovative technologies and tools.

The evolution and revolution of smart contracts on the blockchain network is to bring forth critical advancements in the oil and gas industry. As articulated above, the much-insisted transparency, efficiency, and data security are being elegantly realized through the power of blockchain.

Bibliography

1 Anwar, H. Blockchain in oil and gas industry. https://101blockchains.com/blockchain-in-oil-and-gas-industry/ (accessed 07 January 2021).

2 Ahmad, R.W., Salah, K., Jayaraman, R., et al. Blockchain in oil and gas industry: applications, challenges, and future trends. https://www.techrxiv.org/articles/preprint/Blockchain_in_Oil_and_Gas_Industry_Applications_Challenges_and_Future_Trends/16825696.

3 7 ways blockchain can revamp the oil and gas industry. https://globuc.com/news/7-ways-blockchain-can-revamp-the-oil-and-gas-industry/

4 Where blockchain technology can disrupt the oil and gas industry. https://www.forbes.com/sites/forbesbusinesscouncil/2021/09/27/where-blockchain-technology-can-disrupt-the-oil-and-gas-industry/?sh=597d96f114ab.

5 Blockchain in oil and gas industry: applications, challenges, and future trends. https://www.sciencedirect.com/science/article/abs/pii/S0160791X22000823.

6 Unleashing innovation in oil & gas industry with blockchain: 11 top use cases & benefits. https://www.birlasoft.com/articles/blockchain-oil-gas-industry-use-cases-applications.

19

AI-Inspired Digital Twins for the Oil and Gas Domain

Digital twin (DT), as the name implies, is a digital representation of a real factory or a refinery or simply an asset. For example, an oil and gas asset consists of pipes, valves, oil rigs, pumps, sumps, and many more parts. These parts work in tandem to produce products. However, when there is a need to maintain the components or parts to see if they really are working the way they should, the parts can be modeled digitally to see it as a simulation. These can be used to answer hypothetical questions and can intuitively present these answers. A DT can be used to answer what-if questions by running simulations, and the results should provide insightful and intuitive answers.

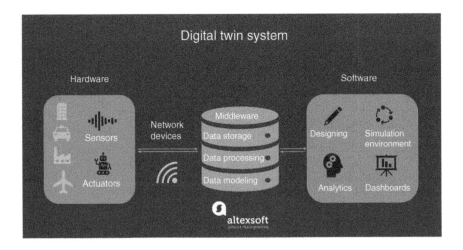

The Power of Artificial Intelligence for the Next-Generation Oil and Gas Industry: Envisaging AI-Inspired Intelligent Energy Systems and Environments, First Edition. Pethuru Raj Chelliah, Venkatraman Jayasankar, Mats Agerstam, B. Sundaravadivazhagan, and Robin Cyriac.
© 2024 The Institute of Electrical and Electronics Engineers, Inc.
Published 2024 by John Wiley & Sons, Inc.

A DT can be used for asset lifecycle management, predictive maintenance preventive maintenance, and operations management.

To better understand how a DT is used, the following key dimensions are needed:

Data: Data are key for a DT to produce useful results. A digital model of many real components will become useful once the data are fed into it. Data can be pressure, temperature, thickness, quantity, material, and many more. Once the data are fed into the digital model, the digital model will be able to replicate the exact behavior of the real part. In other words, the digital model will come to life. Data can be from operations, assets, and can be structured and non-structured.

Visualization: DTs deliver value by providing insights that help in decision-making. DT presents the data in a manner that is easy to understand and consume. The data when visualized create the life-like view of the asset using various techniques such as photogrammetry and 3D modeling.

Modeling: Another critical functionality of DTs is their ability to provide predictive analytics and modeling capabilities. DTs should be able to use the data available to understand the past behavior of the asset being modeled and to predict the future behavior of the asset. This can help in predictive maintenance and prevention of failures.

DTs can create a virtual simulation of the what-if scenarios without investing in creating one in real. Simply put, DTs can simulate the entire production with precise details based on the data feed.

Combining DTs with AI can help create better process designs and in turn help preventive maintenance and improve production.

Machine Learning (ML) and Artificial Intelligence (AI) have been regarded as one of the promising technologies in the manufacturing industry. However, ML methods need a huge amount of high-quality datasets to be trained for decision making and action. DT can generate and provide such clean data for ML. Many times, manual inputs are needed to label data for ML. By using a DT model, training of ML can be accelerated since data comes with the correct metadata from DT.

Few of the areas where DT can be used in oil and gas industry are

1) Asset lifecycle management

The main principles that are the foundations for proper asset-related decision-making within an AM system are the following

□ Life cycle orientation, leading to incorporate long-term objectives and performances to drive decision-making.
□ System orientation, motivating the relevance of a holistic consideration of asset systems in their entirety, and not merely of the individual components.

    ▫ Risk orientation, relating to the relevance to consider risk as assets normally suffer from uncertainties in achieving stated objectives; thus, there is a subsequent need for risk management approaches.

    ▫ Asset-centric orientation, leading to focus on asset data and information of the assets, to make sound business decisions.

Based these principles, every time an asset-related decision must be made, the company should consider two main aspects:

1) The asset lifecycle and the impact of a specific decision that is taken in the asset, that could impact the long-term asset lifecycle
2) At which level the decisions are taken – strategic, tactical, or operative level – within the company and the impact of such decisions

In both the decisions needed to be taken for asset management, DT efficiently combines different assets and their information in one place, thus enabling informed decision-making. Viewing the assets virtually with all the relevant information ensures all the relevant data are available to the stakeholders of the assets. With DT, we can simulate operations and predict the future changes and impacts.

DT can give engineers a virtual representation of what could happen if a maintenance activity is performed even before performing the activity. The advantage of this process is that engineers can save time and money and plan better.

DT needs data and information to produce useful results. One of the data sources can be OSIsoft PI data. PI data comes as time-series tags. Based on contextual data, PI can add meaning to the tag data which are received from sensors. In the fig below, DT gets data from equipment via sensors and the model is created in the tool. Based on the multiple parameters and changes in the input parameters, the lifecycle of the asset can be monitored and planned.

Information management during Asset Lifecycle – The insight gained during asset operations and data from other sources like weather will help gain great knowledge about the asset operations in the oil and gas industry. One of the key challenges is that asset operators have huge information but unable to make use of all the information that is available. This huge cache of information from DT if fed into an ML model can help the models learn quickly and apply intelligence from an AI standpoint. Without the right data, the AI model will not be able to predict accurately.

For DT to help in asset management, we need to

a) Understand the assets owned by the organization. This is key for planning the financing, resourcing, and procurement strategies. Data are gathered from the organizations existing systems including pre-existing systems that are digitalized, paper-based information
b) Cleanse and sort the information to make it useful and remove obsolete details

c) Validate the remaining records with the SMEs and make sure they are current. This could be a manual effort

d) Once the data are cleansed and verified, the data must be made meaningful, in other words, references to the data must be added

e) Once the metadata or references are added, the asset details are understood. The amount of information needed for the assets are decided by the organization based on corporate strategy and objectives

f) Once the details are all updated, onsite inspection using measuring equipment is conducted to make sure the end-to-end data are captured to have precise information about the assets

g) This means the asset is ready for maintenance using a DT system. Users can be presented a completely digital representation of the asset all in one place

Where DT can be used in the energy industry?

The DT application base would be Electricity Grids, Transportation, Greenhouse Production via smart sensors, actuators, RFID tags, IoT sensors, and external data. The sensors and data feeds provide details like pressure, temperature, viscosity, and flow rates. DT can be used in predictive maintenance, preventive maintenance, fault diagnosis, Lifecycle Management, Usage analysis, etc. in operational assets and asset modeling and simulation, viewing the end-to-end asset details before installation of the asset to get a clearer idea of asset behavior after asset installation.

Once the data are ready to be fed into a DT model, the following can be done to ensure a complete asset maintenance and optimization

1) **Asset health**: Here, the existing assets' health is considered and optimized for maintenance costs and effort. The longevity or the remaining lifetime of the system is also assessed based on information received from the existing asset information

Asset reliability:

Production analysis: Production forecast and analysis based on existing demand, capacity, and running time of the assets. Production forecast can also be done by actual loading analysis compared to actual delivery of the products. Production analysis can also be done with checking the demand management and planning the production and supply, which in turn would relate to more asset management processes. The asset management will give details about how much capacity the asset has, how many new upgrades need to be done, and the downtime due to planned maintenance. This would serve as an input to production analysis

Demand management: Based on production forecast, the asset production planning can plan to service the demand and optimize it. Asset details are captured, cleansed, and referenced so that the entire asset information is up-to-date and optimized

2) **Predictive maintenance**

Being able to predict the remaining useful life of an asset is when a particular asset or pump or motor needs maintenance is another use of DT. By just viewing the DT on a screen, an engineer can see the bill of materials used in the pump, prior scheduled maintenance work done, the previous work package for this job, updated diagrams and new additions, reliability study and inspection data on the pump, and monitoring data. He can also check the 3D model to see where the pump sits in the unit, measure the access, and see the other work currently permitted in that area.

A remote asset which needs someone to travel needs maintenance. Using a DT, there is a constant flow of information from the sensors which, after adding context data and analyzing using DT, can produce a reasonable accurate result, which could save a lot of costs and travel to the asset. With the right amount of data, AI can predict asset maintenance or failures without manual intervention

3) **Preventive maintenance**

The DT model can run in parallel to real-world assets and can identify mismatches as and when they occurs. Smart sensors in the assets send near real-time data to a DT, which will constantly compare the operations with the preferred operations. Any difference in the operational parameters can set up an alert, which then helps perform a just-in-time maintenance activity

4) **Operations management**

Using multiple variables such as weather, asset type, and other operation measurement points or set points, engineers can trigger simulations or what-if scenarios to evaluate if the asset is working optimally or if needed adjustments can be made to ensure asset readiness for optimal operations

Benefits for DT for Asset Management:

- Get the full end-to-end picture of the asset
- Predict or anticipate the events before they happen
- Minimize costs and efforts in prototyping and testing – with the right data and asset modeling, multiple scenarios can be tested before making a change
- Plan to implement new technologies and tools which can also by simulated in the DT to show the value added or improvements

19.1 How to Ensure Certainty Using DT for AI

DT provides the maintenance and operations teams with reliable information needed to do their tasks. By bringing together data from multiple sources, risks are minimized and identified early on, data verification and checking is eliminated, and a culture of collaboration is created among the various teams.

A single hub for all the key data and asset information laid out in a simple easy-to-use interface showing what needs to be done to maintain the assets as per performance requirements.

The key value that differentiates between a data hub and DT is that the DT provides a platform to visualize multiple data into a single hub, enabling engineers to take corrective action.

19.1.1 Capture

Data that are captured from different systems must be connected to their contextual data or metadata. One of the key challenges is siloed applications. Data are available in all these applications, but they are siloed and not completely usable. Accessing the critical information is essential for good decision-making. For a single physical asset, there exists multiple digital representations like 3D models, Asset Schematics, Planning spreadsheets, and Asset Performance Information. These types of information are stored in different databases and typically do not work together. Hence, getting the right quality information without looking for it helps in faster decision-making. Also, data access issues could pose challenges in getting the right information.

19.1.2 Interpret

Capturing is one key aspect of getting the data, but getting the right access to data with the right DQ and completeness from different source systems is the key. Interpreting the captured data is the key next step. For instance, we have huge amounts of p&id (piping and instrumentation diagrams) finance data, asset data, PI tags, etc. The OPS team is interested in viewing the live operations data; however, to take actionable insights, the data must be contextualized for visualization. Applying context and connecting tags as the reference value will be needed for creating a good, structured data flow for DT.

19.1.3 Consolidate

Consolidating data from multiple data sources and making sense of them is another key task for a DT implementation.

DT is a core for getting information together and visualizing. Remember DT cannot be a system of record but can use a system for record to get the required outputs. DT is used as an integration and visualization tool. The data must be curated before feeding into DT from an EDW. So how does the different types of data flow into a DT software.

Most datasets are transmitted using json files.

19.1.4   The Types of Data that a DT gets

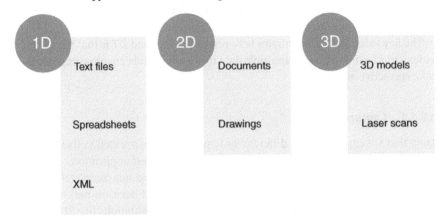

However, 3D files are transmitted using multiple industry standard formats.

3D models come as dxf or dwg, vsd, vue, etc. which are later converted to the native file format as required by the DT tool, or the connector to the DT tool does the conversion. The most used file formats are JSON, CSV, dwg, vsd, etc. The file formats vary between different DT tools.

19.2   Tools Needed to Develop Digital Twins

19.2.1   Engineering Modeling/Drawing Tools

Companies these days have great 3D modeling tools. The 3D designs are exported to other tools via a pipeline. The advantages of this approach are that if a drawing changes at source, it gets updated in the target – here the DT tool. Similarly, the speed of overall development increases significantly as the DT tool will use the latest version of the truth, hence reducing rework or data quality issues.

19.2.2   IoT Devices and Tools

IoT devices and tools can provide real-time data to DT. For instance, PI tags can provide near real-time sensor data from the assets.

19.3   Digital Twin Implementation Approach at a High Level

Ideally, one can build a DT for almost everything, but the challenge lies in creating the DT for every component. So, to implement a DT, its best to choose a small component or process and do a pilot/PoC. Once we understand the component that we are going to twin in the first place, the next steps may be the following:

1) **Involve the team/stakeholders**: The first step in any implementation is to get the key stakeholders and users onboard to the new change. DT models will be likely used by many users, and there is a need to let the users be aware that during construction of the DT model, the team will be involved for gathering information and ideas. The other benefit of engaging the stakeholders is that some key objectives may be missed and that can be pointed by the team. Additionally, we can start using a concise language for the implementation of DT, which can be received from the team/stakeholders.

2) **Establish the data sources**: The foundation of every DT is proper data. Hence, to implement a good DT, it is necessary to establish data sources. Data sources can be different based on different uses. There could be three types of uses

   a) **Standalone asset**: These assets are single components or machines which work pretty much stand alone and need to be modeled digitally for analyzing maintenance or similar activity. The historical data from these machines are fed into the DT system and modeling done specifically for these components.

   b) **Connected assets**: Connected assets are the assets that have a conveyor belt and make use of many machines like just-in-time manufacturing process and downstream or integrated gas production facilities. These facilities will have to be viewed both from an individual component viewpoint and to an overall global asset viewpoint. The way, DT can be implemented here to get the overall process steps and the machines/components involved. The next step is to map the overall machine to data. For specific use cases, it is good to have the details of the manufactured component and map the flow. This information can be used when modeling an end-to-end DT workflow to see impacts if specific changes are introduced in a DT setup.

3) **Choose the right infrastructure**: DTs require real-world infrastructure to work. The following main pieces are needed to make a DT set up a reality

   a) **Sensors**: Most DTs use sensors to collect data from the assets. The exact nature of each device will vary depending on the DT's purpose – a manufacturing enterprise might use motion sensors while an oil and gas refinery will install viscosity and seismic monitors.

   b) **Centralized platform**: Each DT requires a platform that stores all relevant information. Since DT systems need to have a large amount of data to process and produce the models for analysis, it is best to get a cloud-based storage for DT systems to enable massive data processing and reporting.

   c) **Data transmission medium**: Data from the sensors can be transmitted by Wi-Fi to the servers. It will be good to decide on the medium of data transmission for better planning. Wi-Fi works in most cases, but at times there could be obstructions for Wi-Fi, which makes it important to use alternate methods like cable.

4) **Creating a tag register**: Identifying and creating tags for DT models is a key step. A tag register comprises all the tags that will be provided by the IoT sensors as time-series data, which allows for better predictive maintenance.

5) **Identify the use cases**: Use cases need to be identified based on the requirements of the assets and prioritize the low hanging fruits. Based on use cases, it must be decided on what information will be needed to develop the DT models. Based on the information required, the tags that will provide that information will be identified and prioritized for DT setup.

   a) **Data quality issues for DT**: One of the key challenges a DT setup can face is data quality. The main criteria for data quality are their completeness, reliability, accuracy, consistency, accessibility, and timeliness. First, it is necessary to formulate criteria for assessing the quality of data to understand which data are qualitative and which are not. To do this, you can identify five basic ways, the use of which reduces the quality of the data. The table below provides a list of these methods that were identified during the project to manage the quality of the source data.

№	Ways to reduce data quality	Examples
1	Miss a fact	The flow rate was decreased last week due to the shutdown of pumps.
		The fact that last time was problems with electricity has been missed. This omission gives the impression that the pumps are broken.
2	Distort a sequence of events	In order to intensify oil production and increase the interval between wells' repairs a number of geological and technical measures was taken. After carrying out these measures, their economic efficiency was calculated. Economic analysis of geological and technical measures showed a negative result.
		Obviously, by changing the sequence of events, the situation looks extremely illogical.
3	Do not indicate time	A number of wells failed due to breakdowns of submersible equipment.
		Such a message can create prerequisites for urgent repairs of submersible equipment if you do not know that this message dates from last year.
4	Add false information	The pressure of gas supply to the main gas pipeline was different from the established norms because the operator changed the operatig mode of the booster compressor station (BCS).
		The information that the operator changed the BCS operating model is false. Because of it, this operator looks incompetent.
5	Change importance	The oil recovery coefficient (CIN) of the asset was decreased. Therefore, it is necessary to check the reliability of the ground infrastructure of this field. That is an order.
		Actually, this was just a recommendation.

In modern realities, a transition from the generation of a huge amount of useless data to decision support systems is necessary. DTs and decision support systems require high quality of the initial digital data. Unreliable data lead to incorrect decisions and nullify the effect of using intelligent systems such as data mining, machine learning, and predictive analytics. Data quality management ensures the usefulness of digital data and the value of each bit. The key success factor of companies' digital transformation is approaches that ensure the quality of corporate data. For this reason, companies are implementing new business processes, appointing responsible people, and training personnel in data quality management since data reliability, their availability, and consistency in various IT systems require an integrated approach.

It is convincingly proved that the initial data quality is the key success factor of Smart Fields and DTs technologies, which ensures the effectiveness of decision-making systems and, as a result, the investment attractiveness of oil assets.

As per Wikipedia, **Machine learning** (**ML**) is a field of inquiry devoted to understanding and building methods that 'learn', that is, methods that leverage data to improve performance on some set of tasks. It is seen as a part of AI. Machine learning algorithms build a model based on sample data, known as training data, in order to make predictions or decisions without being explicitly programmed to do so.

The different techniques of ML fall under Supervised Learning, Unsupervised Learning, Reinforcement Learning, and Representation Learning.

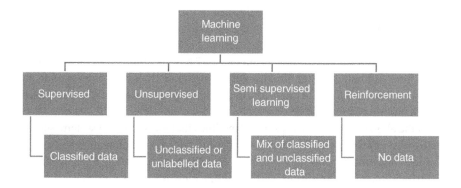

Supervised Learning, as the name indicates, is with the presence of a supervisor. We basically feed the model with known information or data, hence helping the ML model predict the future instances based on the data that are fed. Supervised learning is a technique where input and output patterns are provided to the system. The model is guided with the help of loading the model with the knowledge to facilitate the prediction of future instances. It uses labeled datasets for the training.

Unsupervised learning, as the name implies, is that the model does not have a target output and no training is provided. It is more like the system must learn by algorithms and draw conclusions based on unlabeled or unclassified data.

Semi-supervised learning is a technique of training the algorithm with labeled or unlabeled data. The algorithm is created using labeled data with bare minimum possible, and then the unlabeled data are clustered along with the labeled data.

Reinforcement Learning is where the model output determines the next steps like a game of chess. For example, the input for a particular step will depend on the output of the previous step and so on.

So how does AI benefit from DT and vice versa?

Nowadays, the manufacturing process moves from conventional knowledge-based to data-driven manufacturing. The gap between design and manufacturing is filled by virtual models of manufactured products. These virtual models help mirror the real and virtual world. Due to DT technology, it is easy to do quick virtual verification. One can perform rapid changes in the design when viewing a DT of the asset, thereby improving design and optimizing the process simultaneously.

To enhance oil production rates and deal with emergencies in a timely manner, oil and gas assets must strengthen the management and control measures of each cell in the asset to improve its ability to control the production process. Therefore, in the manufacturing process, it has been a big challenge to decide on how to effectively receive the required feedback from the assets to improve their performance either in real-time or near real-time.

DT provides maintenance and management of assets, providing a digital clone of the assets by considering its properties. In a complete setup of DT along with AI/ML tools, the emphasis must be on data. DT has the ability to receive the right data from different sources, build a virtual model of the assets, and provide an insight to the maintenance engineers to check the performance of the asset based on the information received.

In the above figure, the setups of DT modules include a physical system, sensor networks, virtual model, AI-based analytics module, data visualization tools, continuous machine monitoring procedure, decision support system, and a feedback loop for automated improvements. Combining these technologies makes sure that a digital clone of the physical machine is always available to the assets engineering team, which will help throughout the management of the production life cycle. We can design and implement the DT concept at the unit, system, and system-level in the assets. To implement each submodule, multiple technology options are available.

As shown in the figure below, the on-ground implementation of the industry DT is divided into layers connected via the bidirectional data flow. Each layer exhibits a specific task and is executed via a specific set of either software or protocols or data stores. In manufacturing, DT extends the capacity to replicate, simulate, and minimize the operation and production system, which helps with a proper visualization of every process from manufacturing a part until its assembly.

Now, when the data are aggregated to be visualized in a DT tool, it can learn based on the data fed and whether there needs to be supervised or unsupervised learning. For instance, if DT is used in manufacturing or in the oil and gas domain, the most common areas of learning are virtual manufacturing cell, carbon emission forecast, and fault early warning. To enhance productivity and deal with emergencies in the production process, companies must strengthen the management of each manufacturing cell by predicting failures based on the DT models of the assets in the cell. This can be achieved by creating the DT models, feeding the right amount of data, and creating the ML algorithm based on DT data. The ML algorithm will work with more efficiency if it is fed with the massive data that a DT system can consume for visualizing the virtual models of the assets.

To maximize the benefit from a DT setup, Xu et al. (2019) proposed a two-phase DT-aided fault diagnosis method using deep transfer learning. The proposed method could find potential problems not considered in equipment design by running an ultra-high-fidelity model in a virtual space in DT. It used deep transfer learning to migrate the previously trained diagnostic model from virtual to physical space for real-time monitoring and predictive maintenance. This model could potentially be fed into an ML system to better learn and apply the intelligence.

DT cells are the core component of intelligent manufacturing. The cells are composed of a physical manufacturing cell and a virtual manufacturing cell. The virtual workshop is constructed using data from the existing physical environment virtual cell (DT) to evaluate, analyze, predict, and optimize the production process and create a what-if scenario. Sometimes, when an engineer creates a digital model of a physical asset, it takes weeks for the software to complete the simulations using the sensory inputs to a DT tool. To bypass this issue, engineers create something known as the surrogate model of the physical asset. The surrogate model is a reduced model which does the simulation using lesser computing to quickly show the results. This is done by taking the most important data and come up with ML-based algorithms to predict failures. For example, one of the key issues faced by the oil industry is the change in pump pressure when there is an external event like an earthquake occurring in the vicinity. To predict the possible issues, the important input is the pressure from the pump and weather predictions for this kind of ML output. Hence, the engineer focusses primarily on these parameters to come out with a prediction on the well pressure reduction. The analyst trains each of the models with one or a combination of data types, including simulation, experimentation, field-test, and product-operational data. Model training calibrates the simulation outputs so that their predictions are more accurate.

In the actual production process, equipment failures occur from time to time, which affects the production schedule and costs. If repairs are carried out after the fault occurs, it is often difficult and requires a lot of resources to perform maintenance or repairs. Therefore, it is important to predict the failure of the equipment and the service life of the equipment. The flowchart below shows the process steps.

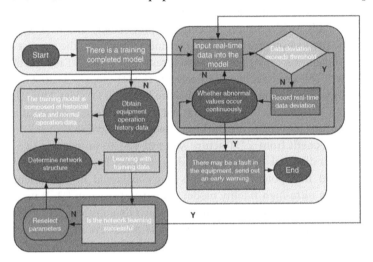

So how does a DT setup make support in predictive maintenance and how does DT add value to the existing AI of machines to save costs. The example below can be applied for different fields like healthcare, manufacturing, and financial modeling.

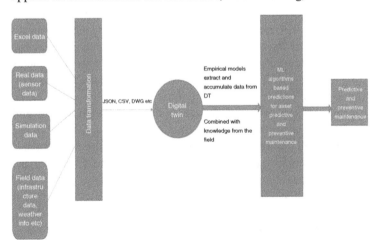

19.4   Digital Twin of Oil and Gas Production

DT of oil and gas production is not only a three-dimensional model, physical entity, virtual entity, and their connection but also the combination of new technology, considering different links and different businesses. So, it comprehensively shows the dual drive of data and physical model, which means dual fusion as the main line, and flexible application to form a comprehensive DT of oil and gas production. First, oil and gas production links are many and complex, involving a wide range of disciplines, so the corresponding dynamic multidimensional multitemporal scale model DT model is created according to different application objects and business needs; the model not only needs to build the geometric and physical dimensional model but also can reflect the geometry, physics, behavior, rules, and constraints of the physical entity (Figure below). At the same time, it describes the evolution process, real-time dynamic operation process, external environment, and interference influence of physical entities from different time scales and integrates different dimensions. A DT model of oil and gas production is the same spatial scale and different time scale model. Second, the physical layer of oil and gas production is constructed by deep integration of new information technology and virtual reality interconnection and integration based on Internet of things. Based on cloud mode DT data storage and sharing service, the transmission layer and data layer of DT of oil and gas production are formed. Based on data analysis, fusion and intelligent decision-making of Big Data and Artificial Intelligence, the intelligent PaaS of oil and gas production DT is built. Based on the virtual reality (VR) and augmented reality (AR) mapping and visualization display, supporting the service application layer, the integration of information physical system and the collection, transmission, and storage of multisource heterogeneous data can be realized; the integration of information physical data, the bidirectional connection, and real-time interaction of virtual and real data can be realized; and the real-time process simulation and optimization can be carried out to provide various on-demand applications and intelligent services. Finally, for data-driven fusion of physical information, data are regarded as the core driving force of DT which emphasizes physical data fusion; in the process of oil and gas production, real-time operation status of equipment, sudden disturbance data, transient abnormal small data, and other physical real-time data are also indispensable parts of the system, reflecting the real state of physical equipment in the system operation. On the other hand, the real-time operational status of the equipment in the system is reflected. In general, the data of oil and gas production DT system include real-time dynamic production data; historical statistical data; some historical scheme design, experiment, test, and research data; and even some knowledge chain and knowledge base, which are constructed into large data of real significance, which can strongly support the analysis and decision-making of oil and gas production. Finally, the information data and physical data are fully interactive and deeply integrated to improve the physical consistency and synchronization of information and improve the real-time accuracy of the results. In the operation stage of DT, how to realize the online parallel controlling in the cyber model and feedback the adjustment instruction to physical system is a key enabling technology.

So how to implement a well-rounded ML solution with DT providing the necessary inputs. We first start with the concept or identification of the use case. We interview the subject matter expert (SME) and come with a use case to conceptualize the need. After identifying the need to create an ML solution, the next step is the preparation phase. In this phase, the infrastructure, tools needed, organizational support needed, business case, and selection of the implementation approach are all defined and finalized.

ML Use Cases: The key starting point of this phase is the identification of ML use cases. ML should not be introduced because it is currently a trend technology with a high potential. Therefore, the first step is to define strategic goals for maintenance improvement that can only be reached with data mining and ML techniques economically. Usually, machines and components fail frequently, and it becomes almost impossible to predict their failure on time. Though there are a plethora of tools and techniques for predictive maintenance, it is still not possible to predict machine failure with the accuracy needed. So, an ML algorithm predicting machine failure needs to be used in the most vulnerable and important business process.

19.5   Solution Approach

The most important part where DT and ML can converge to provide great value is the solution design. The key components needed for an AI/ML algorithm to work is data and the way it is ingested and used. The figure below depicts the application of artificial intelligence in DT in simulation analysis. The whole process starts with the data input, which can be obtained from multiple source systems, spreadsheets, manual entry, etc. However, DT adds much more value from the different data that are entered into the system.

DT systems can process much of important data to help in predictive maintenance. The key challenge would be to ensure the right data with the quality at the right time. One of the key possibilities is to have multiple DT setups for the same physical assets. For example, if there is a pump which must be digitally cloned, we can create multiple DT instances so that the DT instances can be prioritized based on data needed. Pump A may require pressure and temperature data as the primary input. While the additional non-primary data like weather and humidity maybe fed into the DT, the twin model will rely on the primary datasets as input to produce the needed results. So, when the pump does not get the correct primary data, say temperature, there are other sources like the weather, surrounding temperature, wind velocity, and sunrise and sunset time which can produce a rough temperature range which the DT instance can use. One can also aggregate multiple DT instances to produce a large DT instance, say, like a full-fledged refinery to get the desired results for making key decisions. Besides relying on specific reports to measure the performance of assets or components, DT will be a valuable addendum to make much more informed decisions with confidence. DT can also be used to plan installation of large-scale assets, run a full simulation inputting multiple parameters and changing the parameters to see the failure points, and then modify on change the entire design based on the feedback.

19.6 Future of Digital Twins

DTs could be one of the most used technologies in the future. It could become a complete game changer if the right data are made available. We will have smart cities having meta DT. Meta DT is aggregating multiple DT instances to produce a single view of the required physical asset. A smart city may need DT instances of roads, bridges, buildings, buses, etc. which needs to be aggregated. There will be hundreds of DTs, both small and big, and there would be a need for these twins to interact with each other, and this would require enormous computing power, which in turn may spin up and the entire industry all together. There can be VR models which would be completely based on DT. The O&G industry can create massive virtual assets based on DT, and engineers doing maintenance can prepare, train, and then go to the maintenance site to do the asset maintenance. This will really help save lot of time and money.

One example of how DT can combine with XR is by enabling maintenance engineers to operate in a virtual environment/digital environment, and the virtual avatar can replicate the movement of the engineer in a real environment. As shown in the figure below, an engineer sees a virtual/DT of a component from the factory and can analyze and perform maintenance. The movement of the engineer can be replicated by a digital humanoid twin in the factory. One benefit from using safety and other risks can be easily mitigated.

In the future, we may also see people not traveling for any business or other meetings and would virtually attend by projecting themselves in the conference room or may come to the office as a virtual avatar and be able to interact with colleagues. One can even imagine a virtual avatar driving a car and based on feedback can respond. However, the foundation is high-quality, real-time data.

Also, checking how things will work after installing a particular part of the component by simulating the asset using a DT will add great value and reduce repeating the same. There could be a combination of DT with AI-/ML-based bots. This combination could change the way the world sees and does things. For instance, the engineer could sit in his office and a bot can go up to the site where maintenance is needed and can perform a preprogramed maintenance activity. This maintenance activity will be done by the bot based on a program or set of instructions created by the engineer. The engineer would be able to create an accurate set of instructions based on the DT model he sees. Prior to finalizing the maintenance activity, the engineer can simulate and complete a virtual maintenance of the component and arrive at the correct maintenance activity needed to get the optimum result. A complete robotic surgery may be performed with great precision by AI/ML tools guiding the bot. The doctor could create the entire surgery simulation using a DT of the body part to be operated upon, which can then be followed by the bot. The DT of the future can completely take over the human interaction and make human interaction nil if not controlled properly. For instance, a DT of your manager may be sitting in the same room with a DT of you while both would be thousands of miles apart.

If DT can be viewed as a racing car or a 6th Gen Fighter Jet capable of changing the way things are operated, proper data of the right quality are akin to the fuel to the flight or the racing car. One of the key challenges of the DT to grow in leaps and bounds is availability of the right data, less or zero DQ issues, and availability of infrastructure to process the massive amount of data that a DT will need even to perform small simulations for the needed precision. Software systems must be designed to provide real-time data to get the most value of a DT, which itself is a challenge because of network and infrastructure challenges. For instance, in an offshore oil well, huge amounts of data were produced by fiber optic cables, which would produce pulses when there are minute changes in the external environment. These data were so huge that transferring them to the office for analysis was a nightmare. Similarly, DT will require a large amount of data from multiple sources like EDGE devices, sensors, 3D drawings, p&id diagrams, strength of materials information, bill of materials, and weather. Some of the data could be changing quite frequently, and hence DT systems need the feed often. Looking at the benefits and the challenges, it appears that the benefits far outweigh the challenges, and hence there is a very good reason for companies to invest in creating the infrastructure and ecosystem needed for success of the concept of DT – a game changing technology.

Bibliography

1 Digital twins: components, use cases, and implementation tips. https://www.altexsoft.com/blog/digital-twins/

2 Macchi, M., Roda, I., Negri, E., and Fumagalli, L. (2018). Exploring the role of digital twin for asset lifecycle management. *IFAC-PapersOnLine* 51 (11): 790–795, ISSN 2405-8963, https://doi.org/10.1016/j.ifacol.2018.08.415. https://www.sciencedirect.com/science/article/pii/S2405896318315416.

3 Boss, B., Malakuti, S., Lin, S.-W., et al. Digital Twin and Asset Administration Shell Concepts and Application in the Industrial Internet and Industrie 4.0. An Industrial Internet Consortium and Plattform Industrie 4.0 Joint Whitepaper. https://www.iiconsortium.org/pdf/Digital-Twin-and-Asset-Administration-Shell-Concepts-and-Application-Joint-Whitepaper.pdf.

4 Pavlovich, T. and Dron, E. (2020). Data quality and digital twins in decision support systems of oil and gas companies. https://doi.org/10.2991/aisr.k.201029.028. https://www.atlantis-press.com/proceedings/itids-20/125946065.

5 Rosu, L. (2019). Traffic sign recognition on android mobile platforms using computational intelligence methods. https://www.researchgate.net/publication/337318012_Traffic_sign_recognition_on_Android_mobile_platforms_using_computational_intelligence_methods.

6 Wikipedia Contributors (2023). *Artificial Intelligence*. Wikipedia, The Free Encyclopedia https://en.wikipedia.org/w/index.php?title=Artificial_intelligence&oldid=1174569912. revision history statistics. 1174569912 (accessed 11 September 2023).

7 Wikipedia contributors (2023). *Training, Validation, and Test Data Sets*. Wikipedia, The Free Encyclopedia https://en.wikipedia.org/w/index.php?title=Training,_validation,_and_test_data_sets&oldid=1162801019.revision history statistics.1162801019 (accessed 11 September 2023).

8 Kumar, S., Patil, s., Bongale, A., et al. (2020). Demystifying artificial intelligence based digital twins in manufacturing-a bibliometric analysis of trends and techniques. https://digitalcommons.unl.edu/cgi/viewcontent.cgi?article=8543&context=libphilprac.

9 Lv, Z. and Xie, S. Artificial intelligence in the digital twins: State of the art, challenges, and future research topics [version 2; peer review: 2 approved]. https://digitaltwin1.org/articles/1-12#:~:text=Application%20of%20AI%20in%20digital,digital%20twin%20in%20virtual%20space.

10 Shen, Fei; Ren, Shuang; Zhang, Xiang; Luo, Hong; Feng, Chao. (2021). A digital twin-based approach for optimization and prediction of oil and gas production. *Mathematical Problems in Engineering.* 1–8. https://doi.org/10.1155/2021/3062841. https://www.researchgate.net/publication/354345163_A_Digital_Twin-Based_Approach_for_Optimization_and_Prediction_of_Oil_and_Gas_Production.

11 https://www.altexsoft.com/blog/digital-twins/.

12 https://www.atlantis-press.com/proceedings/itids-20/125946065.

13 https://digitalcommons.unl.edu/cgi/viewcontent.cgi?article=8543&context=li bphilprac.

20

Future Directions of Green Hydrogen and Other Fueling Sources

20.1 Introduction

As the current pace of global industrial development continues to increase, so does the demand for energy. This need is often met by burning fossil fuels, which has led to a wide range of environmentally damaging consequences, such as air pollution, global warming, and resource depletion. To address these issues, alternative fuels such as hydrogen are becoming increasingly attractive for use in next-generation industrial applications. The most common substance in the universe, hydrogen, can be produced using several methods, including nuclear energy, natural gas, and renewable energy sources. It is also relatively easy to store and transport, making it an ideal fuel for industrial applications. Furthermore, when used as a fuel, hydrogen produces zero emissions, making it an environmentally friendly and clean substitute for fossil fuels. Hydrogen is already being used as a commercial fuel. For instance, many businesses are already using hydrogen fuel cells to power their vehicles, and hydrogen-powered turbines and generators are being used to provide electricity in some countries. In addition, industrial operations may become less expensive if hydrogen is used as a fuel because it is more efficient than other fuels like diesel. In the future, hydrogen could be used to power a variety of industrial applications, from manufacturing plants and factories to chemical plants and oil refineries. In particular, fossil fuels could be replaced by hydrogen in the creation of chemicals, foods, and other materials. The widespread use of hydrogen as an industrial fuel has the potential to have a significant impact on the global economy. For instance, it could reduce the demand for fossil fuels and help address the environmental consequences of industrial development. Furthermore, the use of hydrogen could spur innovation in the

The Power of Artificial Intelligence for the Next-Generation Oil and Gas Industry: Envisaging AI-Inspired Intelligent Energy Systems and Environments, First Edition. Pethuru Raj Chelliah, Venkatraman Jayasankar, Mats Agerstam, B. Sundaravadivazhagan, and Robin Cyriac.
© 2024 The Institute of Electrical and Electronics Engineers, Inc.
Published 2024 by John Wiley & Sons, Inc.

development of new industrial processes and products. Overall, the widespread use of hydrogen as an industrial fuel has the potential to improve the global economy and address some of the most pressing environmental issues of our time. It is, therefore, essential that governments, businesses, and individuals alike consider investing in the development of hydrogen-based technologies to ensure a sustainable future for the global economy.

20.2 Green Hydrogen Technologies

Hydrogen as an alternative fuel technology is becoming increasingly popular. Hydrogen fuel cells are efficient, clean, and renewable sources of energy. They can be used in a variety of applications, including powering homes, vehicles, and even portable electronic devices. The technology is based on electrochemical reactions between hydrogen and oxygen, producing electricity, heat, and water as the only byproducts. The reaction occurs in a fuel cell, which consists of a stack of cells connected in series producing an electrical current. Hydrogen fuel cell technology is becoming more cost-effective than ever before. It is cheaper than traditional fuel sources such as gasoline and diesel and emits no emissions, making it the ideal choice for eco-friendly transportation. Additionally, fuel cell technology can be used to produce electricity, eliminating the need for grid power. Hydrogen fuel cells are also being used in a variety of other applications, such as powering laptop computers, providing backup power for servers, and providing on-site power for schools and businesses. As technology continues to progress, it is becoming increasingly cost-effective and reliable. Fuel cells are being researched to see if they can be used to supply energy to the entire grid, rather than just the local grid. This could have huge implications for energy production and consumption in the future. As the technology develops, hydrogen fuel cells will become more popular and efficient. They are likely to become the energy source of choice for many people in the near future.

Hydrogen as an alternative fuel technology has many potential benefits, including zero emissions and high efficiency. Possible technologies for hydrogen fuel include the following:

1) **Fuel cells that use hydrogen**: A fuel cell is a mechanism that transforms chemical energy into electrical energy. In a hydrogen fuel cell, the reaction of hydrogen and oxygen yields energy, along with water and heat as byproducts.

2) **Hydrogen combustion engines**: Hydrogen combustion engines work the same way as traditional gasoline engines, but they use hydrogen as the fuel instead of gasoline. The hydrogen undergoes combustion, and the resulting chemical reaction produces motion and heat to power the engine.

3) **Hydrogen fuel storage**: Storing hydrogen in tanks and other portable containers is a key step in utilizing hydrogen as an efficient fuel. Different types of storage methods exist, including pressurized and cryogenic tanks as well as metal hydrides and other advanced technologies.

4) **Hydrogen refueling stations**: In order to make hydrogen fuel available for public use, a network of refueling stations must be established. This is a costly endeavor, but one that is necessary in order to make hydrogen fuel a viable alternative to traditional fossil fuels

5) **Fuel efficiency systems**: Hydrogen fuel can be combined with other fuel efficiency systems, such as hybrid and electric cars, in order to maximize efficiency. By adding hydrogen fuel technology to existing fuel-efficient systems, the fuel efficiency of vehicles can be improved even further.

20.3 Current and Future Industrial Applications of Hydrogen

Hydrogen is being used extensively in a variety of industries today. Not only is it widely used as a fuel, but it also has many different applications in the industrial sector. It is a versatile element that can be used in a variety of ways. In this essay, we will look at some of the current and future applications of hydrogen in the industry.

Currently, hydrogen is of significant importance in nuclear power production. It is used as a coolant in nuclear reactors and also as a means of converting uranium into a usable form of energy. Hydrogen can also be used in fuel cells to produce electricity. Fuel cells are useful for generating electricity from clean sources such as solar, wind, and water.

In addition, hydrogen is used in the petroleum refining process, where it helps remove sulfur from crude oil, making it more suitable for fuel use. In manufacturing, hydrogen is primarily used for welding and cutting metals in factory settings. It is also used for hardening or strengthening certain materials or components. Hydrogen can be used as a means of creating a protective coating on metals, reducing corrosion, and improving their lifespan. In the future, hydrogen will become even more widely used in energy production. It is expected that hydrogen fuel cells will make up a significant portion of the energy grid in the future.

The element will also be used to help reduce pollution in many industries. For example, it is becoming increasingly common for vehicles to run on hydrogen fuel, helping reduce overall emissions and make the transportation sector more efficient and environmentally friendly. Hydrogen is also set to play a pivotal role in the aviation industry in the future. Earlier this year, the first composite hydrogen aircraft was unveiled, paving the way for cleaner and more efficient forms of travel.

Hydrogen is one of the most abundant elements on Earth. Over the years, scientists and engineers have uncovered many uses of hydrogen, from fuel cells to rocket fuel. While the use of hydrogen for energy storage and industrial applications is growing, it has yet to become a widely adopted technology.

The current industrial applications for hydrogen are mainly found within the automobile industry. Hydrogen fuel cells are used in cars, buses, planes, and ships to generate electricity. These fuel cells use the energy released from a reaction between hydrogen and oxygen to generate electricity.

This electricity is then used to power the vehicle's electric motor. While fuel cell technology can be expensive, it offers an alternative to burning fossil fuels.

Another current application of hydrogen is as a feedstock for industrial processes. Because hydrogen is abundant and inexpensive, it can be used in a variety of chemical reactions to produce various chemicals, plastics, and other products. Hydrogen is also used in metal production to reduce waste and the environmental impact.

In the future, hydrogen-based applications could provide an additional source of clean energy. Hydrogen has the potential to store energy from sources like solar or wind, making it a versatile energy storage solution. With the right technologies, hydrogen can also provide an efficient means of long-distance transport of electricity.

Other potential uses for hydrogen include its ability to be used for power generation, desalination and water purification, and synthetic fuel production. Hydrogen could also be used to store and transport energy in off-grid, distributed energy systems. In these applications, hydrogen can be created using excess energy from a renewable energy source and stored and then released later when needed.

Overall, there is a lot of potential for hydrogen in industrial applications. It offers an alternative to fossil fuels and the ability to store and transport energy in a variety of forms. While there are some hurdles that need to be tackled before hydrogen is widely adopted, the current and potential future uses of this element make it an exciting and promising technology.

20.4 The Exploitation of Hydrogen Fuel in a Future System

Due to its many advantages as a fuel, hydrogen has the potential to be a major part of the energy systems of the future. The following are some possible uses for hydrogen fuel:

Transportation: Hydrogen fuel cells can be used to power vehicles, offering a clean and efficient alternative to traditional gasoline and diesel engines. Fuel cell vehicles produce only water as a byproduct and have a much longer driving range than battery-electric vehicles.

Power generation: Hydrogen can be used to generate electricity in fuel cell power plants, offering a clean and efficient alternative to traditional power generation methods. Hydrogen can also be used in combined heat and power (CHP) systems to generate both electricity and heat for buildings and industries.

Energy storage: Renewable energy surpluses can be stored for later use using hydrogen as a type of energy storage. In the absence of solar or wind power, this can help balance the erratic nature of renewable energy sources and offer a steady supply of energy

Industrial processes: In commercial operations like the creation of fertilizers, chemicals, and the refining of petroleum, hydrogen can be used as a fuel. These methods could use hydrogen to decrease pollution and increase energy effectiveness.

Space exploration: Hydrogen is used as a fuel in rockets and spacecraft due to its high energy content and ability to be stored in a compact form.

These are a few possible uses for hydrogen fuel in the future energy infrastructure. The creation of an efficient hydrogen infrastructure, which includes hydrogen production, storage, transportation, and distribution, is necessary for the establishment and use of hydrogen fuel.

20.5 Green Hydrogen: Fuel of the Future

Green hydrogen is being hailed as the fuel of the future. The production and use of green hydrogen have the potential to reduce global carbon emissions and to deliver a clean, renewable, and efficient source of energy.

Green hydrogen is derived from the electrolysis of water. In this process, electricity is used to split the bonds between hydrogen atoms and oxygen molecules in water; the resulting hydrogen can then be used to generate energy. The source of the electricity used in the process is key to its sustainability; when generated from renewable sources such as solar, wind, hydropower, geothermal, and/or tidal energy, green hydrogen is a zero-emission energy source. As such, green hydrogen can be used to decarbonize many sectors of the society, including the industrial sector responsible for large greenhouse gas emissions.

In addition to the positive environmental impact green hydrogen can have, it also offers a reliable and cost-competitive source of energy. It can be used for electricity generation, industrial processes, and as a fuel for transportation, heating, and other purposes. Furthermore, hydrogen offers a resilient power system, as it is available in large quantities and can be stored and transported relatively easily.

The European Union has recognized the potential of green hydrogen and has the ambition to be the world leader in the production of this energy source. Key policies are in place to support green hydrogen, and a number of projects at various scales have been launched. These include Power-to-X, which links power markets and mobility, and the European Clean Hydrogen Alliance, a platform to facilitate investment into the sector and accelerate technological development.

In some countries, the cost of green hydrogen is already competitive, and the prices are expected to decrease further in the coming years. Other countries are also investing heavily in green hydrogen projects, particularly Japan, who plans on having 8 gigawatts of electrolyzers by 2030.

With an increasing number of countries and industry leaders investing in green hydrogen, it is set to become an invaluable part of our low-carbon energy future. As the production costs continue to decline and technologies become available on a global scale, green hydrogen could become a mainstream energy source and the fuel of the future.

In recent years, hydrogen has been gaining traction as an alternative, clean energy source. Today, different types of hydrogen production and use have been developed, with the most prominent being green hydrogen. Green hydrogen holds immense potential as the fuel of the future and offers the possibility of greatly reducing our dependence on fossil fuels.

Green hydrogen is hydrogen that is produced through electrolysis, an energy-based process powered by renewable sources such as water, solar, and wind. This type of hydrogen is not only green in terms of its source of energy but also in its production process, as it does not emit any greenhouse gasses or pollutants into the atmosphere. This makes green hydrogen a clean, emission-free energy source.

There are several possibilities for how green hydrogen can be used. It can be burned as a fuel in hydrogen fuel cells, leading to the production of electricity and heat with zero emissions. Moreover, green hydrogen can be converted into a gas or liquid, allowing it to be used as a fuel in vehicles or used to generate electricity. This versatility means green hydrogen can be used for industrial applications and to power homes, transportation, and even entire cities.

Another great advantage of green hydrogen is its scalability. It can be used on a small or large scale, meaning it is suitable for both residential and commercial applications. Furthermore, green hydrogen production is still in its early stages of development, so there are many possibilities for cost savings as the technology becomes more efficient.

Due to these advantages, green hydrogen is becoming increasingly popular as a sustainable energy source. It has the potential to reduce our reliance on fossil fuels, lower emissions and pollution, and provide clean and efficient energy for future generations.

Although green hydrogen has numerous advantages, there are still some challenges that need to be addressed. For example, production facilities can be costly and require some energy. Additionally, the current network for storing and distributing hydrogen is still quite limited, and consequently becomes a major hurdle for large-scale use.

Nevertheless, with continued research and development, green hydrogen has the potential to become the fuel of the future. By transitioning away from our dependence on fossil fuels and increasing access to clean energy sources, green hydrogen can help us create a more sustainable future.

20.6 Extraction of Hydrogen with Diagrammatic Representation

Figure 1 shows biomass, water, natural gas, and other resources can all be used to produce hydrogen. The most popular techniques for extracting hydrogen are as follows.

Steam methane reforming (SMR): The most popular technique for creating hydrogen is Steam Methane Reforming. In order to generate carbon dioxide (CO_2) and hydrogen (H_2), methane (CH_4) must react with steam (H_2O) at high pressure (30–100 bar) and temperature (700–1100°C)

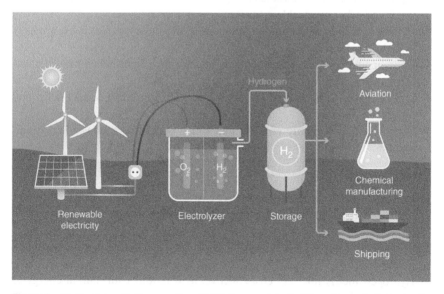

Figure 1   Hydrogen evolution reaction (HER). Source: Adapted from Green hydrogen – A step towards decarbonization (greenportfolio.co).

The reaction can be represented by the following equation:

$$CH_4 + H_2O \rightarrow CO_2 + 3H_2$$

Electrolysis of water: This method involves splitting water into hydrogen and oxygen using an electrical current. The process can be represented as follows:

$$2H_2O \rightarrow 2H_2 + O_2$$

Partial oxidation of methane: In this method, methane is partially oxidized to produce hydrogen and carbon monoxide (CO). The reaction can be represented as follows:

$$CH_4 + 1/2O_2 \rightarrow CO + 2H_2$$

These are some of the common methods for hydrogen extraction, and the method used for hydrogen extraction depends on the available resources and the desired end-use of hydrogen.

20.7 Hydrogen Fuel System Advantages and Disadvantages

Advantages:

1) Hydrogen fuel is a renewable energy source, meaning it can be created from water and other renewable sources.
2) Hydrogen is the most abundant element in the universe, making it a widely available fuel source.
3) Hydrogen fuel produces no greenhouse gases or pollutants when burned, making it a very clean fuel.
4) Hydrogen fuel can be used in internal combustion engines, fuel cells, and other technologies.
5) Hydrogen fuel is highly efficient and can produce more power than traditional fuels.

Disadvantages:

1) Hydrogen fuel is expensive to produce and difficult to store and transport.
2) Hydrogen fuel cells are not yet widely available, making them difficult to implement.
3) Hydrogen fuel is highly flammable and can be dangerous if not handled properly.
4) Hydrogen fuel is not very cost-effective, as it requires substantial energy to produce and store.
5) Hydrogen fuel can be difficult to obtain in certain parts of the world.

20.8 AI-Based Approach for Emerging Green Hydrogen Technologies for Sustainability

The emergence of green hydrogen technologies has provided an opportunity for the world to transition to a sustainable energy system. The development of green hydrogen technologies is an important step toward reducing carbon emissions and achieving sustainable development goals. However, current green hydrogen technologies face several challenges, including cost, scalability, safety, and reliability. To address these challenges, AI-based approaches can be used to develop and optimize green hydrogen technologies. AI can be used to analyze large datasets to identify patterns and trends, which can then be used to develop more efficient and cost-effective green hydrogen technologies. AI can also be used to develop predictive models to forecast the performance of green hydrogen technologies. Additionally, AI can be used to optimize processes and operations of green hydrogen technologies, leading to improved efficiency and cost savings. AI can also support the development of smart energy systems, allowing for more efficient integration of green hydrogen technologies into existing energy systems. By harnessing the power of AI, it is possible to develop and optimize green hydrogen technologies for sustainability.

Green hydrogen is an emerging technology that has the potential to revolutionize the energy sector by providing a clean and sustainable alternative to traditional fossil fuels. AI-based approaches can play a crucial role in advancing and optimizing green hydrogen technologies for maximum efficiency and sustainability. Figure 2 shows here are some ways AI can be used in this context:

Design and optimization of electrolysis systems: Electrolysis is the primary process used to produce green hydrogen from water. AI can be used to optimize the design and operation of electrolysis systems to reduce energy consumption, increase efficiency, and minimize the use of critical raw materials.

Monitoring and control of production processes: AI can be used to monitor and control the production of green hydrogen in real-time, enabling operators to identify and resolve issues quickly, optimize performance, and ensure the production of high-quality hydrogen.

Energy management and optimization: AI can be used to optimize the energy consumption of green hydrogen production processes by predicting demand, identifying energy-saving opportunities, and minimizing waste.

Predictive maintenance: AI can be used to monitor and predict the performance of green hydrogen production equipment, enabling operators to identify potential issues before they occur and perform maintenance proactively, reducing downtime and maintenance costs.

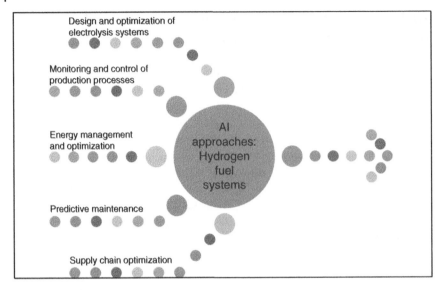

Figure 2 Role of AI in hydrogen fuel systems.

Supply chain optimization: AI can be used to optimize the supply chain for green hydrogen production, from the sourcing of raw materials to the delivery of finished products, minimizing waste, reducing costs, and ensuring timely delivery.

Overall, AI-based approaches can provide significant benefits for the emerging green hydrogen technology, helping accelerate its adoption, reduce costs, and increase sustainability.

The chapter also covers the use of artificial intelligence optimization algorithms in renewable energy systems.

One of the key challenges in renewable energy systems is the variability and uncertainty of renewable energy resources. By using artificial intelligence methods such as machine learning and neural networks, it is possible to predict the availability of renewable energy resources with greater accuracy. This can help renewable energy system operators plan and optimize their operations more effectively.

In addition to predicting renewable energy system resources, the chapter also discusses the use of artificial intelligence methods for predicting output power, load demand, and terminal electricity price. By predicting these factors, renewable energy system operators can optimize their operations to maximize efficiency and profitability.

The use of artificial intelligence planning methods in green energy systems is also covered in this chapter. These methods can be used to maximize energy output and reduce expenses by maximizing the location and scaling of green energy assets, such as wind generators and solar arrays. They can also be used to optimize the scheduling and dispatch of renewable energy assets to meet load demand and minimize curtailment.

An overview of the present state of the art in the use of artificial intelligence techniques and optimization algorithms in green energy systems is given in this chapter. It demonstrates how these methods have the potential to increase the efficiency, reliability, and sustainability of green energy systems and indicates future research paths in this field.

To find the locations where green hydrogen is being produced or has the potential to be produced using AI, you can follow these steps:

AI can be used to help find green hydrogen locations by analyzing data sources such as satellite imagery, population density, and local weather patterns. Using machine learning algorithms, AI can identify potential sites that have the necessary conditions for green hydrogen production, such as access to renewable energy sources, ample water supply, and favorable local climate. Additionally, AI can be used to assess the economic viability of different green hydrogen production sites and optimize the efficiency of the production process.

Gather data: Collect data related to potential locations where green hydrogen could be produced. This may include information on renewable energy sources, such as wind and solar, as well as locations with high demand for hydrogen, such as industrial hubs.

Train an AI model: Train an AI model using machine learning techniques using the gathered data. This may include self-learning, where the model finds trends and connections in the data without specific direction, or controlled learning, where the model receives instruction on data that have been annotated.

Analyze the data: Once the model has been trained, use it to analyze the data and identify potential locations for green hydrogen production. The model may use a combination of factors, such as renewable energy availability, hydrogen demand, and proximity to existing infrastructure, to determine the most promising locations.

Refine the model: Continuously refine the AI model by incorporating new data and feedback from experts in the field of green hydrogen production. This will help improve the accuracy and relevance of the model's recommendations over time.

Evaluate the results: Evaluate the results of the AI model's recommendations by comparing them with existing knowledge about green hydrogen production and identifying any potential discrepancies or areas for further investigation.

Overall, using AI to identify green hydrogen locations can help streamline the process of identifying and evaluating potential production sites, saving time and resources while also promoting more sustainable energy practices.

20.9   Challenges of Hydrogen with AI Technologies

While the use of AI technologies can offer significant benefits in the development and implementation of hydrogen as an energy source, there are several challenges that must be addressed. These include the following:

Limited data availability: AI models require large amounts of data to be trained effectively, and there may be limited data available on hydrogen production, transportation, and storage. This can make it difficult to develop accurate models and limit the effectiveness of AI technologies.

Complex data structures: The data used to train AI models related to hydrogen can be complex and diverse, including data from various sources such as energy grids, fuel cells, and transportation systems. As a result, AI models must be able to handle this complexity to provide meaningful insights.

Safety issues: Since hydrogen is an extremely flammable gas, its production, transportation, and storage raise safety issues. AI technologies must be able to identify and mitigate these risks to ensure the safe use of hydrogen.

Cost considerations: AI technologies can be expensive to develop and implement, and the costs associated with hydrogen production, transportation, and storage can also be significant. These costs must be carefully considered when developing and implementing AI technologies for hydrogen.

Regulatory challenges: The development and implementation of hydrogen technologies may be subject to various regulatory frameworks, which can be complex and vary by region. AI technologies must be able to account for these regulatory considerations to ensure compliance with relevant laws and regulations.

Inflexibility: AI technologies are designed to be used in specific contexts and may not be able to adapt to different scenarios. This can limit the usefulness of the technology and make it difficult to be applied in different areas.

Overall, while the use of AI technologies can offer significant benefits in the development and implementation of hydrogen as an energy source, addressing these challenges will be essential to ensure their effectiveness and successful adoption. effectiveness and successful ad

20.10   The Expected Use and Forecast for Hydrogen Fuel Cells in the Future

The utilization and prediction of hydrogen fuel cells in the future is an exciting prospect. Hydrogen fuel cells are a type of clean energy technology that converts hydrogen into electricity and heat. The technology is widely used in the transportation sector but is also gaining traction in the residential, commercial, and industrial sectors. The future of hydrogen fuel cells is bright. There is a growing demand for clean energy, and hydrogen fuel cells can provide a reliable and affordable source of energy. Hydrogen fuel cells are also more efficient than traditional energy sources, making them an attractive option for businesses and homeowners. The most important benefit of hydrogen fuel cells is that they produce no emissions or pollutants. This makes them a much cleaner alternative to traditional energy sources and helps reduce our impact on the environment. The utilization and prediction of hydrogen fuel cells in the future will depend on the development of new technologies and advances in existing ones. New materials and manufacturing processes will be needed to make hydrogen fuel cells more efficient and cost-effective. Advances in storage and transportation technology will also be necessary to make hydrogen fuel cells a viable option for a wider range of applications. In addition, new policies and regulations from governments and other organizations will be needed to encourage the adoption of hydrogen fuel cells.

20.11 Conclusion

Hydrogen is a versatile element that has significant applications within the industrial and energy sectors. Currently, hydrogen is already widely used for a variety of different purposes, and its usage is expected to increase in the near future. The element will become increasingly important in fields such as energy production, manufacturing, and transportation. Ultimately, hydrogen is a key element that will be instrumental in producing a cleaner, more efficient form of energy in the future.

Green hydrogen and AI have the potential to completely transform the energy sector and make a major contribution to a more sustainable future. Green hydrogen can potentially replace fossil fuels in a number of industries and transportation areas by electrolyzing water using renewable energy sources like wind or solar power.by analyzing vast quantities of data, forecasting demand, and identifying possible ineffectiveness, AI can play an important role in maximizing the production, storage, and distribution of green hydrogen. AI can aid in the creation of novel technologies for producing and storing hydrogen. Additionally, by streamlining processes and reducing waste, AI can help sectors that use hydrogen as a feedstock lower their carbon impact. In general, the combination of AI and green hydrogen may hasten the transition to a low-carbon economy and lessen the impacts of climate change. However, the development of AI and green hydrogen must be done in a manner that is long-lasting, fair, and advantageous to all.

Bibliography

1 Qazi, U.Y. (2022). Future of hydrogen as an alternative fuel for next-generation industrial applications; challenges and expected Opportunities. *Energies* 15: 4741. https://doi.org/10.3390/en15134741.

2 İnci, M. (2022). Future vision of hydrogen fuel cells: a statistical review and research on applications, socio-economic impacts and forecasting prospects, 102739. *Sustainable Energy Technologies and Assessments* 53 (Part C), ISSN 2213-1388, https://doi.org/10.1016/j.seta.2022.102739.

3 Reddy, S.N., Nanda, S., Vo, D.-V.N. et al. (2020). 1 - Hydrogen: fuel of the near future. In: *New Dimensions in Production and Utilization of Hydrogen* (ed. S. Nanda, D.-V.N. Vo, and P. Nguyen-Tri), 1–20. Elsevier. ISBN: 9780128195536 https://doi.org/10.1016/B978-0-12-819553-6.00001-5.

4 Su, S., Yan, X., Agbossou, K. et al. (2022). Artificial intelligence for hydrogen-based hybrid renewable energy systems: a review with case study. *Journal of Physics: Conference Series* 2208 (1): 012013. IOP Publishing.

Index

*The Power of Artificial Intelligence for the Next-Generation Oil and Gas Industry: Envisaging
AI-inspired Intelligent Energy Systems and Environments*, First Edition. Pethuru Raj Chelliah,
Venkatraman Jayasankar, Mats Agerstam, B. Sundaravadivazhagan, and Robin Cyriac.
© 2024 The Institute of Electrical and Electronics Engineers, Inc.
Published 2024 by John Wiley & Sons, Inc.

IEEE Press Series on Power and Energy Systems

Series Editor: Ganesh Kumar Venayagamoorthy, Clemson University, Clemson, South Carolina, USA.

The mission of the IEEE Press Series on Power and Energy Systems is to publish leading-edge books that cover a broad spectrum of current and forward-looking technologies in the fast-moving area of power and energy systems including smart grid, renewable energy systems, electric vehicles and related areas. Our target audi- ence includes power and energy systems professionals from academia, industry and government who are interested in enhancing their knowledge and perspectives in their areas of interest.

1. *Electric Power Systems: Design and Analysis, Revised Printing*
   Mohamed E. El-Hawary

2. *Power System Stability*
   Edward W. Kimbark

3. *Analysis of Faulted Power Systems*
   Paul M. Anderson

4. *Inspection of Large Synchronous Machines: Checklists, Failure Identification, and Troubleshooting*
   Isidor Kerszenbaum

5. *Electric Power Applications of Fuzzy Systems*
   Mohamed E. El-Hawary

6. *Power System Protection*
   Paul M. Anderson

7. *Subsynchronous Resonance in Power Systems*
   Paul M. Anderson, B.L. Agrawal, J.E. Van Ness

8. *Understanding Power Quality Problems: Voltage Sags and Interruptions*
   Math H. Bollen

Printed and bound by CPI Group (UK) Ltd, Croydon, CR0 4YY

27/10/2024

14580125-0005